TELECOMMUNICATIONS:
AN
INTERDISCIPLINARY
SURVEY

TELECOMMUNICATIONS:
AN
INTERDISCIPLINARY
SURVEY

Edited by
L.Lewin

CONTENTS

BIOGRAPHIES OF CONTRIBUTORS

George A. Codding, Jr. is a professor of political science at the University of Colorado. He attended the University of Washington where he earned the degrees of Bachelor of Arts and Master of Arts in 1943 and 1948, respectively; and the Graduate Institute of International Studies of the University of Geneva, Switzerland, where he earned the degree Docteur ès Sciences Politiques in 1952. The next year he joined the faculty of the University of Pennsylvania, where he taught until he took up his present position in 1961. Codding was one of the founders of the University of Colorado's Master of Science in Telecommunications Program. He has worked in the Secretariat of the International Telecommunication Union and acted as a consultant to one of its Secretary Generals. His books include *The International Telecommunication Union: An Experiment in International Cooperation; The Universal Postal Union: Coordinator of the International Mails;* and *Broadcasting Without Barriers.*

G. Gail Crotts is a communications policy analyst in the Office of Plans and Policy at the Federal Communications Commission. She obtained her B.A. from Baylor University. She received her M.S. in journalism in 1969 and her Ph.D. in communications in 1974 from the University of Illinois, Urbana-Champaign. Her thesis was on the public information function of the FCC. She joined the agency in 1974, and helped to establish the Consumer Assistance Office in 1976. In 1977 she joined the Office of Plans and Policy.

A. Terrence Easton is Director of the Graduate School's Department of Telecommunications Management at Golden Gate University in San Francisco, and President of International Communications Management Inc., a San Francisco based telecommunications consultancy firm. He obtained his undergraduate degree in Electrical Engineering from the Johns Hopkins University and received a dual-major Masters Degree in Computer Systems and International Business from the American University (Washington D.C.) in 1972. He joined the graduate faculty of the Center for Technology of Administration at the American University in 1972, and was elected to the faculty of Golden Gate University in 1976. Prior to joining the American University, Mr. Easton held positions with NASA's Apollo Project, Bendix Radio Corporation, GE's Information Services Division and Strager College. Mr. Easton is Professor of Telecommunications and director of the MBA Program at Golden Gate University. He has published numerous monographs and articles and two books on the subject of international communications.

Warren L. Flock is a Professor of electrical engineering at the University of Colorado. He received a B.S. in electrical engineering from the University of Washington, and from 1942 to 1945 he was a Staff Member of the Radiation Laboratory of M.I.T. He received an M.S. in E.E. from the University of California, Berkeley and a Ph.D. in Engineering from the University of California, Los Angeles in 1960. From 1960 to 1964, he was with the Geophysical Institute, University of Alaska, and he then accepted his present position. His interests include telecommunications, remote sensing, radar systems, and solar energy.

Dale N. Hatfield is a Program Policy Specialist with the National Telecommunications and Information Administration, U.S. Department of Commerce. He received a B.S. in electrical engineering from Case Institute of Technology in 1960 and an M.S. in Industrial Management from Purdue University in 1961. He has held telecommunications policy research positions with various government agencies and was formerly the Chief of the Office of Plans and Policy, Federal Communications Commission.

Howard Higman is Professor of Sociology at the University of Colorado where he served as Chairman of the Department of Sociology. He is Chairman of the University of Colorado Conference on World Affairs and Director of the Center for Action Research. His primary field of inquiry is in cross-cultural communication and in the impact of changing technology on social organizations.

Harold E. Hill is Professor and Chairman, Department of Communication, University of Colorado-Boulder. He has B.S. and M.S. degrees from the University of Illinois-Urbana, and his experience includes: several years (at all levels from announcer to program director) in both commercial and public broadcasting; service as a Communication Officer (terminated as Major) during World War II; Executive Vice President of the National Association of Educational Broadcasters in Washington, D.C. for 13 years; has written some 50 articles and made over 100 presentations on communication, particularly media; served as consultant for many broadcasting projects, e.g., helped design the educational television system in American Samoa; 18 years teaching experience at the Universities of Illinois and Colorado; recently received the Distinguished Service Award from the National Association for Educational Communication and Technology.

Baylen Kaskey is head of the Billing systems Department at Bell Telephone Laboratories in Columbus, Ohio. He attended the University of Pennsylvania, where he obtained his bachelor's degree in Mechanical Engineering in 1950, and he did his graduate work at Kansas State College where he was a faculty member from 1950 to 1951. Kaskey joined the technical staff of Bell Laboratories in 1951, where he designed military electronic equipment. He served in U.S. Army Ordnance from 1954 to 1956, working on design of solid propellant rockets. In 1962 he joined Bellcomm, Inc., in Washington D.C., where he headed the Mission Assignments Department. He returned in 1967 to Bell Laboratories to head the Physical Design and Computer Applications Department. On March 1, 1978 he assumed his present responsibilities as head of the Billing Systems Department.

Irving J. Kerner is an attorney in Boulder, Colorado, engaged in the private practice of law. Mr. Kerner's practice is limited to communications matters, primarily involving electronic communications. He obtained his law degree from the University of Colorado in 1966. During the past four years he taught mass communication law at the University of Colorado.

Leonard Lewin is a professor of Electrical Engineering at the University of Colorado at Boulder, where he is Coordinator of the M.S. Interdisciplinary Program in Telecommunications. He was formerly head of the Microwave Department, and later Assistant Manager of the Telecommunications Research Division of Standard Telecommunication Laboratories in the United Kingdom. Earlier, he had worked on radar at the British Admiralty during World War II. His research has focused on antennas, waveguides and propagation, and he was awarded the 1963 International Microwave Prize and also the IEEE W. G. Baker award for a research paper on waveguides. He joined the Boulder faculty in 1966 and was awarded an honorary D.Sc. the following year. Lewin is interested in many aspects of the educational process and in the development of meaningful teaching techniques at all grade levels.

S. W. Maley is Professor of Electrical Engineering at the University of Colorado where he has been teaching telecommunications since 1962. He also conducts research in waveguiding systems and in electromagnetic propagation. He organized and taught courses for the Bell Telephone System Regional Communication Engineering School at the University of Colorado between 1962 and 1970.

Lawrence M. Mead received his B.A. in political science form Amherst College in 1966 and his Ph.D. in government from Harvard University in 1973. He served as a policy analyst in the Department of Health, Education, and Welfare and as a speechwriter for Secretary Kissinger at the Department of State. In 1975, he joined a new research group at the Urban Institute studying implementation problems in Federal social programs. There he participated in a study comparing the decision processes of the FCC and the Environmental Protection Agency for the National Science Foundation. He has published several studies on implementation, bureaucratic reform, and health policy. He recently became Deputy Director of Research for the Republican National Committee.

Robert J. Williams is professor of engineering design and economic evaluation. He attended Michigan State University where he received his bachelors and masters degrees in mechanical engineering. He has taught engineering economy courses in the regular curriculum and to a variety of special groups including the Regional Communications Engineering School at CU and the Telecommunications Masters Degree program also at CU. During 1961-63 he was a full-time consultant to the Government of India on industrial productivity. He served as senior industrial engineer at the Boeing Company in Seattle from 1951-1954. Williams was the Chairman of the mechanical engineering department at the University of Colorado from 1956-1961.

Wesley J. Yordon is professor of economics at the University of Colorado where he teaches courses in microeconomic theory and industrial organization. He attended Wesleyan University from 1949-51, served in the U.S. Army, then received his B.A. from the University of Colorado in 1956. His Ph.D. is from Harvard in 1960, and he has been a Fulbright lecturer in Argentina and Mexico. His research interests have ranged from the mechanics of inflation to the problems of cost analysis in the regulated industries.

PREFACE

The main purpose of this volume is to make generally available
the basic material constituting the MS interdisciplinary program
in telecommunications that has been offered at the University of
Colorado at Boulder since 1971. Initially all of the courses had to
be put together for the program and until now have existed only
in note form. With the publication of this book, both the essential
courses and some valuable ancillary material are made available in
collected form for the first time. It is, of course, not possible in a
single volume to cover in an adequate way the entire field of telecommunications; neither is it possible to teach it in a 12 month
university program. What has been attempted is to provide a
balanced survey with representative material selected to provide
a wide and reasonable coverage. About the only essential course
not explicitly covered here is the one on telephone traffic theory,
since this had already received prior publication and is currently
readily available.* This apart, it is felt that the present work is
complete and self-contained and can be usefully recommended
as a basis for an interdisciplinary study of the large and growing
field of telecommunications.

Leonard Lewin
May 1978

* Beckmann, Petr, *Elementary Queueing Theory and Telephone Traffic* (Geneva,
Illinois: Lee's abc of the Telephone; Traffic Series, 1977.)

INTRODUCTION

THE UNIVERSITIES AND TELECOMMUNICATIONS
L. Lewin, University of Colorado

1. THE INTERDISCIPLINARY NATURE OF TELECOMMUNICATIONS

It would be a great mistake to see telecommunications too narrowly as only the technical aspect of the design and operation of the communication network. Features concerned with government regulation at the international, national and local level, the economics, the management, the legal and social impact; these are but a few of the important aspects that determine the shape and growth of the industry. A good example is CATV which started as a local community development for TV signal acquisition. The long incubation and subsequent form of the FCC regulations have determined in large measure the current structure of this branch of the industry. Copyright laws and federal taxation policy have affected its programming and finance; current development in optical fiber technology will no doubt determine its future growth and social impact.

Although many of the features of the telecommunications industry are covered in one or another discipline in the various departments of universities, the approach is necessarily that of the discipline concerned. Thus, the electrical engineering department of a university or college will be concerned with many things electrical. Some aspects of the light current courses will bear on signal processing. There may even be courses that are specifically designated as telecommunications oriented. But there will be virtually no collaboration with, say, the Political Science department, where a discussion of regulation of public utilities is covered.

The initiation and development of an integrated *interdisciplinary* program in telecommunications is something quite recent to the American university. The earliest of these, and the one I am most familiar with, was undertaken at the University of Colorado at Boulder at the beginning of the seventies, and an account of it forms a substantial part of this introduction. Several other programs, each somewhat different from the others, will also be briefly discussed. There are, of course, many high quality courses in telecommunications offered by various universities, but these are not part of integrated interdisciplinary programs.

2. TEXAS A AND M

With one exception, these programs have all been developed at the Master's level. The exception is at Texas A and M. Initiated in 1975 with encouragement and also financial assistance from the ICA, this program is set at the Bachelor's level, and leads to a B.S. in Engineering Technology with a specialty in Telecommunications. The Telecommunications Technology Specialty is intended to develop the applied scientific knowledge and basic technical skills necessary for a graduate to effectively communicate with employees and apply this knowledge and skill in the telecommunication industry. Emphasis is placed on preparation of the student for understanding of the manufacturing, construction, operation, and design of telecommunications systems, as well as providing the leadership expected of the Telecommunications Technologist. The curriculum comprises a number of courses in various disciplines, including several courses in electrical engineering specifically tailored to the program. A total of 134 semester hours is required for graduation. The emphasis in the curriculum appears to be mainly technical, but, as it evolves, other features may become evident. It is probably too soon to say much at this time about graduates of this program, and precisely what kinds of jobs or further training may be involved for them.

3. GEORGE WASHINGTON UNIVERSITY

An off-campus program in Telecommunications Operations is offered by the Graduate School of Arts and Sciences of George Washington University. It is an interdisciplinary program oriented to operation and management of telecommunications systems.

The degree requires 36 credit hours of coursework taken from the disciplines of Economics, Electrical Engineering, Management, Political Science, Psychology, Sociology, and a Telecommunications Seminar on policy and regulation. Through the School of Engineering and Applied Science and the Continuing Education Program, George Washington University also offers a sequence of five-day symposia on a number of germane subjects such as Telecommunications Policy, Digital Communications, Microprocessors, etc. Although mainly domestic, many features deal with international aspects, including satellites, international telephone, CCITT standards, international Radio Conferences, and much else. The university is greatly aided in this endeavor by its proximity to the U.S. capital, and the availability of government and other experts. Donald Jansky, Assistant Director at NTIA, together with Bud Paul, is largely responsible for much of this program, and he also teaches as an adjunct assistant professor of telecommunications at the University of Pennsylvania. A recent summer session at George Washington University included a course dealing with international and domestic telecommunications institutions, and international policy issues which arise from the interaction between scientific and technological development and activity. Something of the flavor of the program can be gleaned from this very brief description.

4. SOUTHERN METHODIST UNIVERSITY

An interdisciplinary program of a somewhat different character is offered at Southern Methodist University at Dallas-Fort Worth. Their program draws together courses from different schools, enabling the student to concentrate on telecommunications while earning a Master's degree in any of the disciplines of Electrical Engineering, Computer Science, or Operations Research. Degree requirements are 30 hours at the Master's level. Courses on satellites, packet switching, etc., and on policy and regulation cover international features. This program also provides scope for pursuing work beyond the Master's level, working in the respective department. In 1977 they had two Ph.D. students working in this way. One was doing his dissertation in Europe studying the engineering use of broad-band networks with interest on the parameters and issues affecting policy decisions, including legal, social and economic matters. The other was studying the social, economic, political and management issues in international information systems, from the point of view of the user company or institution. The study dealt with developing countries, including Latin America.

This departmental approach, which at the doctoral level is also followed on the Boulder campus, appears to be the only method, at the time of this writing, whereby a doctor's degree in Telecommunications can be awarded in universities in the U.S.

5. GOLDEN GATE UNIVERSITY

The Telecommunications Management Program at Golden Gate University offers professional training in the cost-effective design, utilization and management of currently available telecommunications capabilities (including voice and data communications) in relation to the needs of business and governmental organizations. Developed with the assistance of an Advisory Committee of professionals in the field of Telecommunications, the program was the first of its kind on the West Coast.

The program is in two parts. Students who satisfactorily complete a concentrated 18-unit curriculum of six specified graduate seminars are awarded the Certificate in Telecommunications Management. Designed primarily for the practicing telecommunications professional, the Certificate Program is for those who by the nature of their professional functions may not need the highly specialized management topics covered by the complete M.B.A. curriculum. It is also developed for the specialist whose formal education does not inlcude a baccalaureate degree and who feels that certification of completion of graduate-level studies in his or her field would help to further career objectives.

The Certificate Program curriculum consists of the following 3-unit seminars:

- Organizational Behavior & Management Principles
- Statistical Analysis for Managers
- Introduction to Telecommunications Management

- Managing Corporate Telecommunications
- Introduction to Data Communications Systems
- Design & Management of International Corporate Communications

For the M.B.A. degree in General Management, students must complete both a Foundation Program and an Advanced Program. The former consists of eight 3-unit courses which serve as the necessary background for advanced, graduate-level studies in management. The 30-unit Advanced Program requirement consists of two specified General Management seminars, one seminar selected from each of five specified subject areas (Economics, Finance, Human Resources Management, Marketing, and Quantitative Analysis), and three of the four Telecommunications Management seminars. The Golden Gate program is directed by Professor Easton, who currently heads U.S. operations of International Communications Limited. The courses are offered off-campus, mainly for evening enrollment, and serve to provide training in currently available telecommunication capabilities.

6. SYRACUSE UNIVERSITY

The most recent, at the time of this writing, of the interdisciplinary programs to come to my attention is the Master of Science degree program in Telecommunications (engineering) sponsored by the Department of Electrical and Computer Engineering at Syracuse University. The program is designed to meet the need for professional personnel in the telecommunication field who will be working on problems related to the generation, storage, transmission, retrieval and display of information. The program is interdisciplinary in nature, as it includes a number of integrated courses which cover technical and non-technical aspects, such as historical development, economics, political considerations, and social impact. These integrated courses, which constitute one of the novel features of the program, are primarily technical courses in which the evolution of the discipline, as affected by historical, economic and other forces, is built into the description of the material under study. A minimum of 30 credits is required for graduation.

The program, which accepted students in the fall of 1977, is currently under development as the form the courses will take continues to evolve. There were initially some 50 students enrolled, both on campus and at three off-campus sites. Many of the students come from local companies with telecommunications concerns, so that the program, in fact, acts to supplement and reinforce on-the-job training of graduates in the area.

7. NEW YORK UNIVERSITY

Under the auspices of Dr. Elton of the Alternate Media Center of the New York University School of the Arts, a new interdisciplinary program is being planned, to come into full operation in the fall of 1979. It will involve teaching how to set up new telecommunications applications, with special interest in such areas as local community services, telemedicine, teleconferencing, and legal, business and educational uses. At the time of this writing, the program is planned for a two year, 60 unit syllabus consisting of required courses, elective courses, field experience, seminar participation and a thesis.

The core courses will be required of all students and, as currently planned, are a) Communications Laboratory, b) Interpersonal Communications, c) Organizational Systems, d) Communications and Information Technology, e) The Structure and Regulation of the Telecommunications Establishment, and f) Introduction to Applied Research. There will be one required course each semester in which themes of importance in the program will be explored in a series of seminars. Considerable attention is to be devoted to student advisement, with a segment of the program in the second year devoted to a course of study tailored to the student's interests, and including elective courses selected from other schools of the university. It will be interesting to see how this distinctive program develops over the years.

8. UNIVERSITY OF COLORADO AT BOULDER

The Telecommunications Master's Degree at the University of Colorado is an interdisciplinary program, started in 1971 after more than a year of preparation. The initial idea of developing a Master's Program that was neither an engineering program nor a social science program but helped to bridge the gap between the technology and the social structure in which it is governed and operated, resulted from conversations between some members of the U.S. Department of Commerce and representatives from the university's departments of Political Science and Electrical Engineering.

The original objectives of the program included providing a curriculum in which students without necessarily any formal technical background but with an interest in communications could learn some of the vocabulary and concepts which underlie the technology of telecommunications systems while also providing detailed information on the economics and political structures which govern the industry. At the same time, we wanted to develop a program in which engineers could learn about the social and economic structures while improving their knowledge of the technology. We also hoped that by placing these two groups of students in many of the same classes and by having them work together in teams of two's and three's on projects, that both groups would obtain a better understanding of the complex social, political, economic and technical problems which govern the field.

We saw potential needs for manpower with this broad communications background in government policy and regulating bodies — both at the national and local levels. There also appeared to be a need for broadly trained operating personnel by major users of communications services. For example, it appeared that Cable TV was going to require significant numbers of trained people to write franchises for local governments and regulate the operations of the cable companies.

As the program evolved, some of these original ideas were modified along with the program itself. First, the military surprised us by being extremely interested in the program and sending us many able students. Secondly, the market for graduates from the program appeared to be shifting away from local government and regulatory agencies towards companies who were heavy users of communications equipment, and to interconnect suppliers.

The program at Colorado is now in its seventh full year, and it is possible to see in perspective how some of the various forces have molded it to its present shape. Currently it involves the areas covered by the departments of Electrical Engineering, Political Science, Sociology, Law, Engineering Economy, Business, and Communication. This breadth has been found to be necessary to provide the coverage needed for an adequately trained tele-communications manager, a type of person increasingly in demand in the rapidly expanding field of telecommunications in the modern world. Although the present program administration has been coordinated from the department of electrical engineering, the very broad coverage ensures its fully interdisciplinary nature; administratively it reports directly to the Graduate School of the University.

9. PROGRAM EVOLUTION AT BOULDER

Program Structure.

The program was initially designed around a twelve-month enrollment and the earning of a minimum of 30 credit hours. It commenced with a core structure consisting of two courses from the Engineering Department, one from Political Science, one from Business and one from Sociology. Although not obligatory, most students were expected to take the core courses, together with a number of electives.

The courses earned 12 credit hours each semester. In addition there was the telecommunication seminar, of which more later, and the summer term project or thesis, which earned 6 credit hours for a total of 30 credit hours for the full program. This was normally completed in one full year.

The electives were chosen by the students themselves from a wide variety of graduate programs from the several departments, and more will be said on this subject later. Meanwhile it will be noted that, at the beginning, the engineering core courses constituted only a modest contribution to the total, being only one course in each semester.

Experience with the first year indicated that although some engineering principles which were essential to an understanding of the technical side of the telecommunications industry could be conveyed to students without bringing in much by way of mathematics, a great deal was lost nevertheless by restricting the engineering exposition in this way. Probability, which underlies traffic and queueing theory, requires a certain minimum in mathematics, as does an understanding of both electromagnetic theory and the finer points of modulation theory. To this end, a new core course entitled "Introduction to Communication System Theory" was added and was intended to provide an exposition in an engineering context of the necessary mathematical basis for the ensuing spring courses. In addition, it was found that the curriculum was wanting in another important respect, namely, that a great deal of essential information and understanding pertaining to the nature of transmission of information from one location to another was insufficiently covered in the pre-existing courses. Accordingly, a core course on "Electromagnetic Transmission" was introduced to cover this range of topics. An additional course known as the "Law Seminar" was added to introduce some aspects of legal case history, insofar as they affected a number of important and critical issues facing the telecommunications industry.

Subsequent curriculum evolution was, in part, a logical development of the preceding, and in part a reflection of internal changes at the university. As regards the former, it was soon realized that the students, both those without a technical background and also those who had been engaged, in one way or another, in engineering or communication projects, were often woefully inadequate at even relatively simple mathematical concepts. The introductory course was therefore modified somewhat to ensure that the basics were covered in a way that could be handled usefully by the class. Moreover, a minimum facility with computers was discovered to be absent in a substantial proportion of students, and a one-hour class, "Computing for Telecommunications", was provided to help make up this deficit. It was felt that the introductory course in particular was making up, in part, for a deficient earlier training in mathematics; even those students who had taken a B.S. in engineering at some earlier time found that the material was useful in clarifying ideas that had not been fully understood at the time or had been subsequently forgotten.

Insofar as internal university changes were concerned, the Sociology Department, after participating for two years, was unable to provide an instructor for subsequent courses and the involvement of that department temporarily ceased. Their contribution was covered, at least in part, by the Department of Communication, particularly by a course entitled "Mass Media and Society."

The law professor who had conducted the law seminar later left the university, and no one in that department expressed any interest to cover the particular needs of the Telecommunications Program. The need was met from outside the university, with a new course, "The Legal Structure and Regulatory Scheme", under the auspices of the Political Science Department.

Although these latter changes do not really affect the curriculum structure, they do highlight a specific feature of a program of this type, namely, the dependence of the curriculum integrity on the participation of experts in particular subjects, some of whom are probably few and far between. Proximity to a major government agency or other like organizations is of immense help in making available experts who may not normally constitute part of the regular university faculty.

More recently, several further substantial changes have been introduced. Commencing in the spring of 1976 a "Telecommunications Laboratory" course was started which provides for the roughly 50% of the students with an inadequate contact with the electrical side of the subject. Made possible with a National Science Foundation grant, it is aimed to expose the student in a practical setting to a number of features including such matters as:

i) Antenna patterns and gain
ii) Modulation methods
iii) Time and frequency multiplexing

iv) Noise and interference

v) Queueing, as simulated by a computer

vi) A complete meterological satellite receiving system

The computer course has been extended to three credit hours in recognition of the growing importance of the role of the computer in the telecommunications network, and those students with the necessary background are encouraged to take more advanced courses in programming language and mini- and micro-computers.

The emergence of Cable TV as an important adjunct to the nation's network is recognized by a new course on "Cable Television". This course was the first of three to be scheduled for the summer term. A second involves the Sociology Department, with a course on "The Social Impact of Information". A former FCC specialist, now with the U.S. Department of Commerce, gives the third on "Current Issues in Telecommunications Policy."

With the further development of the curriculum as herein outlined, an additional feature has now become apparent: the number of relevant courses, both core and elective, has grown such that the students are confronted with an almost embarrassing choice. With a 12-hour requirement per semester there are just not enough hours to go around. Very few students voluntarily exceed the minimum, and those that have tried 15 hours per semester, which is the maximum that the Graduate School recognizes, find that the load is excessive. Quite a few students extend their enrollment to 18 months, and where possible, they are now encouraged to do this. The further period thus made available enables quite a few additional courses to be taken, and has the further advantage of helping to prevent the curriculum from being overweighted with engineering-type courses.

In addition to the core courses there are elective courses selected by the student. In principle, the electives can be chosen from anywhere in the graduate bulletin, subject only to the student's advisor's agreement that the choice is relevant to the program. The individual student's interests are thus catered to, but in practice only a relatively small number of courses have been selected in this way. There is some evidence, too, that the outgoing class passes on information as to what are "good" courses to take, so there is a

tendency for the pre-existing choices to persist. The most commonly selected electives are outlined in the program brochure, and
several are included from the Communications Department,
though occasionally students will select from outside this range.

10. TELECOMMUNICATIONS SEMINAR

The Boulder seminar is a weekly event and, although not for credit, it is attended by nearly all the students in the program. There
are a small number of attendees from elsewhere in the university
and also from the public at large. Lasting about two hours, it
usually consists of a lecture-type exposition followed by, or
sometimes interspersed with, questions, comments and discusssion.
The purpose is to expose the students to a wide spectrum of
points of view on a wide range of topics so as to supplement and
also exemplify the in-depth course material.

The program has been extremely fortunate to have available a
number of experts in the field who are available locally. In particular, at the adjacent Department of Commerce facilities (including those of the Office of Telecommunications and the National
Bureau of Standards) there are a number of highly-qualified individuals with first-hand knowledge of aspects of the ITU and its
various subordinate agencies. This has created a unique opportunity to present the student with inside information on the working
of these bodies. Other local experts in various fields cover such
topics as TV and radio programming, frequency spectrum management, local regulatory agencies, the operation of the telephone
system, rate determination, cable-TV operation, satellites, educational communication networks, and many other diverse subjects.

Further afield, we have taken advantage of the occasional presence
in Boulder of national conventions of speakers from across the
nation, as well as inviting speakers from other national universities.
Members of national consultancy corporations in telecommunications have also participated. Both local industry and the Chairman
of the Public Utilities Commission have been particularly helpful
in variously providing financial, documentary and other helpful
assistance to this part of the program.

In association with the seminar program, a number of visits have
been arranged to local facilities of interest. These have included

a field trip to the FAA Center at Longmont, which maintains communications with aircraft throughout a large area centered on Colorado and receives signals from six radars located in four states.

The average student attendance is about the same as the number enrolled in the program, so, allowing for a few outsiders, most students attend most of the time. This is quite significant, for students have little spare time and since the seminar is not for credit it has to compete on a quality and interest basis. After the first year, in fact, it was decided to poll the students after each speaker to ascertain opinions. A short questionnaire was prepared from which a speaker "profile" emerges, which constitutues some guidance as to whom to invite the following year. Some students enthusiastically recommend new speakers from their own knowledge of their home organizations or other personal contacts. These suggestions proved valuable, and also involved the students more intimately with the course.

We feel that this seminar is an essential and valuable part of the total program as it helps to maintain a variety and balance that might be lost from a mere personal selection of elective courses. It provides "real world" exposure of a kind not readily available in a more academically oriented course.

11. THE THESIS

The project or thesis is a substantial part of the student's contribution to the program. The main difference between the two forms is that a thesis must be done by a single student, whereas the project could be a team effort by two, or maybe three, students. Otherwise the same standard is required from each; but as of now only the thesis is accepted by the Graduate School for record and placement in the college library.

Normally the project/thesis is done during the summer months, though most students are preparing for it one way or another from about the beginning of the calendar year.

It is a pleasure to note that well-placed individuals in industry and government have volunteered to help in overseeing certain of these projects, to everyone's benefit.

The choice of subject is left to the students, and almost without exception they turn out to have a prior interest or knowledge that leads them to a suitable choice that is then confirmed by the student's thesis committee. The range of subjects has been extraordinarily wide: from a consideration of the university's telephone exchange to European cable TV; from emergency communications in Colorado to an analysis of telecommunications in developing countries; from the production of TV programs on Public Access to the use of telecommunications by the Boulder Medical Community; from domestic satellites to pay-cable economics.

The thesis length runs about 70 to 380 typed pages, with an average of about 200. Some students take more than the allotted time to finalize after the oral thesis defense is given, and their final graduation is, of course, held up until the work is concluded to the satisfaction of the thesis committee.

Latterly, the existence of these thesis documents has become known outside the university. Some have now been published (by Westview Press, Boulder), and replicated copies of all are available at cost.

12. THE BOULDER STUDENT BODY

The age spread of enrolled students has been from the early twenties to over 50, with an average age of about 30. The military showed a substantial interest in the telecommunications program from its inception. In the first few years it provided about 50% of the enrollment, though this percentage has declined in the last few years. At the same time we have noticed an increase both in applications from abroad, and also in the number of in-state students, who, by and large, have made up only a small part of the class.

Many of the students are "mature" students, with a considerable background in the real world. The military students in particular have a wide range of experience in dealing with large and complex communication systems, and they bring to the class a discipline and a camaraderie which influences the class atmosphere in an unmistakable and beneficial way.

Many students leave their jobs to come to the program and support themselves on savings. Some work part-time , taking two to three years to complete the program. Very small sums are available as grants-in-aid to help these students, but the program does not have a substantial scholarship fund; quite a few would-be students have had to be turned away through inability to provide support.

13. PROGRAM ADMINISTRATION AT BOULDER

Most of the staff for the program were already teaching in various departments, and a re-arrangement of schedules and, in some cases, an increased teaching load, provided the necessary faculty involvement to permit the program to get under way with an initial grant of some $80,000 from the National Science Foundation. Both initial costs of organizing the material and professors' salaries were covered by this. Subsequent grants by the International Communications Association of $6,000 per year for several years enabled some assistance to student fellowships to be made. Currently the program is receiving about $2,500 in donations which are earmarked for student grants-in-aid, but this sum does not go very far, and is quite inadequate to provide even one full-time scholarship.

Mountain Bell in Denver has made a useful contribution which has enabled the speakers' seminar to pay for out-of-state speakers. And as reported earlier, a recent further NSF grant of $38,000 has made it possible to create a telecommunications laboratory to enhance the breadth of the course offerings.

It appears that this program is well known both nationally and also abroad, with interest having been shown from as far away as Australia. An initial mailing of brochures to departments of Electrical Engineering and of Political Science brought in a very large early response. The program is also well-known to the U.S. military and to government agencies. One of the sources of publicity is the students themselves, both in talking to other potential students and also, indirectly, via their projects. The latter often involves them in correspondence with industry, state and national government agencies, and consultancy firms, with a consequent valuable, if unwitting, publicity for the program.

As previously mentioned, the program has been administered mainly from the Department of Electrical Engineering. The question of whether there is an over-emphasis on engineering is an important one in an interdisciplinary program of this character. As mentioned earlier, the number of engineering courses being offered has increased; since the philosophy behind the program has been to emphasize the overwhelming importance for engineers to realize the role played by society and regulating agencies including local, state and national government in the pattern of development and use of telecommunications, the question of balance in the program needs to be continuously monitored. However, the need for the various courses is something which emerged from within the program, so it may be that the form is somehow self-regulating.

Many students, including those from the military, return to their previous jobs after completing the program. About a third of the class seek a new job, either with industry, government or privately. Although a complete record on previous classes has not been kept, it appears that most students have found the job they wanted fairly quickly. There is some indication that previous experience, and, in particular, a good technical background, is a very definite asset, and this encourages us to believe that we currently have the right mix in the program.

14. FUTURE PROGRAM DEVELOPMENTS

The detailed discussion of the Boulder program indicates some of the features constituting an interdisciplinary program in telecommunications. The very brief accounts of other programs at other universities is sufficient to show that each has a somewhat different character, but each is aimed at satisfying an almost insatiable demand by government and industry for competent operators, designers, lawyers, salespersons, economists and managers in the field. At least for some time this demand may be expected to continue and possibly other universities will be led into formulating their own programs. The key to this will certainly be the interdisciplinary character of the program; but the question may be asked as to what disciplines should be included. The more obvious ones are described in the details of the Boulder program, but there is a strong indication that this range may in the long run be insufficient. We are in the early stages of an information explosion made possible by unprecedented technical developments in research, computers and telecommunication network proliferation on a world-wide scale. This is something that

has never happened before in human history, and it opens up awe-
some possibilities both for human progress and also for human
bondage. We are less than a decade away from the period of
George Orwell's prophetic book "1984", and the communication
processes to make this possible are now almost with us. Today the
scientist can *voluntarily* dial into the network. Professor Fest, of
the National Center for Communications Arts and Sciences, fore-
sees a sort of symbiosis between man and his electronic instrument-
ation of a closer sort. He writes, "A Computer, electronically
connected directly to a human brain to provide man with an almost infinite memory capability, is a distinct possibility." Now,
if such a combination were in operation, would it work in only
one direction? Could the individual access the instrument without
its being able to have a form of access directly to him? If it can
give man a much enlarged memory, can it also not be misused by
some dictator for political control? This is still a far-fetched
science-fiction scenario, but so was a landing on the moon as it
appeared only two decades ago. We may not, in fact, be so very
far away from a realistic individual coupling to the network. Dr.
Fletcher, NASA administrator, in a 1976 speech to the Conference
on Satellite Communication and Public Service writes, "Studies
have been made of a Personal Communications System which involves direct broadcast from 'wrist watch radio' to a high capacity,
multibeam satellite for retransmission to ground communications
centrals. The satellite might be 150 ft. in diameter with the capacity of handling up to 25,000 switched channels. The ground transmitter could be small and low power, perhaps 1/25 of a milliwatt.
With large-scale integrated circuit technique, the ground transmitter
need not cost much more than $10. Such a concept allows broad
proliferation of ground stations, bringing practical utilization of
space systems on an everyday basis to the man-in-the street." This
scenario was up-dated early in 1978 when Ivan Bekey, director of
advanced space studies for the Aerospace Corp., appeared before a
Senate commerce subcommittee symposium exploring the opportunities and obstacles associated with space utilization. He passed
a "panic button" mounted on a watch band to three senators and
said such a wrist-mounted alarm transmitter working through a huge
switchboard satellite is one way the space program should be headed. The panic button would send coded calls for help through the

satellite to a local police station. The signal would include a personal identification number, the location of the individual in trouble and perhaps even the nature of the trouble. The same satellite also could serve as a relay station for personal two-way communications anywhere in the United States between people wearing Dick Tracy-style wrist radios. He said that both communications devices could be manufactured for $10 apiece to give millions of Americans tangible benefits from the national space program, which ". . . should provide bold and imaginative services which can be relevant in the everyday lives of millions of citizens, as well as to business and government at all levels."

The satellite that would listen for emergency calls and relay wrist-radio communications would have to be much larger than anything yet placed in orbit. He estimated its antenna would have to be 200 feet in diameter, and it would have to be built in space using the space shuttle orbital transport to haul up equipment and astronaut-builders. The main barriers to building such a massive satellite would be social and political—not technical. It might cost $1 billion to assemble such a switchboard in the sky. But Bekey said it could connect 25 million Americans by radio, and if each paid 10 cents a minute for a call, the entire $1 billion investment could be paid off in a year.

If all these things are technically and economically feasible, we may be fairly sure that they will come to pass. What is much more difficult is to assess some of their outcomes for human well-being. General Sarnoff, former chairman of RCA, once wrote "Every significant new scientific capability has imposed a corresponding responsibility upon those who exercise it. Above all, we must not permit the wonder-working machines to blot out or diminish man. There can never be an electronic substitute for man's private conscience, his sense of justice, compassion and dignity."

Are we able, in fact, to exercise this responsibility? Idries Shah, in his book *Reflections** has a piece entitled, "Behind the Machine" which expresses a concern for this. It reads as follows:

---

*Idries Shah, *Reflections,* Octagon Press 1969, p. 79.

> Man is generally a few paces behind his own inventions.
>
> There are still many people who are revered as figures of authority merely because they can do such things as machines can do more easily. A common example is the awe which people show when faced with someone who has only a good memory or associative capacity, often stuffed with irrelevant facts.
>
> This recalls the refrain 'man is not a machine' frequently used by people whose work and actions tend more and more to convert men into machines.
>
> It is no accident that those cultures which most strongly and often affirm the value and individuality of man are the ones which do most towards automatising him.

It is not merely the knowledge explosion that is of concern here, though it is an important part of the situation. In a highly prophetic moment, Sarnoff once wrote that mankind was on the brink of a communications revolution in which he visualized "high-powered satellites, hovering above the equator, which will broadcast television directly to set owners anywhere in the world." He predicted instantaneous sight, sound and written message communication with anyone, worldwide, and lasers "which in a single beam will produce almost limitless communication channels." These happenings are now almost with us, and with them an enormous increase in the amount of information that we must learn to process. An advertisement by Sylvania suggests that a high school student of today must learn three times the volume of knowledge his father had to at the same age. How much more, we may ask, will be required of children tomorrow! As Margaret Mead puts it, "The information explosion is resulting in more information, being more quickly gathered, more minutely processed, more rapidly diffused—and more overwhelmingly difficult to synthesize."

A further piece from Shah's *Reflections** comments on the individual's response to this growing information environment. Entitled "Drowning" it reads as follows:

*Ibid, p. 24.

To drown in treacle is just as unpleasant as to drown in
mud.

People today are in danger of drowning in information;
but, because they have been taught that information is
useful, they are more willing to drown than they need
be.

If they could handle information, they would not have
to drown at all.

Learning the ability of how to handle information without being
overwhelmed by it is, at least in part, a function of the educational
process to which the universities must contribute. The problem can
probably be managed if there is enough breadth of vision to ensure
that sound judgment and evaluation are developed to the point
where the individual can recognize, validly discriminate and constructively use what he encounters. If the universities can play an
effective role in ensuring the constructive development made
possible by the enormous technical advances that the international
telecommunications network permits, it is by bringing a breadth of
vision to a process which is too narrowly confined if seen *only* in
terms of, say, business management, international law, electrical
engineering, economic efficiency, etc. The importation of value
and purpose is necessary also. This can be done by involving the
curriculum additionally with material taken from philosophy,
anthropology, psychology and the spiritual disciplines. If these
seem strange bedfellows for the more technically oriented concerns
of telecommunications, then we have to realize that knowing *why*
is now becoming just as important as knowing *how*. It is no longer
sufficient to be "climbing Everest, just because it is there," magni-
ficent accomplishment though it may be. We have to come to
grips with the understanding of *why we need to do it*. To say this
is not to belittle great achievements. Rather it is to indicate a shift
in emphasis to a consideration of what *should* be done out of a
great spectrum of possibles. Restraints arising from limitations of
natural resources may eventually determine this anyhow, but it is
much to be preferred that we develop now the wisdom of choice
that will make such eventual forced restrictions unnecessary.

Only time will show whether interdisciplinary telecommunications programs at universities can play an effective role in this. But to be fully effective their breadth will need to be expanded to take in features not traditionally associated with the technical basis on which the growth has until now proceeded.

ACKNOWLEDGEMENTS

I would like to thank Idries Shah and the Octagon Press for permission to reproduce passages from his book *Reflections.* I would also like to acknowledge most sincerely the help and assistance of Mrs. Esther Sparn, the secretary to the Boulder program, for her handling of the manuscripts for the book. And last but not least I would like to express my appreciation to the several contributors to this volume for their help in the provision of material on time, and other assistance in the overall preparation of the book.

Chapter 1

**INTERNATIONAL CONSTRAINTS ON THE USE OF TELE-
COMMUNICATIONS: THE ROLE OF THE INTERNATIONAL
TELECOMMUNICATION UNION**

George A. Codding, Jr.
Department of Political Science
University of Colorado
Boulder, Colorado

1.0 INTRODUCTION

The ultimate use to which technological developments are put depends upon a number of variables, other than the purely technical, including economic, social and political considerations. This is nowhere more true than in the field of telecommunications. From the introduction of the electromagnetic telegraph in the mid-19th Century to the present, telecommunications have provided the life-blood of commerce, government and the military. The importance of telecommunications to these critical areas of human endeavor was such that in a majority of countries telecommunications were made a government monopoly from the beginning , and in all others it was placed under strong government control. When telecommunications crossed national frontiers, as it did almost from the beginning because of the needs of international commerce and international politics, a mechanism was found essential to ensure that it would be used for the maximum benefit of all nations.

The mechanism that was created was the International Telecommunication Union (ITU) which has its headquarters in Geneva, Switzerland. Not only was the ITU the first true international organization among the hundreds now in existence, but over the years it pioneered a number of cooperative procedures which were adopted by the others when they came into being, including the League of Nations and the United Nations itself. Because of these procedures, and the importance of telecommunications, the ITU was able to survive many international crises, including two world wars -- a record few other international organizations have achieved.

To this day the ITU is serving the needs of nations in the ever-changing, ever-expanding world of telecommunications. Basically, the ITU carries out four major tasks: 1) the establishment of binding regulations dealing with certain uses of telecommunications; 2) the creation of non-binding standards; 3) the dissemination of information concerning new developments and uses of telecommunications; and 4) aid to developing nations. This study will investigate all four of these functions, but first there will be a look at the manner in which the ITU has adapted over the years to the continuing telecommunications revolution along with a description of its structure, which has had a major influence on the form that international cooperative action has taken.

Before proceeding, it should be noted that the ITU lays claim to all of the world of telecommunications, including almost any method whatsoever for the transmission of information from one point to another -- telegraph, telephone, radio, and television as well as the transmission of computer data and the like. It also encompasses all techniques of transmission: wire, radio, microwave, and satellites. About the only sphere of the transmission of information in which the ITU does not have primary interest is postal communication; but if certain trends continue, including the ever-rising cost of letter mail, it might well become occupied with that too.

1.1 HISTORICAL BACKGROUND

The manner in which the ITU became involved in the various aspects of international telecommunications is an interesting one. Unlike that which occurred in most of the other UN Specialized Agencies, the subject matter under the competence of the ITU evolved over a long period of time. This section will trace this evolution, concentrating on the international needs that gave rise to each new technological innovation and the manner in which the ITU responded.

1.1.1 The Telegraph

The International Telegraph Union, the direct parent of the International Telecommunication Union, came into being in order to provide governments of Western Europe with a vehicle that would enable them to exploit the international possibilities of the newly developed electromagnetic telegraph. By the second half of the 19th Century, the electromagnetic telegraph was beginning to meet the domestic needs of European governments for a rapid and dependable system of communication. It also had obvious advantages for international communication, from the point of view of both government and commerce; but it was prevented from exercising its true international potential by a number of factors, including differences in equipment and procedure, and a complicated and confusing rate structure.

The response to the problem was the holding of an international conference in Paris in 1865, convened by the French Imperial Government and attended by representatives of twenty interested European countries. This conference had two important products, the first of which was the International Telegraph Convention of Paris (1865) and its appended Telegraph Regulations. In these documents the European nations agreed to interconnect their networks in such a manner as to ensure uninterrupted traffic between their major cities and to use compatible equipment. A single table of tariffs was drawn up and the French franc was chosen as the monetary unit for the payment of international accounts. This new public international telegraph network to be created by the Convention and Regulations was dedicated "to the right of everybody to correspond by means of the international telegraphs."[1]

The second major product of the Paris conference was the International Telegraph Union. While this new international organization, which became the International Telecommunication Union in 1934, was composed in the beginning almost exclusively of European states, it quickly began to take on a truly world wide character as additional states in the world innaugurated telegraph systems and connected them either directly or by cable with those already in existence. The two outstanding holdouts were the United States and Canada. For many years these countries did not sign the Telegraph Convention, nor did they accept the provisions concerning telegraph under the International Telecommunication Union until after World War II, and then only reluctantly. They did, however, send observers to conferences and meetings as did American private operating companies. The United States defended its refusal to accept the telegraph regulations, despite heavy pressure to do otherwise, on the basis of the rather obvious argument that the telegraphs in the United States were operated by private enterprise. Observers have hinted that the U.S. position may actually have been motivated by pressure from the telegraph companies, which did not want any outside interference in their rate structures.

The structure of the International Telegraph Union was quite simple. Prior to World War I, it consisted basically of periodic conferences and a small secretariat, called the Bureau, located in Berne, Switzerland. The functions of the conferences were to provide a place where telegraph experts could meet periodically to discuss telegraph matters and to revise the rules and regulations concerning the international telegraph. The task of the secretariat, which was created at a second international telegraph conference in Vienna in 1868, was primarily to aid in the arrangements for conferences and to act as a central office for the dissemination of information about telegraph matters to the members between conferences. As an example of this latter function, the Vienna Telegraph Conference of 1868 directed that the secretariat gather information relating to the international telegraph, publish regularly a table of telegraph rates, collect general statistics on the international telegraph and publish a journal devoted to telegraph matters. The expenses of the secretariat were borne by the member governments.

The only major change in the structure of the International Telegraph Union occurred in 1925 when it was decided to create the International Telegraph Consultative Committee and the International Telephone Consultative Committee as sub-organs to study various problems affecting the telegraph and telephone between conferences. The International Telegraph Consultative Committee was patterned on the Telephone Committee, which the International Telegraph Union inherited when it took on the regulation of telephones, a subject which will be treated in the next section. In support of his successful proposal to create the Committee, the German delegate at the Paris telegraph conference of 1925 pointed out that there were still many telegraph systems in use, each of which differed from the others in a number of details. Further, there were many questions of operation and procedure that had to be resolved before the international telegraph could be used to its fullest potential. In 1938, in a hotly contested move, it was decided to add rate questions to the competence of the International Telegraph Consultative Committee.

With this basic organization, the International Telegraph Union, and later the International Telecommunication Union, was able to carry out satisfactorily its function regarding the international service, including the setting of appropriate tariffs, and the study of ways and means to make the international telegraph service as rapid and dependable as possible.

1.1.2   The Telephone

The telephone did not come into being until the late 1870's, and its expansion was slow for three reasons. First, in Europe at least, its exploitation was taken over by the already existing postal and telegraph administrations, and was treated as an adjunct to the two older types of communication rather than an important new means of communication, except perhaps in cities. Second, for many years there was no method to connect the telephones of different continents, thus the expansion of telephone took on a national and a regional character and not a truly international one. Third, the international telephone was confronted almost everywhere by the language barrier. For all of these reasons there was less need for international cooperation in telephone matters in its earlier stages than there had been for the telegraph.

The International Telegraph Union accepted the telephone as a subject deserving of its attention, albeit reluctantly. The first proposal for regulation came from the German government at the Telegraph Union's administrative conference in Berlin in 1885. The public had shown its appreciation of telephonic connections between large cities within several countries and were requesting that it be extended to foreign countries. International regulation was therefore in order. Opponents of the German proposal, including the delegations of Austria, Italy and Portugal, argued that it was premature since the service was so new. It was argued that international connections should be left to bilateral agreement until they became more widespread. The compromise solution was to add a short five-paragraph chapter of a very general nature on telephone to the Telegraph Regulations, which left almost complete freedom to administrations, except that it fixed the time unit for charges at five minutes and limited the length of a total call at ten minutes if there were other requests for the use of the line.

It was not until the Administrative Telegraph Conference of London (1903) that it was agreed that there might be enough experience in international telephonic communications to justify the drafting of a set of regulations. Rather than to start from base zero, it was decided to take advantage of the similarities of the two types of wire communication and draft regulations only for those aspects of the telephone service which were different from the telegraph service. These new regulations, consisting of fifteen articles, became the new Chapter 17 of the Telegraph Regulations. One of the articles made all provisions of the Telegraph Regulations which were not inconsistent with Chapter 17 also applicable to the telephone.

The telephone was finally given its own separate set of regulations at the 1932 Madrid Plenipotentiary Telegraph Conference. As mentioned earlier, the Madrid conference was the one where the decision was made to join forces with the International Radiotelegraph Union of which the United States was a member. The United States made it known that its telephone system was run by private enterprise and not by the government. Consequently, the U.S. Government could not be a signatory to an agreement that would result in its regulation. The U.S. was supported in this stand by Canada and a number of Latin American countries. Consequently, the delegates of the Madrid Conference made the new Telephone Regulations apply only to the "European regime." As pointed out in the ITU's centennial publication, the Madrid delegates made it clear the "European regime" meant all of the world except for the Americas. [2]

Although the United States finally agreed to be bound by the Telegraph Regulations and the Telephone Regulations, it still refuses, almost alone, to be bound by the Additional Radio Regulations. [3]

1.1.3   Radio

The second major addition to the subject matter of the International Telecommunication Union, and the one which has had the most revolutionary impact on its operation, was radio. Radio has two major characteristics which made it a likely candidate for international regulation from the beginning: first, if two stations transmit on the same or close to the same frequency they may cause harmful interference; and second, radio waves do not respect national frontiers.

The first few conferences on radio (radiotelegraph in those days) were not necessarily called for this reason, however. Radio had come into practical being close to the end of the 19th Century and was used first commercially for communication with and between ships at sea because it worked well for such communications and it filled an important need. The first company to exploit radio commercially, the Marconi Wireless Telegraph Company, in an attempt to preserve its lead and perhaps even to create for itself a monopolistic position in the field, ordered its operators not to communicate with any other station which did not use the Marconi apparatus. Although there were other motives, the first international radio conferences were called to put an end to this practice and thus enable Marconi's competitors to carve out a place for themselves in this developing, lucrative market. Safety of life at sea was the major excuse used, there having occurred a number of embarrassing if not dangerous incidents when Marconi operators had refused to communicate with others.

Radio was soon used for other services and thus came about the first major task to be performed for radio: the allocation of bands of frequencies to services. The first allocations were for coast stations and ship stations. These were followed in fairly rapid order by allocations for weather reports, time signals, radio beacon stations, broadcasting, aeronautical stations, and on through the myriad of radio services that have been designed for the benefit of mankind including, most recently, those services using satellites.

Two additional developments concerning radio are important to us at this point, namely, the creation of the Radio Consultative Committee and the International Frequency Registration Board. The Radio Consultative Committee had been created by the 1927 Conference of the International Radiotelegraph Union and became a part of the ITU when the Radiotelegraph Union was merged with the International Telegraph Union in

1934. In contrast to what happened with the other two committees, the Radio Committee had a difficult time being approved. The opposition to the Committee at the 1927 conference was especially strong from the United States, France, and the United Kingdom. The delegate from the United States argued that because radio was evolving so rapidly, the Committee might hold back its progress by establishing too rigid principles. The French were afraid that private companies might attempt to obtain the official approval of the committee and use it for commercial gain. The British, who had originally submitted a proposal to establish a technical committee, changed their minds and registered their disapproval on the basis that since no changes could be made to the regulations between conferences such a committee would be useless.

The German and Italian delegations argued that the establishment of such a committee was needed precisely because radio was advancing at such a rapid state and because the committee could relieve the delegates of many of the exhaustive technical studies that they were forced to undertake during conference time. The proponents barely carried the day by a vote of thirty to twenty-six, with eleven abstentions and thirteen delegates absent. The new Committee was to study "technical and related questions" submitted to it by "participating administrations and private enterprises."[4]

Despite their initial opposition to the creation of the International Radio Consultative Committee (CCIR), the United States, France, and the United Kingdom became active participants in its work and the CCIR rapidly gained the genuine respect of all of the members of the Union.

Prior to World War II, with the limited use of radio that existed then, governments rarely had difficulty in finding places for individual radio stations in the allocations that had been made for various radio services at ITU radio conferences. The frequency spectrum is finite, however, and little by little certain bands of frequencies tended to fill up and interference become more prevalent, expecially in the high frequency bands which for a long time were the only bands which could be used for long distance radio-communication.

At the Atlantic City Plenipotentiary Conference in 1947, it was decided to eliminate this problem once and for all by taking two actions: the creation of a master list of frequency assignments for all nations engineered to their present and future needs, and the creation of an International Frequency Registration Board (IFRB) which would have the power to approve or disapprove any additions or changes to that list. In the words of Harold K. Jacobson, the Americans who were responsible for the creation of the IFRB saw it "as something of a cross between the Federal Communications Commission and the International Court of Justice."[5]

As we shall see later, the IFRB never became equivalent of either the FCC or the International Court of Justice, but nevertheless carries out an increasingly valuable function for the members of the ITU.

1.2 STRUCTURE

The structure that has developed over the years to carry out the functions of the ITU may be a little complex when compared to other international organizations, but they are far from complex when compared to any national post and telegraph administration. In essence, the structure of the ITU has five major components: 1) conferences made up of delegates from member states; 2) the Administrative Council, which is in effect a conference of delegates but one of restricted membership; 3) the two international consultative committees, which are similar to conferences except that they do not draft treaties or make recommendations directly to member states; 4) the International Frequency Registration Board; and 5) a secretariat. This section will treat each of these elements of the ITU's structure and, in addition, the extremely important item of financing.

1.2.1 Conferences

There are three types of ITU conferences: plenipotentiary, worldwide administrative conferences, and regional administrative conferences. The plenipotentiary conference is the supreme organ of the ITU and has the exclusive power to amend the basic treaty, the Telecommunication Convention. Plenipotentiary conferences meet at the time and place decided upon by the preceeding plenipotentiary conference, ranging since World War II from five to eight years, and are made up of delegations from all member states wishing to be represented. Specific powers include establishing the budget of the Union and approving its accounts, electing the Administrative Council, electing the IFRB and electing the Secretary General and the Deputy Secretary General. Delegates to plenipotentiary conferences take their duties seriously, and there are always hundreds of proposals at each conference to amend the Convention.

Administrative conferences, which are convened to consider specific telecommunications matters, are of two types, world and regional. The worldwide administrative conference can take up any telecommunications question of a worldwide character including the partial or complete revision of the Radio, Telegraph, Telephone, and Additional Radio Regulations. The specific agenda is determined by the Administrative Council with concurrence of a majority of the members of the Union and must include any question that a plenipotentiary conference wants placed on its agenda. World radio administrative conferences also review the activities of the International Frequency Registration Board and give it instructions concerning its work. A world administrative conference can be convened by the decision of a plenipotentiary conference, at the request of at least one quarter of the members of the Union, or on a proposal of the Administrative Council. Examples of world administrative conferences include the World Administrative Conference for Space Telecommunications held in 1971, the World Administrative Telegraph and Telephone Conference held in 1973, and the World Administrative Radio Conference for Maritime Mobile Telecommunications held in 1974.

The agenda of a regional administrative conference must deal only with specific telecommunications questions of a regional nature. Regional administrative conferences may be convened by a decision of a plenipotentiary conference, on the recommendation of a previous world or regional administrative conference, at the request of one quarter of the members belonging to the region concerned or on a proposal of the Administrative Council.[6]

1.2.2   The Administrative Council

With the failure to make the IFRB the organ intended by its originators, the creation of the Administrative Council was probably the most important addition ever made to the structure of the

ITU. Prior to the creation of the Council the day-to-day work of
the Union was carried out by the Secretary General under the
guidelines laid down by the periodic plenipotentiary conferences.
The Swiss government audited the Union's accounts without charge.
There were a number of reasons behind the successful 1947 proposal to create an Administrative Council to carry on the administrative work of the ITU between meetings of the plenipotentiary conference. The most emphasized reason was the need for continuity,
the need to make decisions between such conferences in the light
of the rapid advances in radio communications and the problems
that ensued. Another was to have a body which could help the
ITU coordinate its work with that of the United Nations and the
other international organizations that were being created immediately after World War II. Along these lines, all the new international organizations which were to become the Specialized Agencies of the United Nations were being supplied with an administrative council, or its equivalent, made up of representatives of states.
Finally, although it was not expressed openly, there was a desire
to eliminate the control of Switzerland over the secretariat.

In effect, the Administrative Council of the ITU is the agent of
the plenipotentiary conference in the relatively long intervals between the meetings of that body. The 1965 convention articu-
lates two major areas of concern for the Council: ensuring the
efficient coordination of the work of the Union and taking steps
to facilitate the implementation of the Convention, Regulations,
and decisions of the various ITU conferences by member countries. While the Convention goes into considerable detail about
these duties, the work of the Council can be summarized in
three major categories: external relations, coordination of the
work of the permanent organs of the Union, and administration. As regards external relations, the Council is the agency which
makes formal contact with the United Nations, the Specialized
Agencies, and other intergovernmental agencies, and can conclude
provisional agreements with them. Coordination includes the review of the annual reports on the activities of the permanent organs of the Union, temporarily filling vacancies among elected officials of the permanent organs, and arranging for the convening
of plenipotentiary and administrative conferences. The most time-
consuming occupation of the Council is that of administration. It

is responsible for drawing up regulations for administrative and financial activities of the Union; supervising the functions of the various organs, including the Secretariat, the Consultative Committees and the IFRB; reviewing and approving the annual budget of the Union in order to ensure "the strictest possible economy;" arranging for the annual audit of the accounts of the Union; and submitting a report on its activities and those of the Union to the plenipotentiary conference for its consideration. The most time-consuming of the administrative tasks is probably that concerning personnel. The Council is responsible for the staff regulations and rules. It sets job classifications and it supervises the application of the UN system of salaries and allowances, the UN Joint Staff Pension Fund and the Union Staff Superannuation and Benevolent Funds. The Administrative Council may also resolve any question not covered by the Convention and which cannot await the next conference for settlement, upon agreement of a majority of the members of the Union.

The Administrative Council meets every spring in Geneva for a three-to four-week session and is composed of thirty-six members (up from the original eighteen of 1947) elected "with due regard to the need for equitable representation of all parts of the world." [7] Each member of the Council has one vote, the Council adopts its own rules of procedure, and the Secretary General acts as its secretary. The Secretary General, his deputy, the Chairman and Vice Chairmen of the IFRB and the Directors of the Consultative Committees all have the right to participate in the deliberations of the Council, without the right to vote, but the Council has the right to hold meetings confined to its own members. Additional meetings may be held if so decided by the Council or by its Chairman at the request of a majority of the members.

1.2.3 The International Consultative Committees

The next element in the structure of the ITU is the International Consultative Committees. The International Radio Consultative Committee (CCIR) and the International Telegraph and Telephone Consultative Committee (CCITT), are close to being international organizations in their own right. The supreme organ in each is the Plenary Assembly which normally meets every three years and which is made up of delegates from all interested administrations and any recognized private operating agency with the approval of a member of the Union. The Plenary Assembly chooses questions that it wishes to study, creates study groups to deal with questions to be examined, and elects its own Director. Questions can also be referred to a Consultative Committee, by an ITU conference, by the IFRB, by the plenary assembly of another Consultative Committee or by any twenty of its members. Both CCI's have their own specialized secretariat, and the CCITT has its own telephone laboratory. Plenary assemblies may submit proposals to administrative conferences and may study and offer advice to member countries on their telecommunication problems.

There are some differences in the mandates of the two Consultative Committees as well as in their method of operation. According to the Telecommunication Convention the duties of the CCIR are "to sudy technical and operating questions relating specifically to radiocommunication and to issue recommendations on them."[8] The duties of the CCITT are similar as regards telegraph and telephone except that in addition to technical and operating questions it can also study "tariff questions." A study group which receives a question to investigate normally prepares a "study program" which is sent out for comments. The comments and the preliminary response are than discussed in a meeting of the whole group and a report, and possibly a recommendation, is drafted which is sent to all participating parties and to the next Plenary Assembly. The Plenary Assembly can accept, reject or modify the draft recommendations. Only the Plenary Assembly has the right to make recommendations to administrations.

In recognition of the fact that the distinctions between telegraph, telephone and radio are fading as a result of new techniques and new technologies, the two Consultative Committees are increasingly forming joint committees and working groups. There is, for example, a Joint Study Group on Noise administered by the CCITT and a Joint Study Group on Transmission of Sound Broadcasting and Television Signals over Long Distances, and a Joint Study Group on Definitions and Symbols administered by the CCIR.

Further, in order to keep a larger perspective of the evolution of telecommunications services, the Montreux Plenipotentiary Conference of 1965 asked the two CCI's to establish a Joint World Plan Committee and regional planning committees as necessary to "develop a General Plan for the international telecommunication network to help in planning international telecommunication services."[9] In pursuance of this request, the CCI's created a World Plan Committee and regional planning committees for Africa, Latin America, Asia and Oceania, and Europe and the Mediterranean Basin.

1.2.4   The International Frequency Registration Board

As mentioned earlier, for a number of reasons the IFRB never achieved the stature that was envisioned for it by the 1947 Atlantic City Plenipotentiary Conference. The most important single reason was the failure of administrations in a series of conferences, called shortly after Atlantic City, to create an overall "engineered" international frequency list. Without such a list, reflecting the current usage and future needs of all the members of the ITU, the IFRB was unable to perform all of its function as arbitrator of the frequency uses of all the nations of the world. Because of the change of function of the IFRB and other reasons including financial ones its size was reduced from eleven to five by the 1965 Montreux Plenipotentiary Conference and might have been abolished had it not been also performing a function that the developing countries considered to be essential to them. (The specific function that the IFRB performs for the members of the ITU will be outlined in the next section.)

As regards structure, the IFRB is made up of five individuals "thoroughly qualified by technical training in the field of radio. . . (who) possess practical experience in the assignment and utilization of frequencies" elected by the Plenipotentiary Conference from candidates sponsored by member countries "in such a way as to ensure equitable distribution amongst the regions of the world." In addition, each member is required to be familiar with geographic, economic and demographic conditions within a particular area of the world."[10]

The members of the IFRB are to be considered as "custodians of
a public trust" and not as representatives of "their respective
countries, or of a region." To make certain that the members of
the IFRB maintain objectivity, the country of which a member
is a citizen agrees to refrain "as far as possible" from recalling
that person during his period of tenure and "must respect the
international character of the Board and of the duties of its
members and shall refrain from any attempt to influence any
of them in the exercise of their duties."[11]

The IFRB elects its own Chairman and Vice-Chairman, who
serve in that capacity for one year, and it has its own specialized secretariat.

1.2.5   The Secretariat and the Coordination Committee

The foundation on which every international organization rests —
and the ITU is no exception — is the secretariat. It is the secretariat that gives an international organization its permanence and
provides the support activities that make it possible for it to exist.
The Telecommunication Convention provides for a General Secretariat to be directed by a Secretary General, who is assisted by a
Deputy Secretary General. Both are elected by the Plenipotentiary
Conference. The Secretary General is the agent of the Administrative Council in that he is responsible to it "for all the administrative and financial aspects of the Union's activities."[12] Thus, while
the Plenipotentiary Conference is the supreme organ of the Union
and sets long-term goals, and the Administrative Concil meets once
a year to make certain that all goes well, the Secretary General is
responsible for the day-to-day operation of the Union. In particular the Secretary General appoints the staff of the Secretariat, or-
ganizes its work, and publishes numerous documents, reports and
studies, the responsibility for which he has accumulated over the
years. The Secretary General may participate in a consultative capacity in all ITU Conferences and meetings. He is also the legal
representative of the Union and is the official welcomer of individuals, both offical and unofficial, who call at the headquarters
of the Union in Geneva.

The general secretariat of the ITU is divided into six departments:
Personnel, Finance, Conferences and Common Services, Computer,
External Relations, and Technical Cooperation. In numbers, the

staff in 1976 was made up of 580 individuals with permanent contracts, 63 with fixed term contracts, and 571 additional employees
on short-term contracts. Officials above the General Service category, for which the principle of geographical distribution applies,
represented some 37 different countries. Among those on the
regular budget, Switzerland had the highest representation, followed by France and then the United Kingdom.[13]

Up to this point, the secretariat of the ITU is quite similar to
that which exists in most other international organizations of
a like nature. But here the similarity ends, and a situation develops which is unique and which has given rise to numerous
criticisms. The Telecommunication Convention provides that
the IFRB, the CCIR and the CCITT shall each have a "specialized
secretariat." The specialized secretariats are not as large as that
of the general secretariat, but impressive nevertheless. The IFRB
in 1976 had a total of 103 employees (excluding the five elected
members), the CCITT had 42 employees, and the CCIR employed
29.

These secretariats are under the direct control of their respective Directors, or the Board in the case of the IFRB, as far as
their duties are concerned, but under the administrative control
of the Secretary General. To further complicate the situation, the
Board and the Directors of the CCITT are given the power to
choose whom they wish as regards the technical and administrative members of their secretariats but this must be done "within the framework of the budget as approved by the Plenipotentiary Conference or the Administrative Council." This is a general budget, applicable to all the organs of the Union including
the General Secretariat. Further, the actual appointment of these
members of the specialized secretariats is done by the Secretary
General, albeit "in agreement with the Director." But: "The final decision for appointment or dismissal rests with the Secretary General."[14]

It was obvious even in 1947 that conflict could arise if members
of the secretariat were under the authority of more than one
head, and it did.[15] In an attempt to straighten out the lines
of authority, at least a little, the Geneva Plenipotentiary Conference of 1959 added the sentence mentioned above which

gave the Secretary General the final decision for appointment or dismissal. Although it would have been welcomed by some, the 1965 Montreux Plenipotentiary Conference would go no further in strengthening the hand of the Secretary General, but did attempt to provide a formal procedure for consultation between the heads of the various organs on matters of mutual interest, especially personnel matters. Article II of the Montreux Convention established an official Coordination Committee to advise the Secretary General on "administrative, financial and technical cooperation matters affecting more than one permanent organ and on external relations and public information." [16] The committee is made up of five individuals: the Secretary General as Chairman, the Deputy Secretary General, the Director of the CCIR, the Director of the CCITT, and the Chairman of the International Frequency Registration Board. The committee is requested to reach conclusions unanimously. If this cannot be done, and the matter is important but not urgent, it must be referred to the next meeting of the Administrative Council. However, if the Secretary General judges the matters in question to be of an urgent nature then he is authorized to make a decision "even when he does not have the support of two or more other members of the Committee". [17]

1.2.6   Finances

The ITU has a very primitive method of financing that remains basically what it was when it was introduced in 1868. The great bulk of the funds necessary to the ITU's operation come from contributions of member governments. The rest comes from such activities as participation in the United Nations Development Program and the sale of documents. The amount that each country contributes to the budget depends upon the contributory class that it selects, ranging from a minimum of 1/2 unit to a maximum of 30 units. The members are free to choose any class of contribution, although a resolution of the Montreux Plenipotentiary Conference suggested that members choose a category "most in keeping with their economic resources" [18] The total of the expenses of the Union not covered by other sources such as the sale of documents is divided by the total number of contributory units to obtain the amount of a single contributory unit. Each member then multiplies this amount by the number of units that it has chosen. For instance, in 1976 there were 41,091,550 Swiss francs out of a total regular budget of 50,296,450 Swiss francs not covered by other sources. This amount was divided by the number 416-1/2, which represented the total number of contributory units subscribed to by all of the members of the ITU in that year. The result was 111.800 Swiss francs, the amount of one contributory unit. Thus, a country such as the Kingdom of Losotho, which subscribed to a 1/2 unit contributory class, paid 55,900 Swiss francs as its share of the 1976 expenses, approximately $22,000 at the exchange rate of the time. In the same year the United States subscribed to the 30-unit class, thus paying 3,354,000 Swiss francs to the 1976 regular budget. The U.S.S.R. was the largest single contributor, paying for 34 units, of which 30 was for itself, 3 for the Ukrainian S.S.R., and one for the Bielorrusian S.S.R. [19]

Until 1973 the ITU Convention did not provide for any sanctions against a state that failed to pay its contributions to the expenses of the Union. As a result there was always a fairly large amount of money owed to the I.T.U. In 1971, for instance, thirty member countries and eight private companies were delinquent for a total of 9,848,574.00 Swiss francs. The largest single amount was the 2,242,543.80 Swiss francs owed by Bolivia. The Torremolinos Conference added a provision to the Convention which provided that members will lose the right to vote in ITU conferences and meetings, and in "consultations carried out by correspondence" if "the amount of arrears equals or exceeds the amount of contribution due from it for the preceding two years."[20] In 1976, although the number of delinquents had risen to thirty-seven countries and twenty-three private companies and agencies, the total amount owed to the ITU was only 3,257,138.19 Swiss francs. The largest amounts in 1976 were the 210,536.65 Swiss francs owed by the Khmer Republic and the 209,745.80 Swiss francs owed by Chad.[21]

1.3 THE PRODUCT

All international organizations perform a number of tasks of importance to the nation states that make up their membership, but few perform more such tasks than the ITU. Indeed, it would be difficult to envision international telecommunications without it. As stated in Article 4 of the ITU Convention, the overall purpose of the ITU is:

"a) to maintain and extend international cooperation for the improvement and rational use of telecommunications of all kinds;

b) to promote the development of technical facilities and their most efficient operation with a view to improving the efficiency of telecommunications services, increasing their usefulness and making them, as far as possible, generally available to the public;

c) to harmonize the actions of nations in the attainment of those ends." [22]

The more specific tasks that the ITU performs in order to achieve those purposes can be summarized under the five rubrics: Regulation, Research and Standards, Information, the International Frequency Register, and Development.

1.3.1   Regulation

The rules for the use of telecommunications that have been agreed upon by the members of the ITU over the years are contained in five main documents: the International Telecommunication Convention, the Telegraph Regulations, the Telephone Regulations, the Radio Regulations, and the Additional Radio Regulations. In signing and ratifying the Convention, the state members of the ITU agree in principle to be bound by the various Regulations. [23] In point of fact, however, many governments including that of the United States treat the Regulations as separate treaties and carry out the ratification process that is provided for in the U.S. Constitution. [24] In addition, administrations can and do refuse to be bound by certain of the Regulations as is the case at present with the United States vis-a-vis the Additional Radio Regulations. While most states do not follow the lead of the United States and consider themselves bound by the Regulations, to be absolutely certain which country is bound to what it is necessary to consult the reservations made by the delegations when signing the Convention and the various Regulations and any which might have been attached to the instrument of ratification. [25]

The Convention.   The International Telecommunication Convention is the ITU's basic document, in effect its constitution. The
Convention gives the ITU its legal existence, sets forth its purposes, establishes its structure, defines its membership[26] and
fixes its relations with the United Nations.[27] It also sets forth
the basic regulations concerning telecommunications in general
and radio in particular.

Since World War II, the Convention has been revised at intervals
ranging from five to eight years. One of the major tasks of recent
plenipotentiary conferences has been to reduce the number of
transitory details in the Convention in order to eliminate the necessity for so many plenipotentiary conferences with all of the
formality and expense that such conferences entail.

The main emphasis of the general rules on telecommunications
in the Convention is on maintaining a truly universal telecommunication network. Members of the ITU recognize the right "of
the public to correspond by means of the international service
of public correspondence" and pledge to establish facilities adequate to meet demands, to use the best technologies, to maintain their parts of the system "in proper operating condition,"
and to use the best operational methods and procedures. Provisions are made for the settlements of accounts between states,
and the gold franc is established as the monetary unit in the
composition of tariffs and the settlement of accounts. Secrecy of
communication is pledged and messages dealing with "safety of
life at sea, on land, in the air or in outer space, as well as epide-
miological telecommunications of exceptional urgency of the
World Health Organization" are given absolute priority. (Government messages are given second priority.)[28]

While the emphasis is on maintaining a smoothly functioning
universal system, members of the ITU reserve the right to make
it less than perfect if they so desire. States retain their right to
stop private telecommunications "which may appear dangerous
to the security of the state or contrary to their laws, to public
order or to decency." In the case of private telegrams, the state
stopping messages must immediately notify the office of origin
except "when such notification may appear dangerous to the

security of the State." [29] As if this was not enough the Convention permits a state to suspend its portion of the international telecommunication service for an indefinite period of time provided only that "it immediately notify such action to each of the other Members through the medium of the Secretary General." [30] Finally, member governments refuse any responsibility to the users of the international telecommunication network, especially as concerns claims for damages.

The section of the Convention dealing with radio is shorter and more specific. Member states agree to limit their use of both the frequency spectrum and geostationary satellite orbits to the minimum, to require the users of mobile radio services to exchange radiocommunications without distinction as to the radio system used by them, and to operate all radio stations in such a manner as to avoid harmful interference to the radio services of stations in other countries. Radio stations of all countries must give absolute priority to distress messages, and the governments of members agree to take all necessary steps to prevent the transmission of false or deceptive distress, urgency, safety or identification signals. Even radio used by the military is not completely exempt from regulation. Although member states "retain their entire freedom with regard to military radio installations of their army, naval and air forces," these installations must "so far as possible" observe all provisions concerning distress messages, measures to prevent harmful interference, and provisions of the Radio Regulations concerning the types of emissions and frequencies to be used according to the nature of the service performed by such stations. Moreover, if military stations take part in the service of public correspondence they must comply with all of the regulations dealing with the public service. [31]

The Radio Regulations.
From a practical point of view the Radio Regulations is probably the most important ITU document. Because of the propagation characteristics of radio and the finite extent of the frequency spectrum, a great deal of coordination of activities of nations and their stations is necessary to ensure that the best interests of all nations are served adequately. This concern is reflected in the breadth and detail of this remarkable document. [32]

The most important single section of the Radio Regulations is Chapter Two which sets forth the conditions for the use of the radio spectrum and allocates sections of it to some two dozen different radio services throughout the world. Some allocations are on a regional basis and others on a universal basis. Unless nations have made a specific reservation concerning a specific service, which many have done, they are obliged to make certain that all of the radio stations operating under their authority use frequency assignments in accordance with the Table of Frequency Allocations. This Chapter is followed by one dealing with notification and registration procedures and another detailing measures to be used to prevent harmful interference between radio stations.

A large portion of the Radio Regulations concerns the manner in which public radio services are to be operated, with special attention to the mobile service. Chapter VI, for instance, includes rules concerning operators' certificates, the class and minimum numbers of operators for ships and aircraft stations, and the working hours of stations in the maritime and aeronautical mobile services. Chapter VII deals with various aspects of working conditions in the mobile services including the frequencies to be used for various purposes and standard procedures. A final section deals with the procedures for handling distress, alarm, urgency and safety signals, and traffic. It is in this section that one finds the international recognition of the familiar distress signals SOS and MAYDAY.

Although the Radio Regulations contain a chapter dealing with radiotelegrams and radiotelephone calls, including such items as priority of messages, routing, and accounting, the details, especially those dealing with tariffs, are found in a document known as the Additional Radio Regulations. It is this document to which the United States has constantly refused to be bound.

The Telegraph and Telephone Regulations. Mainly because of the differences in technologies, the Telegraph and Telephone Regulations are decidely in second place to the Radio Regulations in both depth and scope. In the 1973 Regulations, for instance, the rules concerning the international telegraph are con-

tained in only fifteen articles on fifteen pages and the rules concerning the international telephone system are contained in ten
articles covering fifteen pages. The entire Regulations, including
the Final Protocol (reservations) and Resolutions, Recommendations and Opinions, is exactly 105 pages in length.[33]

Both the Telegraph and Telephone Regulations have major sections detailing services offered and operating methods, including
the keeping of telegram archives. Both also have a major section
concerning rates including accountability and collection of charges,
and both have an Annex dealing with the payment of accounts.

1.3.2   Research and Standards

The second major product of the ITU is the output of the two
Consultative Committees. Few if any international organizations
can boast of a more dedicated and hard-working research complex
than that represented by the CCIR and the CCITT. Their study
groups are composed of some of the finest technical experts in
the governments of the developed countries, supplemented by
experts from industry, and the reports and recommendations
that are produced are universally held in high respect.

The scope of the activities of the Consultative Committees can be
seen from the following listing of the names of their Study Groups:

CCIR Study Groups:

1   Spectrum utilization and monitoring;

2   Space research and radioastronomy;

3   Fixed service at frequencies below about 30 MHz;

4   Fixed service using communication satellites;

5   Propagation in non-ionized media;

6   Ionospheric propagation;

7   Standard frequency and time-signal services;

8   Mobile services;

9   Fixed service using radio-relay systems;

10 Broadcasting service (sound) including audio-recording
    and satellite applications;

11 Broadcasting service (television) including video-recording
    and satellite applications.

CCITT Study Groups:

I	Telegraph operation and tariff (including telex);
II	Telephone operation and tariffs;
III	General tariff principles; lease of telecommunication circuits;
IV	Transmission maintenance of international lines, circuits and chains of circuits;
V	Protection against dangers and disturbances of electro-magnetic origin;
VI	Protection and specifications of cable sheaths and poles;
VII	New networks for data transmission;
VIII	Telegraph and data terminal equipment, local connecting lines;
IX	Telegraph transmission quality; specification of equipment and rules for the maintenance of telegraph channels;
X	Telegraph switching;
XI	Telephone switching and signalling;
XII	Telephone transmission performance and local telephone networks;
XIII	Automatic and semi-automatic telephone networks;
XIV	Facsimile telegraph transmission and equipment;
XV	Transmission systems;
XVI	Telephone circuits;
XVII	Data transmission;
XVIII	Digital networks.

The amount of work that the CCI's produce is also formidable. The XII Plenary Assembly of the CCIR, for instance, produced thirteen volumes of research and recommendations totalling 4,069 pages (excluding the index). The largest was the 530-page volume from Study Group 1, Spectrum Utilization and Monitoring, and the shortest was the 81-page volume produced by Study Group 7, Standard Frequencies and Time-Signal Services. [34]

The 5,400 pages produced by the Fifth Plenary Assembly of the CCITT in 1972 compares favorably with the work of the CCIR. In this nine-volume set the largest single volume, 1,807 pages, was produced by Study Groups XV and XVI (Transmission Systems and Telephone Circuits) and Special Study Groups C and D (Joint CCIR/CCITT Study Group on Noise and the Pulse Code Modulation Study Group). Study Group IV, Transmission Maintenance of International Lines, Circuits and Chains of Circuits, alone produced a volume of research and recommendations of 1,041 pages in length. [35]

The results of the work of the Consultative Committees have only the status of recommendations and are not legally binding on the member states as are the Radio and Telegraph and Telephone Regulations. Nevertheless, because of the high quality of the work of the CCI's their recommendations are taken very seriously by the member states. Some are thus adopted by administrations immediately and others after experimentation. When a recommendation tends to become the universal rule, the subject is raised in a world administrative conference and often then becomes a part of the appropriate regulations. For example, technical sharing criteria for terrestral services and services using space techniques sharing the same frequency band were adopted by the 1971 World Administrative Radio Conference for Space Telecommunications on the basis of CCIR recommendations and incorporated in Article 7 of the Radio Regulations. As another example, Appendix 20D of the Radio Regulations was amended by the 1974 World Maritime Administrative Radio Conference to bring its provisions in line with the relevant CCIR recommendations relating to linked compressor and expander systems. [36]

Because of the dynamic nature of the subject matter with which it deals, for many years the work of the CCIR tended to overshadow that of its sister committee. The CCIR's recommendations in the areas of propagation and spectrum utilization are a case in point. With the development of new technologies, however, the CCITT has taken on a new visibility. Gerd Wallenstein, for example, a participant in the work of the CCITT for a number of years, gives the Committee especially high marks for its recent efforts to achieve common standards which permit inter-

national direct telephone dialing despite "seemingly forbidding obstacles presented by a multitude of national configurations and different signaling systems" and "the strong interests of national supplier industries competing for a share of the world market." Wallenstein credits the hard and painstaking work of the study groups and the fact that those experts "were little encumbered by national political interference, probably because the technical details are too specialized and intricate to be understood by anybody outside a close-knit fraternity."[37]
One of the newer problems that the CCI's are currently investigating is transnational data communication. In order to ensure the effective flow of such communication it is necessary to achieve some sort of system compatibility and common operating methods and, on the commercial level, some common rules concerning payment for services. The 6th CCITT Plenary Assembly gave primary responsibility in the field of data transmission to Study Groups VII and XVII and made provisions for liaison and coordination between these two groups and any other Study Groups where data transmission is invovled.[38]

1.3.3   Information

The making of technical information available to its membership has always been one of the more important functions of international organizations. This exchange occurs in a number of ways: between delegates to conferences and meetings, between officials of national administrations and individuals in the Secretariat, and through the publications of secretariats. The ITU is no exception. The ITU hosts a proportionately high number of conferences and meetings, expecially in the field of radio, and the headquarters of the ITU in Geneva is constantly receiving the visits of officials from telegraph, telephone, and radio administrations as well as officials from private operating companies.

The ITU is especially well known for the quality and the quantity of its publications. The distribution of these publications by the Secretariat to the members of the Union, in the words of the Deputy Secretary General, "guarantees the continued flow of international communications on a world-wide basis."[39]

The Secretariat of the ITU receives its authorization to publish from the Convention, the various service regulations, the Administrative Council, Plenipotentiary Conferences and Administrative Conferences. The Convention, for instance, requires that the Secretariat publish the records of all conferences and meetings, all principal international and regional telecommunication agreements communicated to it by the parties thereto, "data, both national and international, regarding telecommunication throughout the world," technical and administrative information especially useful to the developing countries, information regarding technical methods for the most efficient operation of telecommunication services, and a "journal of general information and documentation concerning telecommunications." [40] The latter is, of course, the prestigious *Telecommunication Journal* which appears monthly in three languages. For the IFRB, the secretariat publishes the weighty *International Frequency List* which contains the particulars of the frequency requirements recorded in the Master International Frequency Register and the weekly, airmail IFRB circulars. The Secretariat also publishes IFRB technical standards, "as well as other data concerning the assignment and utilization of frequencies as are prepared by the IFRB in the discharge of its duties."[41]

The following is a list of other service documents that the ITU secretariat is directed to publish and keep up to date by the Radio Regulations:

1. List of Fixed Stations Operating International Circuits;
2. List of Broadcasting Stations Operating in Bands below 26,100 kHz;
3. List of Coast Stations;
4. List of Ship Stations;
5. List of Radiodetermination and Special Service Stations;
6. Alphabetical List of Call Signs Assigned from the International Series to Stations;
7. List of International Monitoring Stations;
8. Map of Coast Stations which are open to Public Correspondence or which Participate in the Port Operations Service;
9. List of Stations in the Space Service and in the Radio Astronomy Service;

10. Chart in Colors Showing Frequency Allocations.
11. The High Frequency Broadcasting Schedule; and
12. Summary of Monitoring Information Received by the IFRB.

The Telephone Regulations directs the Secretariat to publish: 1) a Yearbook of Common Carrier Telecommunication Statistics; 2) Codes and Abbreviations for the Use of the International Telecommunication Services; 3) a List of International Telephone Routes; 4) a List of Definitions of Essential Telecommunication Terms; and 5) Telecommunication Statistics.

The Telegraph Regulations requires the publication of the following:
  1. Yearbook of Common Carrier Telecommunication Statistics;
  2. Transferred Account Booklet;
  3. International Credit Card for Telegraph Services;
  4. Codes and Abbreviations for the Use of the International Telecommunication Services;
  5. List of Destination Indicators for the Telegram Retransmission System and of Telex Network Identification Codes;
  6. List of Telegraph Offices Open for International Service;
  7. List of Cables Forming the World Submarine Network;
  8. List of Point-to-Point Radio Telegraph Channels;
  9. List of Definitions of Essential Telecommunication Terms;
  10. Telecommunication Statistics;
  11. Routing Table for Offices Connected to the Gentex Service;
  12. Transferred Account Table;
  13. Table of International Telex Relations and Traffic;
  14. Table of Service Restrictions; and
  15. Tables of Telegraph Rates.

The Secretariat also publishes the voluminous reports of the CCIR and the CCITT mentioned above. As a result, the ITU is one of the larger publishers among the international technical organizations. In 1975, for instance, the ITU Secretariat had a publications budget of 6.2 million Swiss francs and it turned out some fifty-two publications amounting to several thousand pages in all. [42]

1.3.4   The Master International Frequency Register

The uniqueness of the task that the International Frequency Board performs merits for it a separate and distinct place in our inventory of the services performed by the ITU for the nations of the world. This task, although far from that envisioned by the delegates to the Atlantic City Plenipotentiary Conference in 1947 who created it, will continue to make an important contribution as long as the radio frequency spectrum remains crowded by radio stations and services. In view of the fact that the radio spectrum is finite and the number of radio stations still continues to increase, this will be a long time indeed.

In essence the IFRB services and maintains a universal register of the frequencies used by radio stations in all of the countries of the world. [43] Articles 9 and 9a of the Radio Regulations provide that all member states must notify the IFRB of frequencies to be used for international communication or which are capable of causing harmful interference to any service of another administration.[44] The Board examines each notification for conformity with the Convention and the Table of Frequency Allocations, and for the possibility that its use will cause interference to the stations of another administration. If the finding is favorable, the Board enters the particular frequency assignment in the Master Register with the date of receipt.

If the notification should not be in conformity with the pertinent regulations, or if there is a probability that it would cause harmful interference with a prior assignment, the Board returns the notification to the administration which sent it, along with its findings and any suggestions it may have to offer with a view to solving the problem. If the administration in question should resubmit the notice with adequate modifications, the assignment is entered by the Board in the Master Register with the date of receipt of the original notification. If the notice is resubmitted in its original form, or with modifications that are still inadequate, the administration in question may nevertheless insist that the frequency be entered into the Register. This the Board will do upon proof that the assignment has been used for at least sixty days without any complaints of harmful interference, and the date recorded will be the original date of notification with certain qualifications.

The Board is also required to survey the Register and with the agreement of the notifying administration to cancel any entry that is not being used.

Through the auspices of the IFRB, the Master International Frequency Register, and the weekly circulars, the telecommunication administrations of the world are provided with a central register of the frequencies in use in the world, a means of obtaining international recognition of national frequency usage, and the expert advice of a highly qualified group of radio frequency specialists.

The task of registering frequency assignments is more complicated than one might expect. In 1976, for example, the IFRB reported the receipt of 81,428 frequency assignment notices. A "full technical examination" was made on 44,072 of them.[45] In addition, the Board investigated some 13,941 notifications to make certain that they were being used as stated in the original notifications, and 9,068 cancellations resulted.[46]

1.3.5   Development Assistance

The fifth area in which the ITU provides a distinct service for its members, and the newest, is development assistance. Before the beginnings of the United Nationals technical assistance program, the ITU had traditionally emphasized activities directed mainly towards the maintenance of a rapid and efficient international telecommunications network and was not overly concerned with what went on within national systems as long as their major cities were a part of the international system. The ITU in effect was dominated by the nations which were the large users of telecommunications and it functioned mainly to meet their requirements and perceived needs.

The new states which achieved independence after World War II and proceeded to join the United Nations and the ITU had different needs. Many of the new states had little in the way of internal communications, and the only foreign links they had were with cities of the former colonial power.

The first actions of the ITU to help meet these new needs of member states came about through outside pressures. In the late 1940's the United Nations became cognizant of the fact that the newly independent states had numerous needs that they could not meet with their own resources, and hence started the UN technical assistance program. After a certain amount of hesitation, the ITU joined the program in 1951 and was allotted a junior partner's share of the available funds. Over the years, the ITU's share of UN-sponsored assistance, now the UNDP, has continued to increase at least in absolute terms. In 1976 the ITU was responsible for a total of $17,170,471 in UNDP funds, $6,250,535 to Africa, $3,392,212 to the Americas, $3,850,211 to Asia and the Pacific, $3,278,250 to Europe and the Middle East, and $399,263 was spent on inter-regional projects. With the addition of $3,071,056 from other sources, the ITU was responsible for arranging for the sending of 493 experts on missions to developing countries, the dispatch of 497 students from developing countries abroad, and the purchase of $4,812,154 in equipment for various field projects. [47]

As more and more developing countries became members of the ITU, more and more emphasis has been placed on development assistance. The 1959 Geneva Plenipotentiary Conference, for instance, formally acknowledged the ITU's commitment to development assistance by adding a paragraph to Article 4 of the ITU Convention, "Purposes of the Union," to the effect that the ITU shall "foster the creation, development and improvement of tele-communication equipment and networks in new or developing countries by every means at its disposal, especially its participation in the appropriate programmes of the United Nations." [48] The Montreux Conference of 1965 saw so many proposals by the developing countries to get the ITU more fully involved in development assistance, such as the organizing of seminars for the telecommunication technicians of developing countries, that the U.S. delegate to that conference was obliged to comment: "It may be stated that if there was one dominant note at the Conference it was the oft-stated belief that the ITU had as one of its basic obligations the requirement to assist the new and developing nations with their individual telecommunication problems." [49] The same trend was also more than evident at the Malaga-Torremolinos Plenipotentiary Conference in 1973.

One proposal that was particularly upsetting to the United States government was that which advocated the establishment of a special ITU development fund supported by contributions of member states to provide assistance for the developing countries to improve their domestic telecommunication networks. While the proposal was defeated at Montreux, it was passed at Malaga-Torremolinos over the strong objections of the United States delegate who declared that if it passed the United States would not contribute to it. The United States still has not contributed to it, and it remains very small.

With the developing nations in a majority in the ITU, and in view of the fact that each state has one vote in ITU conferences and meetings, there is very little possibility that development assistance will not continue to remain one of the ITU's major functions.

1.4 CONCLUSIONS

The International Telecommunication Union has been heavily involved in telecommunications since the development of the electromagnetic telegraph a little over a century ago. This involvement has consisted primarily of establishing binding rules and regulations necessary for the functioning of the international telecommunication system, doing research and setting standards to permit the integration of new technologies into the system, disseminating information necessary for the smooth functioning of the system, and more recently providing assistance to the developing countries to create workable domestic telecommunication systems. All of these functions have met and are meeting a genuine need of the international community.

The machinery has been fairly simple: delegates from member countries meet to revise the rules and regulations as the need has arisen and study committees composed of experts from member countries carry out the necessary research and set standards. The Secretariat provides the infrastructure necessary for holding conferences and meetings and for dissemination of information; the IFRB helps administrations to select and use radio frequencies without causing harmful interference with the stations of other countries. And the cost is relatively small for the services rendered.

All in all, despite some imperfections, the International Telecommunication Union has performed a vital task for the nations
of the world and gives all indications of continuing to respond to
the changing needs of nations well into the future.

REFERENCES

[1]   The early history of the International Telecommunication
       Union, unless otherwise referenced, is from the author's
       *The International Telecommunication Union: An Experi-
       ment in International Cooperation* (Leiden: E.J. Brill,
       1952). This book was republished in 1972 by Arno Press
       of New York.

[2]   ITU, *From Semaphore to Satellite* (Geneva: I.T.U., 1965),
       p. 110.

[3]   ITU, General Secretariat, *International Telecommunication
       Convention, Malaga-Torremolinos, 1973* (Geneva, 1974),
       Final Protocol No. XXXVIII.

[4]   Codding, *op. cit.*, p. 122.

[5]   Jacobson, Harold K., "The International Telecommunica-
       tion Union: ITU's Structure and Functions," in *Global
       Communications in the Space Age: Toward a New ITU*
       (New York: The John and Mary R. Markle Foundation
       and Twentieth Century Fund, 1972), p. 49.

[6]   *ITU Convention, Malaga-Torremolinos, 1973*, Arts. 5, 6,
       7, 53, 54 and 77.

[7]   *Ibid.*, Art. 8, para. 1.

[8]   *Ibid.*, Art. 11, para. 1.

[9]   ITU, General Secretariat, *International Telecommunication
       Convention (Montreux, 1965)* (Geneva, 1965), Art. 14,
       para. 5.

[10]   *ITU Convention, Malaga-Torremolinos, 1973*, Art. 57.

[11]   *Ibid.* Members of the IFRB are forbidden to request or
       receive instructions relating to the exercise of their duties
       from any government, or "any public or private organiza-
       tion or person."

[12]   *ITU Convention, Malaga-Torremolinos, 1973*, Art. 9,
       para. 1.

[13]   ITU, *Report on the Activities of the International Tele-
       communication Union in 1976* (Geneva, 1977), p. 71.

[14]   *ITU Convention, Malaga-Torremolinos, 1973,* Art. 74.

[15]   See Codding, *op. cit.,* pp. 433-435.

[16]   *ITU Convention, Montreux, 1965,* Art. 11, para. 1.

[17]   *ITU Convention, Malaga-Torremolinos, 1973,* Arts. 12 and 59.

[18]   *ITU Convention, Montreux, 1965,* Resolutions, Recommendations and Opinions, No. 15.

[19]   *Report, 1976,* pp. 74 and 82-93.

[20]   *ITU Convention, Malaga-Torremolinos, 1973,* Art. 2, para. 2 and Art. 15, para. 7.

[21]   *Report, 1976,* pp. 130-131.

[22]   *ITU Convention, Malaga-Torremolinos, 1973,* Art. 4.

[23]   *Ibid.* Art. 82.

[24]   See, for instance, the notification in the Report on the Activities of the ITU in 1974 concerning the "deposit of the instrument of ratification of the Radio Regulations" on the part of the United States of America. ITU, *Report on the Activities of the International Telecommunication Union in 1974,* Geneva, 1974, p. 112. See also David M. Leive, *International Telecommunications and International Law: The Regulation of the Radio Spectrum.* (Dobbs Ferry, N.Y.: Oceana Pub., Inc., 1970).

[25]   States can and do attach reservations to particular parts of the Regulations themselves at the time of negotiation or refuse to be bound by the whole Regulation that they helped draft, but this latter is not often the case.

[26]   Any member of the United Nations may become a member of the ITU by formally acceding to the Convention. A non-UN member must apply for membership and, if its application receives a two-thirds affirmative vote, may accede to the Convention. In point of fact the membership provisions no longer have any great importance since almost every independent state is already a member. Membership in the ITU as of May 30, 1978, stood at 154.

[27]   In essence the two organizations pledge their cooperation in matters of mutual concern, agree to exchange documents and other information, and give each other a limited

right to participate in conferences and meetings. See *ITU Convention, Malaga-Torremolinos, 1973,* Art. 39 and Annex 3.

[28] *Ibid.* Arts. 18, 22, 23, 25, 26, 28, 29, and 30.

[29] *Ibid.,* Art. 19.

[30] *Ibid.,* Art. 20.

[31] *Ibid.,* Art. 38.

[32] The latest complete Radio Regulations is that drafted in Geneva in 1959: a conference has been called for September, 1979, to bring it up to date. In the past it was the custom of the ITU to re-work the entire Regulations at one conference, but as mentioned earlier, in more recent times the evolution of the technology has been such that it has been necessary to convene conferences between times to make necessary changes in the regulations concerning particular services. Consequently, the 1959 Regulations are supplemented and brought up to date by the final acts of the various WARCS that have taken place since 1959 such as the Extraordinary Administrative Radio Conference to allocate frequency bands for space radiocommunication purposes, Geneva, 1963; the Extraordinary Administrative Radio Conference for the preparation of a revised allotment plan for the Aeronautical Mobile (R) Service, Geneva, 1966; the World Administrative Radio Conference to deal with matters relating to the Maritime Mobile Service, Geneva, 1967; the World Administrative Radio Conference for Space Telecommunications, Geneva, 1971; the World Administrative Radio Conference for Maritime Mobile Communications, Geneva, 1974; and the World Administrative Radio Conference for the Planning of the Broadcast Satellite Service, Geneva, 1977.

[33] See ITU, *Final Acts of the World Administrative Telegraph and Telephone Conference (Geneva, 1973),* Geneva, 1974. The signatures of the delegates to the conference alone account for 32 pages of the total text.

[34] See CCIR, *XIIIth Plenary Assembly (Geneva, 1974),* Geneva, 1975, Vols, I and VII.

[35] See CCITT, *Green Book — Fifth Plenary Assembly (Geneva, 1972),* Geneva, 1973, Vols. III and IV.

[36] See U.S., Department of State, *Report of the United States Delegation to the World Maritime Administrative Radio Conference of the International Telecommunication Union, Geneva, Switzerland, April 22-June 7, 1974* (TD Serial No. 50), Geneva, 1974 (mimeo), p. 29.

[37] Gerd D. Wallenstein, "Collaboration without Coercion: The I.T.U. as a Model for Worldwide Agreement-Making," in his *International Telecommunication Agreements* (Dobbs Ferry, N.Y.: Oceana Publications, Inc., 1977), Vol. 1, p. 59.

[38] From a speech by Richard E. Butler, Deputy Secretary General of the ITU, entitled "World Telecommunication Development and the Role of the International Telecommunication Union" delivered in Brussels, February 9, 1978, to the International Conference on Trans-National Data Regulation.

[39] *Ibid.*

[40] See *ITU Convention, Malaga-Torremolinos, 1973,* Art. 56.

[41] *Ibid.*

[42] *Report, 1976,* pp. 102 and 108-109.

[43] For the definitive work on the IFRB consult David M. Leive, *International Telecommunications and International Law: The Regulation of the Radio Spectrum* (Dobbs Ferry, N.Y.: Oceana Publications, Inc., 1970).

[44] *Radio Regulations, 1976,* Vol. I, Art. 9, para. 1.

[45] *Report, 1976,* p. 14.

[46] *Ibid.,* p. 18.

[47] *Ibid.,* pp. 50-51.

[48] ITU, General Secretariat, *International Telecommunication Convention, Geneva, 1959,* Geneva, 1960, Art. 4, para. 2, sec. d.

[49] U.S. Department of State, Office of Telecommunications, *Report of the United States Delegation to the Plenipotentiary Conference of the International Telecommunication Union, Montreux, Switzerland, September 14 to November 12, 1965* (TD Serial No. 973), Washington, D.C., December 15, 1965 (mimeo.), p. 10.

Chapter 2

THE FCC AS AN INSTITUTION*

G. Gail Crotts
Federal Communications Commission
Washington, D.C.

and

Lawrence M. Mead
Research Division, Republican National Committee
Washington, D.C.

*The views expressed in this chapter are those of the authors and do not necessarily represent those of the FCC.

THE FCC AS AN INSTITUTION

2.0   INTRODUCTION

We usually do not think of the FCC as as institution at all. We think of it as a group of seven commissioners establishing policy or handing down rulings *ex cathedra.* The literature on the FCC is mostly about its regulations, not about the agency that produces them. Discussions of the FCC as such will usually mention its legal mandates and describe the basic organization of its bureaus rather than analyze how it really functions. Speaking generally, ". . . the literature on the regulatory commissions is replete with formalistic, legalistic, and purely descriptive accounts of how such agencies are structured, what their legal powers and authority are, and what they have done or not done." [1]

The regulatory literature usually assumes that regulations were decided for the reasons stated in the official decisions--or at least by some *intellectual* process (see Cox, 1970, for a typical description of FCC policy during the 1960's in these terms). [2] In fact, the administrative and political process surrounding decision-making matters just as much. Newspaper accounts of major decisions--and common sense--tell us this. Nevertheless, the *process* of regulation has been studied very little. Economists have academic theories of how commissions should regulate and political theories of why they do otherwise Yet we still know very little of why the regulatory commissions do or do not live up to the norms critics propose for them. The answers lie somewhere in the goals and behavior of the commissions as organizations. [3]

Our purpose here is to begin to fill this void. We mean to show that the Federal Communications Commission can be understood not only as a *single actor*, making policy for intellectual reasons (the model mentioned above), but also as an *organization* and as a *political actor* which tend to make policy for reasons of their own. Each of the three models is developed using elements of social, organizational, and political theory, and is meant to suggest hypotheses for concrete research. We will illustrate the models with descriptions of the FCC and its major decisions in order to show that they can be applied usefully to the agency. Our purpose, however, is more to suggest than to verify them.

We are heavily indebted to Graham Allison, who first applied the three models together in an effort to explain a governmental decision. [4] We have made some adjustments to his approach, which are explained below. Allison's work opened up powerful new avenues for research on foreign policy decision-making. In a more modest way, we hope to do the same for the study of regulatory decisions.

Each of the following three sections describes one of the three models, shows how it applies to the FCC, and uses it to analyze several major FCC decisions. Some of the decisions are discussed more than once, since the different models illuminate different aspects of the same case. The fifth section uses the models to analyze and assess major proposals for the reform of the FCC. The conclusion argues for the importance of institutional development in the reform of regulatory policy.

2.1 MODEL 1: THE SINGLE ACTOR

As mentioned above, most writing about the FCC assumes intellectual reasons for the way the FCC behaves. Such accounts treat the FCC as if it were an individual, or a cohesive group of seven individuals, making policy for stated reasons. This "single actor" or "rational actor" model assumes that the FCC, like

"economic man," acts rationally to maximize values of importance to itself. To understand its actions, according to this model,
we need only to imagine the calculations which led the FCC to
its conclusions. In the literal sense, we understand FCC action
by *rationalizing* it. [5]

We use the single actor model here in a more normative form
than Allison does. His treatment assumes that political actors
pursue self-interested objectives; the focus is on the calculations
of interest leading to action. Here we focus on the goals the
actors are presumed to maximize. We take as goals those norms
to which the FCC is commonly held by the professional groups
within the Commission and by external authorities and critics.
In our use, Model 1 says that these norms explain all of what
the FCC *should* do and much of what it *does* do. The reasoning
is that policy is decided as if by individuals; that these actors
favor actions which may be publicly justified; and that the
reasons for the actions derive from the central values pursued
by the Commission.

What values does the Commission serve? The professional
groups and external authorities involved with the FCC tend
to assert different values. Technically-minded people, such
as engineers and economists, generally say the FCC should
obey its *legal mandate* to enforce the public interest on the
industry and should promote technical and economic *efficiency.*
Lawyers and judges say it should follow *due process* in its rule
making, and that its decisions should satisfy norms of *jurispru-*
dence (that is, they should mold past precedents and current
cases into coherent regulatory law). Public interest advocates
say it should follow the *popular will* and achieve the *public*
good in ways not confined to the dictates of either economic
calculation or law. These characterizations are of course only
approximate.

These three normative positions all imply that FCC actions will
have authority to the extent they adhere to the appropriate
values. The three positions correspond quite closely to the
three forms of political authority discussed by social theorist
Max Weber. The following subsections take up the three positions, explain each in terms of Weber's theory, and describe

the influence each appears to have had on the FCC's structure and decisions. Each offers a different version of the single actor view of the Commission.

2.1.1   Rational Authority

The ethic for the FCC which states that it should execute its legal mandates and achieve efficiency seems the most self-evident of the three. These premises are so deeply embedded in our culture that they seem obvious.

They reflect what Weber called the "rational" or "legal" form of authority. In modern government, according to Weber, citizens typically give allegiance to impersonal laws. There is a "government of laws rather than men." Where administrative discretion is necessary, governance is by expert functionaries chosen on the basis of merit, not by political favorites or by others chosen by ascriptive criteria. Allegiance is to an impersonal order composed of laws and bureaucracies ruling with the authority of expertise. [6]

Rational authority has been especially important for the independent regulatory commissions. These agencies were set up by Congressional mandate in order to solve difficult allocation problems by technical means. Much of the rationale for establishing independent regulatory agencies was the assumption that experts acting with delegated authority could regulate business in a less partisan way than politicians would by themselves.

Weber presents rational authority as an "ideal type." No government and no commission fully lives up to it. The ideal has political weight, however, because of its unquestioned legitimacy.

For an FCC official, the norms of rational authority command (1) adherence to the FCC's legal mandate, insofar as it dictates specific decisions, and (2) expert calculation, insofar as the official is left to make decisions himself. Calculation means to make policy in the light of consequences. It means choosing the option that is most advantageous in technical or economical terms.

The formal structure of the FCC reflects the norms of rational authority. The chairman and the six other commissioners are named and confirmed in ways set out by law. They are nominated by the President and confirmed by the Senate. They serve seven-year terms, staggered so one commissioner is replaced each year. The chairman is appointed from among the commissioners by the President and serves at his pleasure. A chairman removed by the President, however, would remain a commissioner to the end of his term unless he chose to resign. The commissioners rule by delegated authority and may be overruled by statute. They make decisions in their own name but with the advice of an expert staff of some 2000 persons organized in bureaus and offices. (The bureaucratic structure is discussed fully in Model 2 below.)

The norms of rational authority, as Model 1 predicts, have had great influence on the FCC's evolution. The legal basis of the FCC's authority is simply assumed by most students of the agency. The history of communications regulation can be understood as a search for the right kind of agency resting on the right kind of legal mandate. The Commission's authority is based on legislation, and deficiencies in its authority can also be attributed to legislation.

The FCC was formed by an amalgamation of earlier agencies and authorities which had proven insufficient for the job. The Commission was created by the Communications Act of 1934 through a merger of the Federal Radio Commission set up in 1927 and those aspects of the Interstate Commerce Commission dealing with telephone and telegraph. [7] An important later enactment is the Satellite Communications Act of 1962, which defines the FCC's authority over satellite communications. [8] Other legislation to clarify, strengthen, or limit the Commission's authority has repeatedly been urged by its friends and critics. All make the assumption--consensual in modern politics--that statute law is the strongest authority an agency can have.

Legislation directly governs some specific FCC decisions. For example, the Communications Act specifies that the FCC

must enforce technical requirements for radiotelephone equipped ships, must approve the extension of lines in the common
carrier service, and must make an annual report to Congress. [9]
Other things that it cannot do, such as censor broadcasting,
are denied it by the Communications Act or the Constitution.
[10] Sometimes the Commission acts or refuses to act because
of a threat of legislation in Congress. When the Commission
appeals to facts like these in justifying its actions or inactions,
it is clearly acting in the manner supposed by Model 1.

The FCC and its decisions have also been molded by the other
aspect of rational authority — expert calculation. The Communications Act in fact leaves the agency much discretion; the Commission seeks to exercise it on a basis of expertise. It has its own
staff of experts in the various communications fields — broadcasting, common carrier, cable television, land mobile radio,
etc. — and it solicits advice form outside experts and interests
before it makes major decisions. Its decisions are usually justified in technical and economic terms, even when bureaucratic
or political influences have been important (see below). That
is, the decision claims to lead to the most favorable technical
or economic consequences.

The expert side of the FCC is most clearly seen in its more
technical decisions, notably those involving spectrum management. Since the radio spectrum has limited range of usable
frequencies, only a limited number of frequencies may be
allocated to the various radio services in an area without interference between them. As the demands for spectrum space
grow and shift, the Commission must decide anew how to divide up a scarce resource. The assumption that it could do this
scientifically was a major rationale given for the FCC at its
founding.

Several examples will show that the rationalist ethic clearly influences spectrum decisions, whatever other motives are involved.
In 1945 the Commission shifted FM radio from the 42-50 megahertz (MHz) band to 88-108 MHz on grounds that interference
from skywaves would be less at the higher frequencies. While the

move inflicted considerable economic cost on established FM stations, the Commission viewed the long-term technical benefits as overriding. [11]

In 1952, after seven years of deliberation, the Commission decided to allocate 70 UHF channels for the use of television in addition to the existing 12 VHF channels. The options of a television system using VHF only or UHF only were rejected for a host of complex reasons. While economic interests favoring VHF or UHF were involved, the issue was heavily technical and was fought out in technical terms. In addition, the decision to use both VHF and UHF was based on the prediction that the UHF stations could develop despite VHF competition. These technical and economic projections committed the Commission to several later actions to foster the development of UHF stations. [12]

In decisions of 1970 and 1974-75, the Commission withdrew the 806-960 megahertz (MHz) band from the use of UHF-TV channels and reallocated it for land mobile radio use. The initial reason was simple — the pressure of increasing demand for land mobile radio services. However, the decision followed a lengthy inquiry by the Commission, a careful consideration of options, and a call for technical studies into the demand for land mobile in the future. [13]

The FCC sometimes listens to outside expert opinion about the best way to achieve effective and efficient policy. In 1971, the Commission authorized the licensing of private companies to compete with AT&T in the specialized common carrier business (that is, wire and microwave communication for computer signals and other specialized uses). [14] The decision, which rested on a series of precedents, broke with the tradition of regulating AT&T as a public utility. Instead, AT&T (or at least its specialized services) would be held accountable to the public interest by increased market competition. The decision reflected, in part, a growing consensus among economists that competition may be the best way to discipline regulated industries, given their ability to capture the regulating commissions and use regulation to protect themselves against innovation. [15]

These decisions show some of the plausibility of Model 1. Only a total cynic would deny that the rationalist values of the FCC importantly influence its decisions. Model 1 argues that the Commission acts as an individual would searching for actions which may be publicly justified. Commissioners know that to act in the name of rational authority confers legitimacy second to none.

2.1.2 Traditional Authority

The second set of norms specified by a Model 1 analysis of the FCC are those of lawyers: decisions should follow due process and be well-argued in the light of precedent.

These values are akin to those Weber associates with the "traditional" form of political authority. If rational authority is impersonal and expert, traditional authority (which usually precedes it historically) is conservative and highly personal. A traditional ruler, such as a feudal chieftain, does not *make* laws; he discovers them in the past. He rules in the light of precedents. And to the extent his own judgment is necessary, he gives it as a judge rather than an expert. He does not calculate; he construes past law to fit present circumstances. He does not make general policy; he issues rulings in particular cases. While the norms of rational authority are adherence to law and expertise, those of the traditional ruler are tradition and discretion. [16]

At the FCC, discretion is necessary because of the very vagueness of the legislative mandate. The Communications Act imposes very few specific policies or actions on the Commission. The most well-known injunction in the Act is that the FCC must regulate communications according to "public interest, convenience and necessity." This frees the Commission to set or change policy on the basis of its own reasoning, as in the specialized common carrier case just mentioned. It also weakens the Commission against the pressure of the regulated interests. Commissioners can rarely say their hands are tied by higher legal authority. They *have* to respond to demands for change, in part, because they have the authority to do so. This, plus changing membership on the Commission, means that stable policies are difficult to sustain. Inevitably, policy comes to mean a sequence of case decisions responsive to precedent and discretion, but without any overall rationale.

Traditional-legal norms appear to have influenced the FCC's structure and procedures at least as much as rational ones. Indeed, at first glance the FCC appears to be a court rather than a modern bureaucracy. Alongside the staff bureaus, which advise the Commission on the basis of expertise, are the administrative law judges, who advise using adversary proceedings in the legal manner. The Commission is required to order a hearing when it confronts "a substantial and material issue of fact," for example in the renewal of a broadcasting license or in a rate making case. [17] The decision of the administrative law judge may be appealed to a Review Board and then to the Commission itself. Commission decisions, in turn, may be appealed to the U.S. Court of Appeals and to the Supreme Court.

The Commission tends to use court-like procedures not only to judge individual cases but to make general rules. On major issues, the commissioners themselves may hold oral arguments and listen to lawyers argue for the various sides. Recently, the Commission began holding public meetings at which anyone may present grievances or proposals for change — a merger of the governing and judicial functions closely akin to a medieval court.

The rational and traditional modes of decision making are to some extent in tension. When hearings are not involved, the FCC's bureau staff may advise the commissioners about policy *ex parte* (off the record) and play the dominant role that rational authority imagines for it. When there are hearings, however, the bureau may only participate and advise the Commission on the record, as if it were a private party. In the specialized common carrier case mentioned above, the Commission could have chosen either route. If it had ordered comparative hearings, pitting AT&T against the specialized common carrier companies, the common carrier bureau, which favored competition, would have been only one party among many. Since it chose the rulemaking procedure, there was no adjudicatory proceeding. The Commission acted in response to written submissions and oral presentation, both formal and informal. Hence, the common carrier bureau was able to summarize the viewpoints in its own terms, remain in control, and persuade the Commission to adopt a major policy change. [18] *Ex parte* contacts have historically been considered helpful to the Commission in rulemaking proceedings, but have recently been challenged as inappropriate. In 1977 the Appeals Court suggested that rules against *ex parte* contacts might have to be applied even in rulemaking proceedings. [19]

A commitment to adversarial methods is one reason the FCC's expert staff capability is not greater than it is. The Commission has lacked the expertise to analyze most highly technical issues on its own. It has relied instead on data and analysis submitted by the affected parties or by outside experts. One instance is regulation of AT&T's common carrier rates. [20] Another is the design of rules governing the certificating of cable television. [21] According to some critics, in both cases heavy reliance on the industry for information may have made policy less effective than otherwise. Until recently, there were virtually no economists at the FCC. [22] In 1972, Chairman Burch appointed a special assistant for planning and policy, and in 1973 established an Office of Plans and Policy, to try to fill this void. Even OPP at first was headed by an attorney or engineer. The appointment of an economist as chief, in the spring of 1978, signaled a change of focus.

The Commission's traditional style of decision making is best seen in the relatively non-technical context of broadcast license renewals. Stations supposedly hold their licenses as a public trust and must renew them every three years. On the one hand there is a history, amounting almost to law, that stations are renewed regardless of how closely they have fulfilled their promises made at the previous renewal about public service broadcasting, attentiveness to community needs, and other obligations. [23] Most renewals are granted routinely by staff using delegated authority. On the other hand, renewal can be capricious. When community groups or rival stations file a petition to deny a renewal application,

the commissioners must personally rule on the petition, sometimes after a hearing to determine fact. In such decisions, the
Commission rules with discretion, as a judge would. The cases
may be used for unexpected decisions that may represent departures in policy.

In 1961, in the KORD, Inc. decision, the FCC announced that
stations would henceforth have to take their public obligations
seriously or have their renewals denied. [24] In 1969, the Commission denied the application of WHDH-TV in Boston in favor
of another applicant on grounds that this would diversify ownership of communications facilities in the area. [25] In 1971, the
Commission rejected a petition to deny the renewal of WMAL-
TV, Washington, D.C., on grounds of employment discrimination, but propounded a "zone of reasonableness" doctrine that
has been evolving ever since. [26] All of these decisions sent
shock waves through the industry, but none has altered the
practice of granting most renewals routinely.

Another force for discretion is the power of the courts to review FCC decisions. Occasionally the Appeals Court has ordered
the FCC to hold a hearing on an application when community
groups have lodged a petition to deny renewal. [27] The courts
will sometimes use FCC cases to launch major policy departures
of their own. For example, when the Appeals Court reviewed
the WLBT-TV, Jackson, Mississippi, decision in 1966, it formally
extended legal standing to contest renewals to community groups
with no economic interest in the outcome other than that of a
consumer. This decision opened the doors to many public interest challenges to renewals and, in a broader sense, fundamentally widened the range of interests represented in FCC decisions.[28]

The essence of the renewal process is that the FCC has never
reduced it to rules, as it would if rational norms were opera-
tive. The Commission has never been able to quantify the standards that stations must meet for renewal nor established the
types of cases it will personally review. [29] An attempt to
do so after the WHDH case, under Congressional pressure, was
overturned by the courts. [30] Rules for review proposed internally by the Broadcast Bureau were accepted by the commissioners in 1973, then ignored.[31] Only after considerable

public pressure were internal guidelines used by the staff in reviewing EEO compliance made public. [32] The Commission wants to preserve its discretion to respond to external pressure or personal whim, and this is sanctioned by traditional norms of authority.

The prevalence of traditional norms at the FCC is not surprising About half the commissioners and much of the staff have routinely been lawyers, a background attuned to precedent, adversary proceedings, and judicial discretion as a basis for policy making. [33] If, as Model 1 argues, agency action will be driven toward actions which individual commissioners can justify in terms of their dominant values, traditional norms will color much of what the FCC is and does.

2.1.3   Charismatic Authority

The third set of values useful for a Model 1 analysis are those of the public interest advocates. They assert that the FCC should follow the popular will in its decisions and should serve the public in ways not confined by legal precedent or economic calculation.

We would call these norms populist. They appeal in a general way to what Weber called charismatic authority. The charismatic commands allegiance through the display of remarkable personal gifts rooted in a sense of higher calling. Weber himself speaks of charisma in connection with religious prophets. We use the term here in a looser sense to describe any political leader who commands authority more on the basis of his personal qualities and his emotional appeal to "the people" than on his specific actions or policies. [34]

The commissioners are driven toward a populist style by the collegial nature of the FCC. No one of them can make decisions without the support of the others. Each has to build a personal constituency inside and outside the FCC if he or she wants to lead. This is true even of the chairman, although he controls the agenda and manages the routine administration of the Commission. The chairman's power, like that of the

President within the larger political system, is mainly the power of persuasion. In large part, his authority is merely personal, and potentially charismatic.

Commissioners easily see their job as political. Most have been associated with politics before going to the FCC. For some, to appeal over the head of industry to the people is second nature. To do this gives them personal identity in an otherwise collective leadership. Politically, charisma has the advantage that it centers authority in the individual, not in bureaucratic or judicial structure.

The populist appeal is also endorsed by the larger political system. Democratic political theory supports the idea that the meeting of the leader and the led yields a more basic form of authority than bureaucratic or judicial institutions. If commissioners seek actions they can justify publicly, as Model 1 predicts, democratic values will tend to drive them toward populist behavior.

Populist norms have had increasing influence over FCC action in recent years. From the 1960's on, the FCC has been under growing pressure from public interest lawyers, community groups, and popular forces generally to be more and more responsive. Individual chairmen and commissioners have tried to respond by making direct personal appeals to the public.

The clearest example is Nicholas Johnson, a highly controversial activist lawyer who served on the Commission from 1966 to 1973. Johnson was viewed by the industry as a subversive and a zealot. He made himself spokesman for the public interest advocates and minority and community groups who were seeking to make broadcasting and FCC regulation more accountable to the public. He actively encouraged these elements to take the public service obligations of broadcasters seriously and to challenge the license renewals of unresponsive stations. [35] While still a commissioner, he wrote a book on how to mobilize popular opinion against the broadcasting establishment. [36] After he left the Commission, he became head of the National Citizens Committee for Broadcasting, one of the major public interest groups involved in FCC proceedings.

Johnson was typically charismatic in his isolation from established institutions. He sat on the Commission only in order to reform it. He did not enjoy or seek great influence among the FCC staff, and he refused to engage in political compromise with the industry. His influence among the other commissioners and outside groups was real, but it rested almost entirely on his personal ability, integrity, and eloquence — that is, on charismatic qualities.

Richard Wiley, chairman in 1974-77, was also charismatic, but he used his ability to master the institutions rather than attack them. His unusual energy and control of detail gave him, perhaps, more power over the Commission staff than any other chairman. Because of his political ability, he usually had the support of the other commissioners and even the regulated industries. For instance, he encouraged the networks and the NAB Television Code to institute the Family Viewing Hour, as a way of limiting violence on television, without having to make it mandatory.

Some chairmen have made explicitly populist appeals to the people as a way of balancing the dominant power of the communications industry. Newton Minow, chairman 1961-63, dealt with what he called the "hostile environment" of his position by articulating the resentments of people against broadcasting. His "vast wasteland" speech voiced the disillusionment of many with the commercial banality of television and has provided a slogan for critics ever since. [37] Wiley instituted a series of regional open meetings throughout the country in 1974-76, to hear the grievances of people firsthand.

Populist or charismatic authority appeals to the feelings of ordinary people, not the support of established interests. Despite their own influence over the Commission, broadcasters often perceive the FCC as a radical juggernaut which reifies every popular resentment toward broadcasting into some coercive policy. [38] And, indeed, many of the Commission's more important recent decisions affecting broadcasting have sprung directly from perceived popular distaste for excessive sex, violence, and commercialism on television. The 1974

Policy Statement on Children's Television expressed the Commission's intention to limit commercialism in children's programming. [39] In its 1975 Pacific Foundation WBAI or "seven dirty words" decision, the FCC sought to restrict the use of obscene language in broadcasting. [40] And in 1976, Chairman Wiley facilitated the Family Viewing Hour with the networks in order to exclude gratuitous violence from the early portion of prime time television.

However, charismatic authority also tends to be unmeasured. When leader and followers connect in a direct, populist way, the one tends to promise and the other to demand everything they feel. The agenda often goes beyond what is allowed by other forms of authority--specifically, by law or tradition. Conflicts can arise. Popular resentments toward broadcasters tend to demand of the FCC a power to censor programming which, in fact, the Communications Act and the First Amendment do not allow. [41] Hence, for example, the Children's Television policy seems to commit the Commission to programming policies that it could never enforce. Initially, the WBAI decision was struck down in appellate court as an act of censorship; the Supreme Court, however, overturned the lower court's decision. [42] The Family Viewing Hour also has been disallowed because Wiley did not observe normal due process in reaching agreement with the networks (this decision has been appealed). [43]

Other times, popular decisions offend not law but reason. That is, they are not expert in the terms of rational authority. They lead to unfortunate consequences. This happens because charismatic authority resists restraint by the norms of economic rationality. [44] The prime time access rule, 1970, limited network programming to three hours nightly in an attempt to make more evening time available to community-oriented local programs. [45] In fact, such programming was not forthcoming, and the dominance of the networks probably increased. [46]

Although Chairman Burch dissented from this ruling, he pushed through another with a similar rationale. The cable television rules of 1972 were supposed to foster CATV development. Cable promised to diversify control over broadcasting and had become a symbol for those seeking to break the power of the networks over television. The rules met fierce industry resistance. However, cable has not developed as rapidly as expected. The reasons, besides the recent economic recession, may include the explicit public service obligations which were initially laid on cable in order to win the support of Commissioners Johnson and H. Rex Lee.[47]

Charismatic values are in some ways the most—and in other ways the least—usable norms for a Model 1 analysis of the FCC. No others trace political action so clearly to individual actors rather than to institutions. On the other hand no others make action so unpredictable. The populist leader must follow the popular whim of the moment. Because he has little connection to structure, his actions do not follow clear patterns, as do those of the expert civil servant or judge. Hence, it is difficult to state hypotheses about what action will occur.

2.1.4   Summary

To sum up, the single actor model has obvious explanatory power for the FCC. It is true that Commission decisions proceed importantly from the mind of the individual commissioners. Policy is clearly affected by the individuals who happen to sit on the Commission at a given time. [48] It is also true that commissioners approach their decisions in part as intellectual and moral problems. Options which can be rationalized in terms of the dominant values of the Commission do clearly have an advantage in the decision process.

A Model 1 view would trace the complexity of FCC behavior, not so much to the organizational and political influences considered below, as to the complexity of the Commission's norms. The values of economists and engineers, lawyers and judges, and populist leaders and followers, are all available and, to some degree, in tension. Any or all the norms may play the rationalizing role that is central to the Model 1 explanation of official behavior. It seems that different norms tend to come to the fore for different kinds of decisions. Rational norms are

dominant for heavily technical decisions (spectrum management, common carrier); traditional-legal ones for broadcast
license renewal; and charismatic ones for broadcasting policy
decisions.

Methodologically, Model 1 has the advantage that the analyst's
imagination can do much of his research for him. Since action
is attributed to individual official's rationalizing their behavior,
the student of the FCC need only imagine how the effort to
maximize relevant values would have led him to the action in
question. There is no need to pick apart the "black box" of the
FCC to understand its intricacies; one can intuit its workings
from the outside.

However, this advantage implies a danger. With Model 1, the
analyst seeks understanding by projecting his own reasonings
on the phenomenon he wants to explain. What assurance can
he have that the commissioners really acted for the apparent
reasons, even if they claim these reasons publicly? There may
be a plausible *correlation* between the apparent reasoning and
the agency action, but how can we know that the connection
is *causal*? [49]

To know this, there is no substitute for on-the-spot research
inside the agency to see how it actually functions — the kind
of research we noted in our introduction to be lacking. The
moment we look at actual decisions close up, it is clear that
individual officials rationalizing their actions do not dictate
policy by themselves. Bureaucratic and political factors not
discussed in Model 1 now come to the fore. These influences
are emphasized, respectively, in our second and third analytic
models.

2.2 MODEL 2: ORGANIZATIONAL PROCESS

A second approach to understanding the FCC is to look at
it as an organization. Bureaucracy came into the picture often under Model 1, but not as an independent influence. Rather, official action was attributed to the reasoning of individual officials, and organization was merely the place where

they worked. In this second approach, FCC decisions are attributed to the influence of organization *as such*. This model argues that organization has inherent strengths and weaknesses which affect decisions quite apart from the choices of individual actors. Often, this model contends, policy is not *chosen* at all. It is simply an *output* of a bureaucratic process with dynamics of its own.

According to theories of public administration, bureaucracy should have no influence of its own. Civil servants are supposed to carry out laws and policies with perfect obedience and efficiency. Experience clearly tells us they do not. Organization theory is an attempt to formulate why not. Major concepts drawn from organization theory generate Model 2's hypotheses about what causes FCC behavior.

2.2.1 The Formal View of Bureaucracy

In the formal view, bureaucracy behaves according to the norms of rational authority already considered above. In Weber's theory of bureaucracy, an organization of officials, just like any one individual official, is accountable to two standards: (1) obedience to policy embodied in laws and the decisions of higher officials, and (2) expertise in serving the public interest where discretion is allowed. Bureaucrats follow rules laid down for them by higher authority, and they are trained, recruited, and promoted on the basis of merit. As Weber puts it, bureaucracy means "A continuous organization of official functions bound by rules" and "the exercise of control on the basis of knowledge." [50]

Weber adds a third norm for bureaucracy peculiar to it as an organization which is very significant. It is that the officials should have no independence of higher authority. In Weber's terms, they must not be able to "appropriate" their positions. Officials must not have a right to their jobs or anything they work with. Just as a modern economy depends, in Weber's view, on the total expropriation of the worker by the capitalist, so modern government depends on the expropriation of

office holders by the state. Political modernization means the
end of rule by semi-autonomous potentates, as under feudal-
ism. Civil servants then become totally dependent on their
superiors. The superiors can hire, fire, promote them, and
give them orders at pleasure, subject only to rationalist norms
that law and efficiency be served. [51]

Because of its obedience to rational norms and its inability to
resist form within, Weber's bureaucracy has no identity of its
own. It is the shadow of the Model 1 policy maker acting on
rational values. It has the faceless and efficient character of a
machine:

> "Precision, speed, unambiguity, . . . continuity, discre-
> tion, unity, strict subordination, reduction of friction
> and of material and personal costs — these are raised to
> the optimum point in the strictly bureaucratic admin-
> istration. . ." [52]

The formal theory might be called the "organization chart view of
bureaucracy." If reality followed the theory, the FCC would function exactly the way it appears on the Organization Chart (see page
60). On paper, the commissioners sit over the entire FCC. All are
equal in voting on decisions, but the chairman has the main administrative authority over the staff. The bureaucracy consists of four
bureaus covering the main areas of Commission responsibility
(Broadcast, Common Carrier, Cable Television, and Safety and Special Radio Services), plus the Field Operations Bureau and seven
offices with cross-cutting functions (General Counsel, Plans and
Policy, Opinions and Review, Review Board, Administrative Law
Judges, Chief Engineer, and Executive Director). The distinction
between the bureaus and offices is close to that between "line"
and "staff."

In theory, communication lines among the units should be fairly
simple, much as they appear on the chart. The bureaus and offices
are functionally differentiated. The main overlaps should occur
between "line" and "staff" offices dealing with the same issue.

Most communication should be, not among units, but between them and the commissioners whom they advise on decisions and who in turn tell them what to do.

In theory, communication between the commissioners and staff is mostly about rules and the making of rules. Regulations are essential to formal bureaucracy. While FCC Rules are meant, of course, to govern the industry, the FCC also needs them to govern itself. Through regulations, superiors tell subordinates how to make decisions. The rules tell officials what information they are to consider and what standards to apply in making judgments. [53] Guidance is essential when staff have to make thousands of routine decisions each year. For example, over 100,000 complaints are received annually about radio or television interference alone. Formalism based on rules is the only way to avoid arbitrary or inconsistent decisions. [54] When there are no clear rules, administration is not properly bureaucratic. Decisions are governed instead by usage and discretion — the norms of traditional authority.

The FCC has unusual latitude to make rules because of its quasi-legislative mandate under the Communications Act. The Commission's rules occupied 734 pages in 1956 and 1347 pages in 1976, a growth of 84% in 20 years.

The Commission also has rules for the making of rules. The Administrative Procedure and Judicial Review statutes prescribe in detail how to go about rulemaking. [55] The Commission issues a Notice of Inquiry (NOI) to seek expert testimony from affected interests on a policy issue. It issues a Notice of Proposed Rule Making (NPRM) when it seeks comment on a proposed new regulation. After reviewing the responses, the Commission may, in the case of an NOI, issue an NPRM, setting forth its response to comments, its own view of the policy issues, and proposed rules. In the case of an NPRM, the Commission will issue a Report and Order which sets out the final

Federal Communications Commission
Organization Chart   November 1977

Office of Plans & Policy

Office of Opinions & Review

Office of General Counsel

Administrative Rules &
  Procedures Div.
Research &
  Trial Div.
Legislation Div.
Litigation Div.
Public Access Div.

Industry Equal Employment
  Opportunity Unit

Office of Chief Engineer

International &
  Operations Div.
Laboratory Div.
Researchs &
  Standards Div.
Spectrum Allocations Div.

Planning & Coordination Staff

Broadcast Bureau

Broadcast Facilities Div.
Complaints &
  Compliance Div.
Hearing Div.
License Div.
Office of Network Study
Policy and Rules Div.
Renewal & Transfer Div.

Cable Television Bureau

Certificates of
  Compliance Div.
Policy Review &
  Development Div.
Research Div.
Special Relief &
  Microwave Div.
Records & Systems
  Management Div.

The Commissioners

CHARLES D. FERRIS   CHAIRMAN

ROBERT E. LEE   JAMES H. QUELLO
ABBOTT M. WASHBURN   JOSEPH R. FOGARTY
MARGITA E. WHITE   TYRONE BROWN

Review Board

Office of Administrative
Law Judges

Office of Executive Director

Administrative Services Div.
Consumer Assistance Office
Data Automation Div.
Emergency Communications Div.
Financial Management Div.
Internal Review & Security Div.
Management Systems Div.
Personnel Div.
Procurement Div.
Public Information Officer
Records Management Div.
The Secretary

Field Operations Bureau

Enforcement Div.
Engineering Div.
Regional Div.
Violations Div.

Field Installations

Common Carrier Bureau

Accounting & Audits Div.
Economics Div.
Facilities & Services Div.
Hearing Div.
Mobile Services Div.
Policy & Rules Div.
Tariff Div.

Compliance & Litigation
  Task Force

International Programs Staff

Program Evaluation Staff

Field Office

Safety and Special Radio
Services Bureau

Aviation & Marine Div.
Industrial & Public
  Safety Facilities Div.
Industrial & Public
  Safety Rules Div.
Land Mobile Spectrum
  Management Div.
Legal Advisory &
  Enforcement Div.
Personal Radio Div.

Regional Branch

rules to be instituted. Parties opposing the decision may petition the Commission to reconsider on grounds that it lacked necessary information or that new information has come to light.

In 1974, for example, the FCC faced an overload on the existing 23 Citizens Band radio channels because of the exploding popularity of CB. The Commission issued a Notice of Proposed Rule Making to increase the number of CB channels to 70, tighten technical standards for CB equipment, and simplify CB operating procedures. After comments, the Commission issued a First Report and Order adopting many of the technical and operating standards but postponing the more sensitive issue of new channels. Then in March, 1976, it issued a combined Notice of Inquiry and Further Notice of Proposed Rule Making in which additional technical issues were raised and a proposal for 40 rather than 70 channels was made. After comments were evaluated, a Second Report and Order was issued adopting these proposals. Petitons to reconsider the decision were filed by the Association of Maximum Service Telecasters, Inc., and the American Broadcasting Companies. Although the Commission did not alter its position on the technical standards or the number of new channels, it did amend its Second Report and Order to specify the date beyond which the manufacture and sale of 23-channel CB sets would be prohibited. [56]

While these procedures emerge from a judicial tradition, as we saw earlier, they also provide needed structure for a bureaucratic staff. They tell the large number of people involved how to go about making policy. The Notice of Inquiry and Notice of Proposed Rulemaking serve to focus the rulemaking process for the FCC as well as outside parties. The staff prepares all Commission documents at the direction of the Commission and summarizes the comments received from outside parties.

Other procedures, purely internal to the agency, govern consultation among bureaus on issues of common concern. One unit, for instance, will be given the "lead" in preparing a decision to be recommended to the Commission, but the document, called "an agenda item," will be routed to the others for concurrence before going to the commissioners. In the CB case, the Safety and Special Radio Services Bureau initiated and guided the rulemaking, but the Office of the Chief Engineer was consulted on technical standards, the Field Operations Bureau on enforcement, the Broadcast Bureau about possible television interference, the Office of Plans and Policy because of the long-range planning implications, and the Office of the Executive Director because of the added personnel and funds necessary to implement the decision.

Rules dealing with substance and procedure work to coordinate officials by limiting the options they may consider in decisions. The rules exclude some options as improper or impolitic and steer action toward others. In the CB case, standard routines for rulemaking caused policy to be made one way and not another, and the new regulations in turn set norms for subsequent regulation of CB radio. Rules define the common patterns of behavior in terms of which officials are organized. In this sense, the FCC as an organization is literally defined by its rules. [57]

The formal norms of bureaucracy clearly influence the FCC's structure and functioning. A Model 2 analysis of the FCC, which attributes outcomes to organizational process, must clearly pay attention to them. Yet the formal view of bureaucracy, like the ideal of rational authority, is a norm never fully realized. Organizations may be driven toward it by efforts to manage more efficiently, but they never fully attain it. As every citizen knows, the Post Office is simply *not* as orderly, obedient, and efficient as theories of public administration say it could and should be.

Nor is the FCC. The economics and reform literature on the Commission expresses great impatience with it. Why *is* it so reluctant to do its job? Why is it so suspicious of new technology in communications? Why is it so heavily influenced by the industry? The Commission was set up to solve regulatory problems on a rational basis. Why so often does it fail to do so? [58]

2.2.2   Bounded Rationality

Much of modern organization theory is an effort to explain
why formal norms of bureaucracy are never fully achieved.
The dominant school, that of Herbert Simon, suggests that the
formal view of bureaucracy is utopian. Officials could not live
up to it even if they wanted to. The formal view tends to assume that perfect rationality is possible. The norm of perfect
adherence to law and policy would be feasible, however, only
if superiors could clearly communicate policy downward and
verify compliance, and if subordinates could clearly apprehend
it . The norm of efficiency or expertise would be feasible only
if officials could communicate the necessary information to
each other and craft optimal solutions to every problem. In
fact, they cannot meet either norm because of fundamental
limits to their ability to absorb information. Human reason
cannot integrate an unlimited number of facts or variables. Its
reach extends to a certain horizon and stops. It can only proceed further through rules, habits, or other conventions that
simplify reality. In Simon's term, it is "bounded." [59]

Control Problems.   Cognitive limits place severe curbs on the
ability of bureaucratic superiors to control what goes on in
their organization. They are few, and their staff are many. It
is mentally impossible for them to direct or monitor what
staff members do in detail. The staff, for their part, are absorbed in the organization. They are immersed in the countless interactions with other staff and offices required to solve
daily problems. Communication lines are not primarily vertical, to and from superiors and the Commission, as the organization chart shows, but heavily horizontal among bureaucratic
units; and this is more true the more the issues are technical
and non-routine.[60] For the staff, directions from the Commission are merely one input among many. Even with the best
will in the world, it is hard for them to *hear* what the commissioners want, just as it is hard for the commissioners to speak
clearly to them.

The control problem gets worse the further down in the organization one goes. The commissioners can usually get the cooperation of the bureau chiefs and other senior staff who directly

advise them, since many of these are political appointees directly accountable to them, and all tend to be highly dedicated. More junior officials, however, must be supervised and monitored through intervening bureaucratic layers. At each level orders are subject to distortion and evasion. There is "authority leakage" as orders come down the line. Leaders often have the sense that their directives disappear into an abyss, without any effect on the organization. Lower-level officials, who chiefly implement Commission decisions, may spend very little of their time actually doing what the Commission wants. Bureaucratic sub-units may drift apart and follow policies of their own. [61]

As an example, some notices of rulemakings take the staff months to prepare — even when the commissioners have expressed great interest in them. Commission decisions on personnel, budget, and administrative matters are often delayed or not implemented until another alternative can be "sold" to a majority of the commissioners. For example, Chairman Wiley publicly announced in 1976 that the Commission planned to produce a CB handbook to explain CB rules in layman's language to literally millions of new CB-ers. But at the time, the staff had already decided to rewrite the CB rules into plain English, eliminating the need for an explanatory handbook. Although periodically the chairman would inquire about the handbook, the lower echelon staff never took it as a serious assignment. In 1978, the Commission adopted the plain English version of the CB rules; it never issued a handbook. [62]

For Model 2, the control problem suggests the hypothesis that some FCC actions may not have been chosen by the Commission at all. Rather, they may result from decisions by individual employees or bureaus whom the Commission does not fully control. Some decisions by staff may become known to the commissioners and hence controllable only when they are elevated to public view by the press or Congress. The leadership of the FCC, like other agencies, spends a lot of time and energy just reacting to actions by their own personnel.

The publicity surrounding a "bootlegged" copy of the Roberts
report on children's television shows how press coverage of
actions (in this case recommendations) can compel Commission response. Roberts was the head of the task force on
children's television which Chairman Burch set up in 1971.
Less than two weeks after Richard Wiley became chairman in
1973, Jack Anderson ran a column about Roberts' recommendations to curtail commercials on children's programs, among
other proposals. Cole and Oettinger point out that the "ex-
plosive" details in the report were in fact known to members
of Congress and had been discussed in the trade press; it was
the forceful publicity given to the report that demanded a
public response. The column appeared in the *Washington
Post* the morning that the commissioners appeared before the
Senate Subcommittee on Communications for their annual
oversight hearings, compelling them to answer questions about
it. [63]

Such examples are not unusual, although most often the publicity occurs in the trade press. In fact, internal memos between
bureaus, and from bureaus to commissioners, are often written
with the full expectation that they will be highlighted in the
trade press.

Information Overload.   Related to the control problem is information overload. Because the commissioners are overbur-
dened, they cannot give full attention to every decision they
face. The weekly Commission agenda may include up to fifty
or more separate items. A single, day-long Commission meeting
may take up 25 or more of these--many of them highly techni-
cal--for which commissioners have had less than a week to prepare. [64] The press of business means that FCC staff have to
compete with each other to get the time and attention of the
Commission. Sometimes they cannot break through the log
jam with urgent information. Hence, the Commission may be
unadvised about the full implications of a decision or the fact
that resources are unavailable to implement it. The basic problem is that the FCC, like other agencies, is hierarchical, and
coordination depends on the action of a handful of leaders
at the top. [65]

Because of the amount of information to be processed, commissioners are heavily dependent upon the staff. They cannot attend to all the information coming up from below. What items staff members decide to show them heavily influences what they do. According to Cole and Oettinger, the staff usually give the Commission only one recommendation on each issue, rather than offering a range of options. When faced with an "option paper," commissioners will inevitably ask the staff to recommend one course of action. They simply have no time to do otherwise. Obviously, this gives the staff great power over decisions. [66]

The bureau chiefs are the most powerful gatekeepers of information. It may be that junior staff have knowledge or opinions that differ from those of the chief, but a bureau chief can keep them from getting through to the Commission. For example, the head of the Safety and Special Radio Services Bureau wanted to discontinue annual equipment measurements for land mobile transmitters. His staff wrote him memos opposing the idea--in part because parties who filed comments in the case had opposed it. The bureau chief, however, held to his view and finally produced a document arguing for discontinuation. With the staff sitting quietly in the meeting, he presented his view to the commissioners and they accepted it. [67] Accordingly, when Chairman Wiley shortly after appeared before the Association of Public-Safety Communications Officers (APCO), which had opposed the change, he was greatly surprised by the industry's opposition.

From the information overload problem Model 2 derives the hypothesis that FCC actions may result, not from premeditation, but from the problems or facts the commissioners happen to hear about from the staff below.

The more salient an issue is politically, the more likely the commissioners are to hear about it from outside sources, escape control by staff, define their own options, and make the decision themselves. The cable television rules of 1972, for example, were hammered out by the commissioners among themselves because of heavy pressures from the affected industries, Congress, and the Office of Telecommunications Policy. [68] On the other hand, the more technical and obscure the issue, the less likely the Commission is to act on its own and the more likely to defer to staff. In 1971, the FCC issued new rules authorizing competition with AT&T in the specialized common carrier field almost entirely because the Commission's Common Carrier Bureau recommended this course. The decision was a major policy departure but attracted little public attention at the time, and, therefore, little personal attention from the commissioners. [69]

Satisficing. Another consequence of bounded rationality is that officials do not seek perfectly optimal solutions to every problem, as the rationalist ideal prescribes. To do this would require processing too much information too often. Instead, they "satisfice;" that is, they seek solutions which are not perfect but known to be satisfactory, either because they can easily be changed later or because they build on the agency's established ways of doing things. [70]

The FCC often satisfices by making short term rather than long term decisions. The tentative decision can be revised in the light of experience. To make a final decision would require calculating all of the consequences in advance. The incremental approach allows the consideration of consequences to be spread over time, not concentrated in advance of the decision. [71]

In 1974, the Commission issued a policy statement on children's television based on a voluntary agreement with the broadcasters to restrict commercials in children's programming, but postponed promulgating the policy as rules. In 1970 the Commission reallocated the 806-960 MHz band to land mobile radio but postponed the issue of what to do with existing television translators operating in this band. (Translators are used to transmit television signals beyond the primary service areas of a station.) Eventually the 700 or so translators already in the band were "grandfathered" — allowed to remain there — while all new translators were assigned to channels below 806 MHz. [72]

In 1976, the FCC voted to permit "closed captioning" of television programs for the benefit of the hearing -impaired. ("Closed captioning" is a technique for transmitting caption signals, to be decoded by translators in special TV sets purchased by the hearing-impaired. This decision is sometimes referred to as the "Line 21 Decision" because the coded signal uses Line 21 of the television screen.) There was considerable pressure from the deaf community to require captioning, but the Commission did not go beyond permitting it.[73] Chairman Wiley told representatives from various organizations for the deaf that the Commission could not go further, and the obligation was now on them to lobby broadcasters to provide captioning. Obviously, the decision could be revised if the organizations convinced the commissioners it was insufficient.

A second way to satisfice is to base decisions on past policies which have proved workable. A fresh solution is not calculated for every new problem. Rather, old solutions are adapted incrementally. One reason is that the organization already has heavy investments in established ways of doing things. To change policy for every new issue would involve not only information costs but the learning and psychological costs of shifting the organization to new procedures. [74]

Bureaucratic inertia may explain why some FCC functions seem to go on as they always have regardless of policy changes by the Commission. The Broadcast Bureau has consistently followed a policy of renewing virtually all broadcast licenses, despite occasional Commission decisions appearing to hold stations more accountable for meeting their public service obligations. [75] The Common Carrier Bureau still depends on AT&T for much of the data and analysis necessary to set common carrier rates, even though advancing technology and the advent of competition in specialized common carrier create a greater need for analytic capability within the FCC. [76] These bureaus simply have established ways of doing things that are not likely to change quickly.

The commissioners are each appointed for seven years and serve an average of three. They confront an organization where standard operating procedures are already established. Bureaucratic

inertia favors the power of the career officials, who are permanent, over that of the commissioners, who come and go. Lee
Loevinger, a former FCC commissioner, has written:

> "The middle management of career executives
> know that the announced policies will change as
> top political appointees change, but that the rou-
> tine work of the bureaucracy will go on regardless. . .
> So middle management makes minimum conces-
> sions to policy pronouncements by top manage-
> ment, secure in the knowledge that before any
> very significant change in policy can be fully im-
> plemented it will be changed as a result of a polit-
> ical change in top management." [77]

One career executive, particularly secure through his expertise
and tenure, simply refused to change an internal interpretation
of the the definition of "broadcasting." He retorted to an an-
gered Commissioner Wells, "I was here before you arrived and
I'll be here when you are gone." Commissioner Wells, in fact,
left the Commission a few months later.

At the heart of a Model 2 analysis of the FCC is the idea that
organizational routine seriously constrains the options policy
makers have available for decisions. The bureaucracy is not
infinitely flexible and able to do whatever leaders want, as the
formal view tends to assume. The leaders can only choose among,
and perhaps incrementally adjust, the routines the organization
already knows. The computer is already programmed, so to
speak, and the decision makers can only choose among the programs. [78]

As a result, policy change and bureaucratic change often have
to go hand in hand. The bureaucracy tends to resist innovation.
If the policy maker means to change policy, he has to fight
bureaucratic battles. Sometimes, the agency must be reorganized so that the bureaucratic mold is broken and fresh thinking
can occur. If new options are to be available for decisions, new
units must be created to represent them in the bureaucratic
policy process.

On occasion, the FCC has had to create task forces to explore new policy areas because the established structures were inhospitable. For example, in 1966, under pressure from cable television interests, the Commission established a Cable Television Task Force to look into ways to expedite the development of cable. Previously, responsibility for cable had been lodged in the Broadcast Bureau, which looked upon it as a competitor of broadcasting and tended to stifle its growth. Not by coincidence, when Burch saw the need for new cable rules in 1970, the task force was elevated to bureau status and played the major staff role leading to the 1972 cable regulations. Correspondingly, the Broadcast Bureau played a lesser role than before.[79] In 1971, the Broadcast Bureau was hostile to the idea of an inquiry into children's television. Therefore a task force on the issue was lodged officially in Broadcast but actually attached directly to Chairman Burch's office; it provided the staff work for the 1974 policy statement on children's television.[80]

2.2.3   Social Psychology

The control and information problems of bureaucracy and its satisficing behavior all result ultimately from the limited rationality of officials. But this analysis still assumes that bureaucrats are at least trying to satisfy the formal norms for bureaucracy. The more radical critique, stemming from social psychology, suggests that officials are not trying to be obedient or efficient at all. They are serving social purposes of their own.

This reasoning extends the idea of bureau routines mentioned above. Organizations develop not only standard operating procedures but distinctive "subcultures." These attitudes not only justify standard procedures but come to constitute an entire world view for staff members. In order to operate efficiently, bureaus need an internal consensus on goals. But consensus often becomes as end in itself.[81] The ethos deepens the more the bureau is isolated from a need to change and the more it ages.[82] Eventually, the culture can condition officials to the point where they are positively incapable of dealing rationally with new challenges.[83]

Political appointees who come into an agency at the top discover that they cannot change the bureaucratic subculture, only use it or avoid it.[84] The FCC commissioners sidestepped the entrenched attitude of the Broadcast Bureau by setting up the Cable and Children's Television Task Forces, mentioned above. For more incremental policy changes, they have tried to win over the relevant bureau and encourage it to implement the new policy with the same enthusiasm as the old.

For example, the Commission has tried to persuade the Field Operations Bureau to bolster its public service activities in the field, due to the rapid increase in numbers of citizens band radio operators (from 1 million in 1974 to over 13 million in the spring of 1978) and the national prominence of consumerism. The engineers, who comprise most of the bureau, would much prefer to continue routine tasks of enforcement and licensing. However, recent chairmen have produced gradual change by allocating the bureau more funds for public outreach and publicly praising bureau performance.

Commissioners have also tried to change the Broadcast Bureau's attitude toward petitions to deny broadcast license renewals. Cole and Oettinger report that the Broadcast staff will not generally investigate renewal applications that look weak unless a petition to deny prompts them to. They also usually advise the Commission to deny or set aside such petitions. Commissioners Johnson and Hooks have advocated a more critical attitude by staff, with limited success. Hooks argued that ". . .[The Commission] should not stand behind a procedural barrier on the apparent side of a licensee and let the matter ride simply because the complainant is without the . . . resources or legal acumen to mount a perfect attack." The renewal staff simply does not perceive itself as an investigative arm of the Commission. They do not have the instincts or habits of a prosecutor. [85]

The more basic point is that organizations cannot avoid having to adjust to their members as people. Individuals seek not only to perform well on the job but to find generalized meaning for their lives. They need and want the emotional security afforded by the bureau ideology. They seek to round out their jobs, however narrowly the organization may define them, so that their lives are personally satisfying. [86] They want to use the authority of their positions to participate in the decision-making process. The formal structures of bureaucracy must continually adjust to the social needs and processes of its members. The adjustment may lead to new problems, new structures, then new adjustments, in a continual interplay. In principle, organizations simply *cannot* be as rigid and static as the formal norm supposes. [87]

In theory, staff members are supposed to be subservient to the organization. In practice, the agency must negotiate implicitly for their support. Employees have their own reasons for seeking to work in the organization and cooperate with others. These objectives must be met, or they will be satisfied in ways damaging to the agency mission. [88] Employee concerns have to do with prestige as well as pay. Amenities such as the color of offices, the size of desks, or whether the floor is carpeted can become symbols of esteem that affect people's performance. Such needs are probably more pressing today than in the past because people have become more individualized and may be less well socialized into organizations than previously.[89]

Under Model 1, it was argued that individuals may exercise *authority* at the FCC through charismatic personal qualities. Under Model 3 below, we will argue that they may exercise *power* if they are major stakeholders in the game of bureaucratic politics. Here, the argument is that individuals influence the organization at all levels in a much more mundane sense. Tolstoy argued in his novels that history is determined, not by the leaders known to history, but by the countless actions of myriad ordinary people just living their lives. The same could be said of policy at the FCC. The actions of the Commission are affected significantly by who happens to occupy a given office at a given time — and even how they feel on a given day. Im-

portant functions are often shifted between offices, every bu-
reaucrat knows, not for reasons of high policy (Model 1) or
bureaucratic organization (Model 2), but simply because of the
special competence — or incompetence — of individual staff
members.

One of the authors who works at the FCC could mention a
number of instances where a single, unsung employee used
unusual management ability, legal expertise, creative problem-solving, or negotiation skills to move ahead on a particular decision or program. Equally, individuals who lacked
the skills or attitudes required by their positions have undercut the Commission's performance quite seriously.

Of course, personal influences are random by definition. They
can only be told about through journalism or the gossip of in-
siders. They take no single direction; they could never be described in theory, which must appeal to general causes. For
the same reason, Model 2 says they can never be entirely controlled. To a degree, people are simply *disorderly,* and no amount
of organization can make them otherwise. The limits of organizational rationality lie in ". . . limitations on the power of individuals and groups to influence the actions of other individuals
and groups."[90]

2.2.4   Explicit Constraints

Social psychology says that employees do not, as even bounded
rationality assumes, seek to satisfy the formal norms of bureaucracy. But at least, only their nature as human and social animals
prevents them from doing so. A yet more pessimistic critique
would point to the many ways the very structure of the bureaucracy aids and abets disinterest in doing the job.

In Weber, again, bureaucracy requires that officials have no rights
to their jobs; they must be totally dependent on their superiors.
American bureaucracy fails to meet this standard in several re-
spects. The political culture has always been hostile to hierarchical
authority. In several ways, the dominant pluralism has denied bureaucratic managers effective authority over their staff.

Before the advent of civil service with the Pendleton Act of 1883,
American public administration at all levels of government was
staffed through patronage. Public jobs were viewed as "spoils"

to be distributed to the followers of the victorious party. This accorded with populist attitudes which were suspicious of governance by experts and in favor of rule by amateurs. [91] These attitudes live on in such politicized hiring practices as "political" appointments at the top of bureaus, the politicized nature of some civil service hiring criteria (preferences for veterans and handicapped), and the informal preferences for women and minorities legitimized under affirmative action programs.

Patronage compromised one of Weber's norms for bureaucrats — expertise — but sometimes served the other — obedience to higher authority. Officials obeyed the politicians who could fire them. When civil service systems arose after the Pendleton Act, expertise was enhanced to some extent, but obedience was undercut.

Civil service was supposed to eliminate patronage personnel practices and instead base hiring and promotions on "merit." In practice, personnel decisions have come to be based on a combination of seniority, test results, and experience ratings that have limited connection to on-the-job performance. The system has taken away most of the supervisors' discretion to hire and fire for their own reasons, and with it most of their power to motivate their staff to perform. While the Federal civil service is more meritocratic and less rigid than most state and local systems, [92] it does represent one of the major problems facing FCC managers.

This constraint has increasingly been compounded by another — public service unionism. Unions have organized much of local and state government bureaucracy and are making progress in Federal agencies. They tend to support civil service rules strongly just because they tend to shield employees from supervisory authority. Though unions rarely question merit selection and promotion in principle, they want these norms construed in ways that strongly favor seniority and restrict the discretion of managers. Increasingly, pay and work conditions for government employees are set wholly or partly through collective bargaining. [93]

Civil service is a reality at the FCC, and unionism has recently
come. An Employee Representatives Board has long existed to
bring employees' grievances to the attention of management. A
union attempt to organize the Commission failed in 1971. In
June, 1978, employees voted for the National Treasury Employees Union as their bargaining agent for non-supervisory personnel. Before, personnel grievances had taken collective form.
In 1977, the Broadcast Bureau license examiners caused a work
slowdown because they were not paid as much as other examiners doing, in their view, equivalent work. The current move to
unionize the Commission arose from widespread opposition to a
shift to later working hours decreed by Chairman Ferris in November, 1977.

All of these constraints have the effect of denying bureaucratic
superiors control over their staff. They give employees exactly
the entitlement to their jobs that they should not have in formal bureaucratic theory. The result is that in the FCC, as in
all government agencies, there are "deadwood" employees who
do little work and whom supervisors must simply work around.
The Carter Administration has recently proposed civil service
reforms to restore somewhat greater authority to managers.
Most public service unions have already expressed opposition.
This verifies the view of Weber, who well knew his model was
only an ideal type, that ". . . historical reality involves a continuous, though for the most part latent, conflict between chiefs
and their administrative staffs for appropriation and expropriation in relation to one another." [94]

However, a more important explicit constraint on FCC organization may be a more subtle one — the influence of professional groups. Of the FCC's 2127 staff in February, 1978, 321 (or
15 percent) were lawyers and 515 (25 percent) were engineers
or electronic technicians. Professional people bring to their work
qualities of intellectual and economic independence that may
be contrary to bureaucratic organization. Their ways of thinking
are molded to professional, not organizational, mores. Because
their skills are transferable, they can often leave bureaucratic
settings they find uncongenial. They tend to be a force for innovation in organization, not for order or stability. [95]

At the FCC, lawyers tend to argue for policy options that can be defended in an adversary setting in court, not those that might be effective or efficient in terms of practical consequences. In 1977, for instance, the staff recommended that an unauthorized broadcast antenna be ordered demolished and then rebuilt after a construction permit had been issued, in order to avoid an undesirable legal precedent. The four commissioners who were *not* lawyers carried the vote, 4-3, to let the antenna stand. The news reports read, "common sense won." [96]

Lawyers also insist on obeying the letter of the law. Recent changes in international law have required that every radio operator permit contain a photograph of the holder. Engineers in the Field Operations Bureau recommended that photos be required only for licenses likely to be used abroad — where international law would apply. The lawyers on the Commission could not bring themselves to support what would have been a minor breach of a treaty. They eventually yielded to staff arguments that to require photos on all licenses would inconvenience millions of operators and require funding the fifty FCC field offices to emboss the licenses. The new policy was adopted by consent, rather than in a public meeting, to minimize the publicity. [97]

Engineers, for their part, can be equally rigid on technical issues. Their professional ethic tells them to argue for the solution that is technically the most satisfactory, regardless of the economic, legal or political consequences. Strong arguments from the FCC engineers have sometimes helped push through important technical changes in the face of strong resistance. One example is the decision in 1945 to shift FM radio to a higher frequency in the spectrum. Another is the preparations now underway for the World Administrative Radio Conference due to begin in September 1979. WARC meets every 20 years to settle spectrum management issues on which nations must cooperate, such as which frequencies to use for international aircraft and shipping. FCC engineers are preparing position papers outlining optimal solutions to the technical problems, seemingly unable to recognize that some Third World nations may resist these proposals merely because they come from the United States. For an engineer, the idea that contingency positions should be prepared in deference to political realities is difficult to accept.

Of course, lawyers and engineers serve values — traditional and rational ones, respectively — which we found above to be deep-seated at the FCC. In this sense, these groups are very much part of the organization. The point here is that they are not subject to close *control* as bureaucratic theory says they should be. The ethos each group believes is part of the subculture of the FCC as an organization, not something the leadership has chosen on grounds of policy. A Model 2 analysis of FCC action assumes that it will be colored by these beliefs, whatever the commissioners choose.

In summary, the FCC bureaucracy is altogether less hierarchical, efficient, responsive, and cohesive than the formal theory of it suggests. Not only the inherent limits of organization but explicit constraints on bureaucratic authority limit the ability of the commissioners to obtain what they want from the staff. The FCC, like the rest of the Federal bureaucracy, is substantially a law unto itself. [98] Much of what the Commission does results in some way from this fact.

2.3 MODEL 3: POLITICAL PROCESS

According to Model 1, FCC actions can be understood as if they proceeded from officials acting as individuals. According to Model 2, bureaucracy, which in theory should be the invisible servant of the policy maker, in fact has influence of its own. Our third perspective explains FCC action by looking to political rivalry and compromise among the FCC's leaders within the Commission and its various political masters without. If Model 1 says action results from *individual* choice and Model 2 from *organization,* Model 3 says it results from *collective* choice among the *leaders* of organizations.

The last section used as a touchstone Weber's formal requirement that bureaucratic officials have no independence of their superiors. Organizational influences on decisions all result from breaching this condition in some way. The analogous requirement for Model

3 is that organizations should receive unambiguous political direction. Weber usually assumes that bureaucracy has only a single master. The only political, as opposed to career, official in a bureau is the minister who heads it, and he is the conduit for laws and orders emanating from the larger political system. [99] From him comes the clearcut political will that the bureau needs in order to satisfy the norm of obedience and, also, the norm of efficiency, since optimal means cannot be calculated until specific goals have been set as ends.

2.3.1   A Plural Executive

In public administration theory, unity of command is commonly achieved by placing a single executive at the head of an organization. The FCC has a seven-person plural executive. This is the first of several ways politics enters to make a unified political will unavailable to the agency. Weber says that collective leadership serves the political values of representativeness and extended deliberation before decision. But it compromises the executive's authority over staff and the precision and consistency of decisions. [100]

That assessment is valid for the FCC. A political process, as much as an intellectual or managerial one, must occur before the Commission can make policy. Important FCC decisions tend to be preceded by vigorous horse-trading among the commissioners. A classic example is the cable television rules of 1972. Chairman Burch wanted new regulations that would allow the cable industry to expand, but he could get the support of commissioners H. Rex Lee and Nicholas Johnson only by requiring that the cable operators provide access for educational programming and the general public. [101]

The whole trend of FCC decisions changed around 1969-70, when Republican appointments tipped the Commission majority in a conservative direction. The earlier liberal Democratic majority, symbolized if not led by Nicholas Johnson, had initiated populist, anti-industry actions such as the prime time access rule and rules limiting ownership of newspapers and broadcast stations in the same market by the same company. Though the new chairman, Dean Burch, carried through some of these changes, the Republican majority generally intervened less in broadcasting. [102]

This change helps explain why the Commission in 1969 granted
WHDH-TV to a competing applicant in order to diversify media
ownership in Boston — a landmark liberal decision — and then,
under Congressional pressure in 1970, issued a policy statement
on license renewals very favorable to established stations. (The
Court of Appeals later disallowed the policy.) [103]

When decisions are political compromises, they do not result from
choice in the Model 1 manner. The commissioners all participate
in the political process, but no one of them usually chooses the
outcome. The result often cannot be rationalized by the values of
any one of the players.

2.3.2 Bureaucratic Politics

Even if the FCC were headed by a single executive, however, outcomes would be swayed by bureaucratic politics at lower levels. We
saw above that policy can be influenced by the established mind-
set of bureaus within the FCC. Model 3 adds to this the expectation that units will actively compete for power in a political process. In their drive for security and autonomy, the units will compete more vigorously for "turf" than is rational for the organization as a whole. [104] The competition of the Broadcast Bureau
and the Cable Television Task Force (later bureau) for control
of cable regulation was a struggle of this kind.

For Model 3, the key figures in official action are the political
executives atop the bureaucratic units. Bureaucratic politics is
preeminently bargaining among people in high offices. Bureaucratic leaders tend to act as politicians seeking to build a constituency for themselves and their organizations. They also tend to
be, for Model 2 reasons, captives of the flow of information given
them by their staffs. A steady stream of issues and decisions presented from the bureau's viewpoint tends to orient them to its
viewpoint and wean them away from purely personal action. Both
as politicians and bureaucratic spokesmen they tend to become
advocates of their organizations — and, frequently, of the segments
of their industry they are charged to regulate. "Where you stand
depends on where you sit." Further, the need to bargain forces
them to stake out their views more strongly and with greater confidence than they would if they could make policy without com-

petition. Each can assume responsibility only for *one* viewpoint. The outcome may be in the general interest, but it will be a compromise no one of them has argued for.[105]

Several of the FCC's major decisions show the importance of effective or ineffective leadership in the bureaucratic infighting. Between 1971 and 1974, there was a struggle between the Broadcast Bureau and the Children's Television Task Force over whether the Commission should issue rules to restrict commercials on children's programs and encourage programming designed for children. The task force leaders, who wanted new policies, strengthened their position by effective advocacy within the Commission and leaks of their proposals to the press outside. The Broadcast Bureau, which opposed change, was driven on the defensive despite its greater size and influence. The Bureau was able to persuade the Commission to issue a policy statement rather than rules, but this concession came primarily because of changes in the Commission's membership.[106]

The 800 MHz spectrum management decision pitted the Safety and Special Radio Services Bureau, which favored reallocation of UHF-TV frequencies to land mobile use, against the Broadcast Bureau, which opposed it. Land mobile demands were sufficiently strong to persuade the Commission finally to reallocate the space, but it did so on condition that a spectrum management task force be established to insure that the additional space was used efficiently. The Broadcast Bureau wanted the space used well to prevent future land mobile demands, and hence supported the task force. However, neither Broadcast nor Safety & Special controlled the task force. The Chief Engineer, with a well-timed proposal, got control of it and used it to add some 70 positions to his office.

As the program grew, a second dispute pitted the task force, which wanted to use its own regional offices to assign mobile frequencies, against the Field Operations Bureau, which wanted its field offices used, and the Executive Director, who was concerned about the increasing budgetary demands and wanted the required computer services centralized in Washington. After extended controversy, the Executive Director, with support of the

Office of Plans and Policy, won over the Chairman and Commission, causing the task force to be absorbed in the Safety and Special Bureau and most of the work of its Chicago regional office moved back to Washington.

In 1976, the issue arose whether there was a more efficient way to license CB operators. Established procedure was for the FCC to issue licenses to CB operators, just as for other kinds of land mobile radio. The alternatives were to allow self-licensing, under which new operators would simply mail in applications to be filed by the FCC, or to abolish licenses altogether. Safety and Special supported the existing procedure, because the necessary personnel added resources to the bureau. The other offices and bureaus favored the other options, but their failure to agree clearly on one of them weakened their position. Also, the commissioners, naturally, tend to give extra weight to the opinion of the lead bureau on an issue. For these reasons, Safety and Special was able to dominate the argument before the commissioners, and the decision went in its favor.

Interestingly, the Commission gave the lead on the issue initially to an inter-bureau committee headed by the Deputy Executive Director, exactly to prevent Safety and Special from dominating the decision. This was the first time a task force structure had been used to evaluate an issue relating to only one bureau. The committee was supposed to settle the issue and save the Commission having to make more than the formal decision. But bureaucratic habit was too strong. The committee presented options, rather than a single recommendation, for fear of preempting the decision of the Commission. [107] And the committee members were allowed to argue their separate cases before the commissioners, just as if there had been no committee at all. Hence, the committee generated research on the issue but hardly any effect on the outcome. This shows how organizational routines can mold the environment for bureaucratic politics.

The staff offices of the FCC tend to be closely associated with the Chairman, and sometimes are used by him to counter the weight of the service bureaus. Dean Burch established the Office of Plans and Policy in 1973 to make sure he would get planning advice independent of the bureaus (and sometime ahead of the the the other commissioners). The General Counsel's office did the main staff work on the cable rules for Burch before the cable task force was promoted to bureau status. The General Counsel also supported Burch against Broadcast on the children's television issue by arguing that the Commission had full authority to regulate on the question. [108] This pattern resembles that of larger Federal departments, in which the Secretary commonly has staff units comprising an "Office of the Secretary" to help him control the operating bureaus.

So far, all our models have looked at the FCC itself to explain its behavior. Model 1 looks to the justifications available to individual officials making decisions, Model 2 to organizational structure, and Model 3 to whatever factors — personality, leadership, job pressures — govern the performance of bureaucratic "players" in the internal political contest. [109]

2.3.3   External Politics

In addition, Model 3 looks to political forces impinging on the FCC from the outside. The consideration of external factors is where Model 3 departs most clearly from the other two approaches.

A major influence on the FCC is, of course, the regulated industries, most prominently broadcasting and common carrier. The major broadcasting "players" include the networks and the National Association of Broadcasters (NAB). The major common carrier is the American Telephone and Telegraph Co. (AT&T). Their power is based on a political version of the control of information which we saw, in Model 2, gives bureaucrats power over the commissioners. In theory, the Commission is accountable to Congress and the public, and the industries are defenseless before it. In fact, influence goes to whoever gives the FCC most attention. Congress and the public, whose interests are diffuse, give the Commission only sporadic attention and hence have less power over it than appears. The industries, whose interests are concentrated on what the FCC does, give it a great deal of attention and gain, in return, disproportionate influence. [110] The constant drumfire of their lobbyists, lawyers, and publications tends to overwhelm other inputs to the commissioners and attune them to an industry viewpoint. [111]

Industry influence depends on constant, informal contact with the Commission. An important medium is the Washington communications lawyers who represent businesses before the FCC. They practice "law by telephone," in which they both inquire about policy and seek to influence it. The power of the law firms is enhanced by the fact that many commissioners and staff go to work for them or the regulated industries when they leave the FCC. [112] Of the 33 commissioners who left the FCC between 1945 and 1970, 21 went to work for the communications industry or communications law firms.[113]

As we saw under Model 1, rules against *ex parte* contact with the commissioners are designed to restrain unofficial influence on the FCC. However, they have applied only to adjudication cases. Only recently has the Commission initiated a rulemaking to establish the extent to which rules against *ex parte* should be applied to rulemaking proceedings. In one important instance — the 1975 rules for pay TV — the Appeals Court overturned a decision because of the extensive off-the record lobbying that was allowed. [114]

Industry influence, as well as technical considerations, helps explain the Commission's decision of 1952 to force UHF television to develop against the competition of VHF stations. Heavy lobbying by RCA, owner of the NBC network, precluded the option of an all-UHF system even in some localities. [115] To a lesser extent, a desire to protect AM radio interests may have influenced the FCC's 1945 decision to shift FM radio to a higher spectrum band, a move which made obsolete existing FM equipment and delayed the development of FM stations. [116] On the whole, FCC policy may be said to have delayed the development of cable and pay television as competitors of broadcasting.

These cases all suggest that industry influence is a reason why, as critics have pointed out, the FCC is suspicious of new technology. By definition, new technology is not yet deployed in industry. Industry interests naturally defend the technology they have. The FCC tends to protect the existing economic investments of companies. It hesitates to put at risk a system spoken for by economic interests, in favor of a new system spoken for only by enthusiasts and academics. [117] There seems to be a notion of economic standing parallel to that of legal standing. Ideas are taken seriously only when an industry has given them economic reality "on the ground." Cable television has been able to get even partial attention and support from the Commission only because it has some economic presence as an industry.

A second political influence of growing importance has been consumer groups interested in making broadcasting more responsive to the general public. These groups came into prominence following the landmark 1966 decision of the Court of Appeals in the WLBT case which granted community representatives legal standing in license renewal cases. Consumer groups were inspired by the oratory of Nicholas Johnson when he was on the Commission. They have been represented by, and are barely distinguishable from, public interest law firms, such as the Citizens Communications Center (CCC). They include the National Citizen's Committee for Broadcasting (NCCB), the NAACP, Action for Children's Television (ACT), the National Black Media Coalition and the United Church of Christ (the plaintiff in the WLBT case), among others.

The groups challenge license renewals on grounds that the licensee has not met standards of community responsiveness, public service, or affirmative action. They also contend for general broadcast policies favorable to the general interest as they see it. A notable example has been ACT's campaign since 1968 to restrict commercials and improve programming for children's television. ACT parlayed an attractive populist appeal, strong advocacy, and effective publicity into considerable pressure on the FCC. Counterpressure by the NAB caused the Commission to confine itself to a policy statement stating standards for commercials voluntarily agreed to by the broadcasters. ACT appealled unsuccessfully in the Court of Appeals to require the Commission to set binding rules.[118]

Like the industry, the public interest groups seek close contact with the FCC. A bit of a "revolving door" has already grown up between them and the Commission, parallel to the one between the communications law firms and companies and the Commission. At the end of his tenure in 1973, Nicholas Johnson left the Commission to become the head of NCCB. The former head of the Citizen's Communications Center became administrative assistant to Charles Ferris when he became chairman in November, 1977.

However, the typical weapons of the groups are not influence within the FCC, but advocacy and publicity aimed at a larger public. This indicates a basically different political style. As we saw above, the industry is most able to protect highly concentrated economic interests if it is given a quiet atmosphere in which it can dominate the flow of information reaching the commissioners. The public interest groups, on the other hand, seek to protect the diffuse interests of the general public. To do this, they must break open the closed world of industry influence and force the commissioners to pay attention to a wider constituency. They seek to use as weapons the politicians and the press — the two political forces speaking for the undifferentiated public.

The groups seek to link the closed world of the FCC with the larger public. They operate to a degree like industry lobbies in order to find out about Commission deliberations and seek to influence them, but they use these resources to mobilize outside political forces. They work to give the general public the resource it cannot have by itself — a degree of day-by-day attention to the Commission rivaling that of industry. Hence, the position of the groups is paradoxical. They seek success as lobbyists only in order to undercut lobby politics. Presumably, if they could ever mobilize the public permanently, their function would disappear. [119]

The advent of the public groups has no doubt been salutary for the FCC. They tend to make real what might otherwise be mainly formal — the Commission's accountability to the general public. They give organizational presence to the general interest and hence tend to prevent capture of the agency by the industry. [120]

However, a Model 3 analysis allows no heroes. Good policy cannot to be chosen independently of the political process. No one actor, however well-intentioned, can simply cause FCC policy to serve the general interest. The groups can have influence only by engaging in a political struggle with the industry. To influence the outcome they must sometimes state their views at an extreme that would not serve the public interest if it prevailed. Some public interest advocates sound as if they would destroy the industry if they could. In this sense, political competition corrupts even players with disinterested motives. The outcome may in fact serve the public interest, but despite — perhaps *because* of — the fact that none of the competitors have argued for it.

2.3.4   Multiple Masters

The pressure of interest groups is only one dimension of external politics. A second is competition among the official institutions who have authority over an agency. Here is where the departure from Weber's presumption of a unified political will is most blatant.

The FCC has three official masters — Congress, the Executive, and the courts. Traditionally Congress has been the most powerful. Although Congress has delegated near-total authority to the FCC to deal with communications, it may always revoke it and overrule the Commission by statute. It also has the power to confirm commissioners' appointments, determine the agency's budget, and investigate its policies. The mere threat, as well as the use, of these powers has given Congress pervasive influence over much that the Commission does. Congressional oversight is concentrated in the Communications Subcommittees of the Commerce Committees of both House and Senate, but extends also to the Appropriations Committees and individual Congressmen or staff who seek to influence the Commission.[121]

Traditionally, Congressional influence has been exercised on behalf of broadcasters and against FCC attempts to regulate in a populist spirit. One reason is that Congressmen are heavily dependent on local broadcast stations for the media coverage they need to be reelected. [122] The Senate Communications Subcommittee, long headed by Rhode Island Senator John Pastore, often took this stance, for instance by influencing Dean Burch to revise the proposed cable television rules in the interests of broadcasters and broadcast copyright owners. [123] More recently, however, the House Communications Subcommittee, under San Diego Congressman Lionel Van Deerlin, has pursued a much more populist line, reflecting the increasing influence of public interest lobbies and attitudes. In June 1978, the House Subcommittee produced a bill which could, if adopted, revamp significantly the authority and practices of the FCC, along with the Public Broadcasting Corporation and the National Telecommunications and Information Administration.[124]

Some of the FCC's more adventurous decisions have been taken under threat of Congressional action and have led to negotiations with Congress. In 1961, the Commission proposed to expedite the development of UHF Television by "deintermixing" UHF and VHF stations in eight local markets. It was proposed to give six markets over wholly to UHF, reviving an option rejected for the whole country in the UHF decision of 1952. Alarmed VHF broadcasters mobilized Congressional support to stop the action. However, FCC Chairman Minow obtained in return passage of legislation to require that all new television sets be capable of receiving both VHF and UHF channels — another reform long desired by UHF interests. [125] In 1963-64, broadcasters used Congressional support, plus a weak majority on the Commission itself, to stop an FCC move to limit permissible advertising time on the air. [126] In 1969, the Commission gave the license of WHDH in Boston to a competing applicant in order to diversify media ownership — its first-ever revocation on overtly populist grounds, since the station was innocent of obvious malfeasance. As mentioned above, Congressional pressure, combined with a changed political balance on the Commission, forced the Commission to issue guidelines for comparative hearings involving regular license renewals which were more favorable to established licensees. [127]

Common carrier regulation, which has been too technical to attract much political interest, has received increasing Congressional attention because of FCC efforts to introduce competition into the field. The full import of the 1971 decision allowing specialized common carriers to compete with AT&T was not at first apparent to anyone outside the FCC's Common Carrier Bureau, which sponsored it. However, the decision survived routine court challenges. The Commission followed it up by denying petitions to reconsider, questioning the rate submissions AT&T proposed in response to competition, and favoring competition in domestic satellite and land mobile communications. [128] AT&T and its supporters became alarmed and appealed to Congress. A Consumer Communications Reform Bill, known as the Bell Bill, was introduced in 1976 which would essentially roll back the FCC competition policy and restore AT&T's position as a public utility monopoly. The bill was reintroduced in 1977 and 1978. [129]

The Executive's powers over the FCC are less pervasive than those of Congress but increasingly important. Their basis is the President's authority to nominate the commissioners, to name the FCC chairman (who serves at his pleasure), and to make communications policy through channels which are competitive with the FCC. Also, the President's staff reviews senior staff appointments at the FCC, and the Office of Management and Budget (OMB) reviews its personnel ceilings, budget requests to Congress, and legislative proposals. [130]

Presidents have sometimes used task forces to preempt FCC policy making. President Johnson's Task Force on Communications Policy, 1967-8, held up FCC action on domestic communications satellites. Presidents have also used task forces to evaluate the performance of the FCC and other regulatory commissions. The Landis Committee reported to President-elect Kennedy, and the Ash Council reported to President Nixon.

President Nixon's creation of the Office of Telecommunications Policy (OTP) in 1970 gave the Executive a standing point of leverage on FCC affairs. OTP presided over the final negotiations leading to the Consensus Agreement that resolved the dispute over the Commission's 1971-72 cable television rules. [131] A rivalry eventually grew up between Chairman Burch and Clay Whitehead, the head of OTP. In 1978, the Carter Administration moved OTP to the Commerce Department to become the National Telecommunications and Information Administration, headed by Henry Geller, former General Counsel of the FCC and advisor to Dean Burch during the cable controversy.

A basic source of Executive influence is that communications policy has become too complicated for the FCC to make alone. The issues extend far beyond spectrum management, broadcast licensing, and common carrier rate regulation — the core Commission activities. New technology raises a host of legal and economic issues that bring many other agencies into the political game. OTP has appeared as a party to FCC proceedings. The Justice Department's anti-trust division put pressure on the Commission to extend its rules against multiple ownership of media facilities to cover newspapers and to stop the merger of ABC and ITT. [132] The minute the number of agencies multiplies, the FCC tends to lose its preeminent position in communications policy. Power gravitates to the President or White House staff who have the authority to coordinate all the Executive players.

The most influential external institution for the FCC today, however, may be the Courts of Appeals, particularly the Appeals Court for the District of Columbia. While the courts traditionally have reviewed Commission decisions for adherence to due process and the FCC mandate, today they increasingly decide the merits of cases as well.

To a great extent, legal appeals are an extension of the interest-group politics reviewed earlier. The FCC itself is a quasi-judicial body, as we saw under Model 1. It uses adversary hearings for adjudication cases and solicits comments of an adversary nature before making rules. Hence, to appeal a decision usually results in an adversary proceeding in court not very different from the original proceeding before the FCC. Often, losing parties who appeal are merely trying to win on legal grounds the case they have first lost on substance before the Commission. Legal action is the continuation of politics by other means.

When the courts rule on due process they ask whether the FCC has adequately considered all the interests and viewpoints relevant to a decision, and specifically whether it has complied with the Administrative Procedure Act. Precedents govern the range of interests entitled to "standing" (representation) in cases. The importance of the Appeals Court's 1966 decision in the WLBT case was that it gave community groups standing even though they lacked a clear economic interest in broadcasting. The specialized common carrier decision of 1971 had to survive a legal challenge alleging that the FCC should not have used rule making procedure, which requires only written comments, but adjudication, which would have required lengthy comparative hearings among AT&T and the applicant companies. [133] In 1977, the courts overturned the restrictions on violence in early prime time (the "family viewing hour") which Chairman Wiley had negotiated with the broadcasters, on grounds that the agreement was tantamount to rule making and the Commission had not consulted other interests.[134] In 1977, ACT unsuccessfully challenged the FCC's 1974 Policy Statement to force it to issue binding rules on children's television, rather than simply generating a policy statement.[135]

The courts also hear challenges to the Commission's legal authority to regulate. Such questions arise because the Communications Act is vague and does not clearly give the Commission authority over new types of communications that have arisen since 1934. , The precedents, some of which go back before the Communicaᴸ tions Act, have become in effect part of the FCC's legal mandate. The most vexing recent question has been whether the Commission had the authority to regulate cable television. Since cable is not mentioned in the Communications Act, the courts have held in decisions since 1968 that the FCC may regulate cable only for purposes "ancillary" to the protection and development of broadcasting, for which the Commission's responsibility is clear. But how wide a range of cable regulation this doctrine covers is still unclear.[136]

Federal courts increasingly decide questions of substance as well. Their traditional criteria of due process and statutoriness have become so discretionary that they cannot be applied without, in effect, judging the merits of the case. Once legal standing is extended to groups with no immediate economic interest in a case, private suits for redress of grievance can be turned into wide-ranging, quasi-legislative inquiries. When community interests are defined this broadly, how to balance their interests depends more on judgment than law. Also, the statutory basis for Federal regulation has become less clear as Congress has delegated authority to agencies in increasingly broad and vague ways. [137] The FCC's very vague mandate given in 1934 was an early example of this trend. Therefore, whether a regulation is statutory becomes more a matter of discretion than law. These changes have given the Federal courts an increasing role in regulatory policy, at the expense of both Congress and the agencies. [138]

An uncertain mandate, among other factors, has allowed the Appeals Courts virtually to take over cable television regulation since 1972. Judges have affirmed the FCC's authority to regulate cable but have decided many of the details for themselves. In 1972 the Court of Appeals approved the Commission's mandatory cable origination rules as "reasonable ancillary" because they increase the number of outlets for community expression and augment the public's choice of programs and type of service. [139] In 1977 the court overturned rules preventing cable operators "siphoning" broadcast signals out of the air for cable transmission, on grounds that siphoning had created no real economic threat to broadcasters. [140] In 1978, the court ruled that the imposition of minimum public access and channel capacity standards on cable systems was improper. [141] Also in 1978, the court held that the FCC has the authority to preempt state and local price regulation of special pay cable programming, and that the way the Commission has done this is adequate and effective. [142]

2.3.5   The Pluralist Problem

Model 3 traces FCC outcomes to political bargaining in the absence of a clear political will. The lack of clearcut goals seems initially due to the bargaining itself, among the commissioners, FCC bureaus, outside interests, and other institutions. The players see to it by their competition that the policy is set by a political process, not by any one interest acting alone. However, the politicking is due to a larger cause — the inability of the overall political system to agree closely on goals. Regulatory commissions like the FCC reflect the problem at an extreme. If the FCC had a highly specific legislative mandate, as some newer regulatory bodies do (such as the Environmental Protection Agency), its leaders could base policy on clear obligations to law, in the manner of Weber's rational authority, and pay less attention to political pressures. But the FCC arose precisely from Congress' failure to agree. In fear that politics would infect communications regulation, Congress delegated the area wholesale to the FCC. It has refused to clarify the FCC's mandate even in areas like cable television where the Commission's authority is seriously in doubt.

The uncertain mandate has merely shifted the political pressures from Congress to the FCC, not excluded them. The Commission is subject to political pressures precisely because players know it has the discretion to yield to them. If an agency operates under no authoritative commands, the effort to influence it is worthwhile. Congress itself, as we saw above, seeks to influence the FCC. The Commission's main problems, a vague mandate and undue Congressional pressure, go hand in hand. [143] Only closer collective agreement about goals for communications could eliminate them.

The lack of agreement on policy also means that organizational arrangements are politicized. Traditionally, Congressional committees and interest groups have shaped the structure of Federal agencies to fit their own needs, not the norms of public administration theory. Programs or units are reorganized or their departmental location changed in order to change the interests they serve. [144] The FCC has periodically faced such threats. One or another bill to reorganize the Commission has been pending in Congress in most years since its founding.

The Commission has so far escaped forcible reorganization. Its main structural problem is something more subtle — lack of a clear relationship to the Executive or Legislative Branches. Agencies with a clear mission usually receive it from the Executive or Congress, and are organizationally linked accordingly. Thus, OMB is clearly accountable to the President and the Government Accounting Office (GAO) to Congress. The major departments are accountable to the President through their political appointees and to Congress through their legislation and budgets. The political authorities accept some responsibility for protecting these entities from contrary political pressures. The agencies are viewed as *agents* of government.

The FCC, however, is a kind of subordinate government. It is supposed to be "independent." It has no clear relationship to either the Executive or Congress, or to the courts. Rather, it has some of the nature of all three. It has quasi-legislative functions, but is closely associated with the Executive and behaves like a court. [145] In this position, as we have seen, the FCC gets very little protection. All three branches of government can influence the FCC, but none takes responsibility for shielding it from the others. All can sway it for short-run reasons, but none helps it develop a coherent long-term policy for communications. The FCC has to do this alone.

The FCC's lack of a clear political or organizational "place" opens the door to pluralist pressures. The agency is not part of government in such a way that it knows clearly whom to follow and what to do. Rather, it is one of a number of organizations and interests interacting to determine communications policy without any clear hierarchy. The environment is uncomfortable for the FCC, but endorsed by the larger political system. Pluralism is a deeply American approach to achieving social order, maybe more American than either representative democracy or hierarchical, bureaucratic authority. [146]

To the extent pluralism is a reality, the Model 3 analysis of the FCC becomes plausible, and the viewpoints of Models 1 and 2 implausible. Leaders cannot rationalize actions according to their own values, as Model 1 imagines, if they must continually adjust to the demands of others. Expert calculation of the kind stressed by rationalist values can play a role only insofar as competing interests use it as a weapon to influence decisions. [147] Nor will decision result primarily from bureaucratic dynamics, as Model 2 says, if the major determinants are political pressures from outside the organization. Political executives are not, it turns out, very interested in organization at all. They spend most of their time reacting to external pressures. [148]

In short, the policy process cannot be rationalized and cannot be organized coherently as long as political bargaining is dominant. Leaders can only adjust; they lack the security to calculate or to manage. [149] Bargaining cannot even be understood in as theoretical a way as the dynamics specified by the other two models. The nature of pluralism is to provide, and to reveal, very little structure of its own.

2.4 THE REFORM OF THE FCC

The foregoing analysis can help us understand the proposals
usually made to reform the FCC. We will find that they are
usually based on one or another of the analytic models reviewed
above. The models underlie reformers' views of what is wrong with
the FCC and how it could be made right. Each model explains what
a particular class of reforms can accomplish — and what it cannot.

2.4.1 The Critique of the FCC

The reformers agree closely about the shortcomings of the FCC,
aside from details. In theory, the Commission is supposed to enforce the public interest on the communications industries by requiring responsive and efficient service. Regulation, rather than
the market, has to do this, because the industries must for technical reasons (e.g., the limited radio spectrum, the need for an
integrated national phone system) be oligopolies subject to only
limited competitive pressures. The goal is not simply that the industries should thrive in market terms; it is that they should be
responsive to the public and new technology in ways the market
would not, by itself, require.

Critics have claimed that, in practice, the FCC has been much
more concerned to protect established industry economic interests than to enforce either responsiveness or efficiency. [150]
The Commission has allowed broadcasters to cater to a mass
commercial market rather than serve local communities, and
it has to some extent shielded them from competition from
alternative technologies (e.g., UHF and cable television, FM radio).
It has allowed AT&T to serve some common carrier needs exclusively, excluding until recently competitors who might have
served them better. These outcomes indicate poor performance
by the FCC even if the industries are profitable and, by international standards, technologically advanced. [151] The Commission's public goals have not been fully achieved.

2.4.2 Will and Expertise

Some critics, following Model 1 assumptions, argue that the
problem can be traced to the minds of the commissioners and,
by extension, all FCC personnel. The Commission does not

sincerely will the public interest, or it does not know how to achieve it. If the FCC were more principled and more expert, it could achieve its goals.

We saw above that the single actor viewpoint speaks of maximizing at least three different sets of values. Economist critics of the FCC have been those arguing most forcefully for rational values. They see better FCC performance primarily as a matter of intelligence. They want to improve the quality of policy analysis, planning, and evaluation done at the Commission. All these terms mean to make policy in the light of foreseeable economic consequences. Since economists are those most skilled at economic calculation, they should exercise increased authority at the FCC. One suggestion is that economists sit on the Commission itself, but none ever have. [152] The Commission has gone part way in instituting the Office of Plans and Policy in 1973, which all by itself vastly increased the agency's economic expertise.

Lawyers, the advocates of common-law or traditional values, seek to reform the FCC by improving its internal processes and judicial review. Lawyers were very active in the 1930's and 1940's, when a leading reform issue was how to reconcile administrative law with judicial notions of due process. The Administrative Procedure Act of 1946 codified the legal view of administrative due process for the FCC and most other Federal agencies. [153] Lawyers have argued, further, that active judicial review of FCC actions is an effective way to hold the Commission accountable and force it to perform. This view is shared by some economists. [154] It was endorsed in the Ash Council reorganization proposals, which specified a special Administrative Court to take over review from the Court of Appeals. [155]

Public interest critics, for their part, assume populist values. Their diagnosis of the FCC problem is that the commissioners are coopted by the regulated industries and detached from due allegiance to the general public. "Better men", more able and more principled, should be appointed. Their separation from the industry should be enforced by giving them longer terms than those currently in force and forbidding them from going to work for the industry for an extended period after they leave the Commission. [156] Also, representatives of the general public or consumers could be added to the Commission. [157] The proposed rewrite of the Communications Act reflects this view to the extent that a commissioner's term is changed from seven years with the possibility of reappointment, to ten years without reappointment, and the present one-year prohibition on representing industry interests after leaving the Commission is extended to include supervisory personnel as well as commissioners.

All these proposals are defensible in themselves. But they ignore the reality that the commissioners, however able and principled, must carry out their decisions through an *organizational and political process.* This was the criticism that Models 2 and 3 made of Model 1. As Weber puts it, leaders cannot exercise authority by themselves; they need an administrative staff. [158] Just as much, they need to be politicians.

The economists' critique tends to neglect the importance of organization. They reject, for instance, the Ash Council's contention that the organization of the regulatory commissions is a problem separable from the substance of regulation. [159] This implies that if policy is the product of good intentions and expertise, its implementation and success can be taken for granted. We have seen how important organization and politics actually are for the FCC. Such factors lie outside the normal range of economic reasoning. [160]

Lawyers' suggestions for improved due process and review tend to ignore the adverse institutional consequences. If regulation could be reduced to court-like deliberations about policy, legal approaches could be nothing but salutory. But deliberation has costs in time and effort, and rules once made must be administered. Judicial procedures for rule making and adjudication emphasize precise representation of affected interests over speed and flexibility. They also give the regulated industries a dominant role in proceedings, as against inputs which are tough to represent, such as economic calculations and the interests of the unorganized public. That is, they emphasize traditional over rational or populist values. Critics of the regulatory commissions commonly find that they have been "over judicialized" — tied up in knots by court-like procedures. The Ad-

ministrative Procedure Act was a triumph for the legal values
in policy making, but costly to others. Its structures may have
to be relaxed before a more balanced approach can emerge. [161]

Public interest reform proposals sometimes seek to disrupt the
institutional apparatus they would need to be effective. Reformers
of the Nicholas Johnson or Ralph Nader type tend to campaign
against the institutions rather than seeking to use them for more
enlightened purposes. Populist leaders may be effective as critics
and prophets; however, to bring about change they must come to
terms with organization and politics. They have to compromise
with existing institutions, and as charismatics, this is just what
they resist. [162]

2.4.3   Managerial Reform

A second group of reformers, far from ignoring organization,
makes it central. The perspective is managerial; the FCC's major weakness is that the bureaucracy is not controlled by the
political executives. Even an advocate like Nicholas Johnson
argues, after experience on the Commission, that the commissioners are at the mercy of their agenda and staff. [163]

The answers are to be found in public administration orthodoxy.
The formal principles of bureaucracy must become fact as well
as norm. The commissioners must communicate their will to
staff more clearly through more specific delegations of authority,
and they must devise a management information system to see
that orders are carried out. [164] Better communications channels up and down the hierarchy, in other words, will overcome
the control and information blockages that work to separate subordinates from their managers.

More than this, some reformers have questioned the basic principles of the Commission form. They say that the plural executive inherently prevents decisive management and should be replaced with a single executive atop each agency. Further, the
commissions should be made accountable to the President like
other agencies, since their independence (and, in practice, considerable subservience to Congress) prevents the President from
developing and implementing consistent policy.

A series of official studies of the regulatory commissions has proposed to (1) strengthen Presidential authority over the commissions; (2) absorb them into the executive departments, perhaps with continuing independence for adjudication functions; (3) strengthen the managerial role of the Chairman and Executive Director within each commission; and (4) strengthen policy planning and evaluation capability. The Landis Report of 1961 argued forcefully for the first change, recommending special offices within the White House to coordinate regulatory policy. [165] The Brownlow Report, 1937, and Ash Council Report, 1971, argued aggressively for the second, although Ash specifically left the FCC independent because of the political sensitivity of communications regulation. [166] The Hoover Commission studies of 1949 and 1950 gave detailed attention to the commissions' internal management. [167]

All these recommendations appear to have had some influence on the FCC. The President's role in communications policy was strengthened by the creation of the Office of Telecommunications Policy. OTP, now the new National Telecommunications and Information Administration, also constituted a new administrative authority (under the Executive), separate from adjudication (left independent in the FCC). The chairman's managerial role has been strengthened, and the founding of the Office of Plans and Policy gives him greater planning capability. The new, proposed Communications Act would reduce the number of Commissioners from seven to five, thus increasing the managerial ability of the Commissioners.

The second Hoover Commission, 1950, helped bring about a shift of much of the FCC from professional to functional organization. Offices of Accounting, Engineering, Law, etc., ceded to functional bureaus for broadcasting, common carrier, etc., leaving only the Office of Chief Engineer and a relatively small Office of General Counsel organized on a professional basis. [168] This weakened the grip of the professional castes on the organization and strengthened the hand of higher management.

But we must ask how much has changed. The criticisms of the FCC are almost unchanged from 30 years ago. The commissioners still lack the authority over their organization that managerial

norms say they should have. The FCC remains slow-moving and
set in its ways. [169] The same proposals for reorganization are
made repeatedly. The reason may be that the suggested changes
have not gone far enough. Reform, especially, has barely touched
the explicit limits on supervisor's managerial authority, notably
the civil service.

Two explanations are more fundamental. One is that managerial
reforms are self-limiting. Most of the bureaucratic problems discussed under Model 2 are inherent in organization itself. Efforts
to overcome them through reorganization run into the same problems. Efforts to clarify communications up and down the hierarchy
run into the same distortions of orders and messages they are designed to overcome. [170] Strengthening the formal position of
the leadership can accomplish only so much when the real problem is control of the lower levels. Ultimately, the inherent limitations of coordination through formal rules cannot be overcome
through the imposition of more formal rules. Bureaucracy is not
a panacea; there are hard limits to what it can accomplish. [171]

The other explanation is politics. As we saw under Model 3, the
main problem of bureaucratic managers may not be management
at all. It is, rather, to assemble a consensus for their policies among
diverse players inside and outside the organization who are *not*
clearly subject to their authority. Outcomes apparently due to
ineffective management may really be due to a balance of political forces which executives and administrative reformers cannot
change. We saw that a Model 1 diagnosis of the FCC tends to
leave out organization. Equally, a Model 2 analysis tends to leave
out political influences such as Congress and the regulated industries.

2.4.4   Political Change

A third method of reform seeks to improve the FCC by changing
its political environment. Unlike the other two, this approach has
not had a coherent school behind it. The authors propose it as a
construct to organize random suggestions here and there in the
FCC's history and literature. Rather than reorganize the FCC from
within, the political tactic aims to expose it to pressure to improve
from without.

One variant is to mobilize the public interest groups which already play an increasing role in Commission policy making. The theory behind the idea has been developed by disillusioned reformers of social programs who have given up trying to change bureaucracy from within. If pressure from advocate groups and experts from the outside can induce agencies to change, the argument goes, there will be less need to tinker with the "black box" of agency organization. [172]

Applied to the FCC, organized public pressure seeks to counter the dominant influence of the regulated industries. As discussed above, public interest groups seek to break the industries' grip on the commissioners' attention and information, and hence make politically real to them what is otherwise merely formal — their accountability to the general public. [173] One way to expand the groups' influence on adjudication cases would be to reimburse the expenses of public participants or groups taking part. In June 1978, the Commission adopted a Notice of Inquiry into this possibility. [174] A way to expand their role in rule making would be to appoint their representatives to task forces or advisory bodies exploring specific issues. For example, in the summer of 1977 a special notice about the World Administrative Radio Conference proceedings was mailed to public interest groups to acquaint them with the issues under consideration and solicit their participation.

A second political tactic is to mobilize the institutions that influence the FCC within the Federal establishment, and perhaps create new ones. A standard tactic of bureaucratic politics is to create competitors for the agency one wants to pressure or weaken. To a degree, President Nixon did this when he created the Office of Telecommunications Policy. Congress did it when, in 1962, it divided control of communications satellites between the FCC and the Communications Satellite Corporation. An Administration which seriously wished to reform the FCC might be well advised not to try to persuade the Commission or reorganize it, but to orchestrate pressure on it from the other Executive Branch players in communications policy. To some extent the proposed Communications

Act uses this tactic by specifying the relationship between the FCC and the Public Broadcasting Corporation and National Telecommunications and Information Administration.

The major hope of this approach is probably the courts. We saw above that Federal judges have taken control of some aspects of communications policy away from the FCC. Some critics view this not as a threat to the agency, but as a salutary check likely to lead to improved rule making. Out of judicial case law may come the coherent policies which the FCC seems unable to formulate by itself. [175]

The limitation of political reform is that pluralism tends to allow only incremental change. To add or strengthen one group or institution in the pluralist interaction cannot produce large results when the other players remain unchanged. Hence, political solutions may be more symbolic than real. [176] It is implausible to suggest that public interest groups or the courts can transform the FCC when these forces have had only limited influence on the Commission to date.

A more basic objection is that pluralism is not subject to direction. The nature of it is to deny any one interest control over the outcome. The proposed Act reflects inevitable compromise between the various interests. No reformer will be able to control the FCC political process for his own ends, however enlightened. How could he know that the public interest groups, or rival agencies, or the courts would serve the purposes supposed for them in the reform plan? The reformer cannot shape the process as if he stood apart; he can only join in it as one player among many.

2.5 INSTITUTIONAL DEVELOPMENT

Our review of FCC reform proposals has been sobering. The gains of reform have been limited in practice. The theoretical analysis seems to indicate why. The proposals grounded in one model seem to leave out necessary perspectives drawn from the other two. The organizational and political proposals have limitations specified in their own models. Prospects for change seem *inherently* circumscribed.

The problem, however, is really that reform ideas tend to be piecemeal and short-term. They propose gimmicks which will improve the FCC's performance if applied right now. All will be well if the chairman is an economist, has more authority to manage his staff, or can call on public interest groups to balance the industry. Such suggestions must have limited effect, since each leaves unchanged the other institutional features of the FCC.

But suppose all institutional limitations were eased at once. Suppose there were, not marginal changes, but better appointments to the Commission, better public administration, and better politics. Then changes in one dimension would not be limited by rigidity in another. If the commissioners were more capable and dedicated, they would make better rules. [177] If Congress gave the FCC a more specific legislative mandate, such as the new Act, and a more responsive civil service system, the FCC would perform better as an orgianization.[178] If the public cared enough for regulation so that politicians routinely made an issue of it, regular representative democracy would balance the influence of the industry without a need for public interest groups.[179]

Changes on this scale would transform the nature and performance of the FCC. Such changes have seemed to be beyond the reach of deliberate reform. Even to suggest them seems utopian. And yet, from a more *historical* perspective, they have occurred. What *deliberate change* has not achieved has been achieved partially by a process of *development* over time.

As a group, the commissioners probably are more able and dedicated, their staff more expert, their political environment less dominated by the industry, than they were 10, 20, or 30 years ago. Certainly, the Commission has regulated in a more activist and probably more effective manner since about 1960 than it did before then. The same could probably be said of American government and politics in general. The institutions really *do* perform better today than they did 50 or 100 years ago. Of course, progress may be outweighed by higher expectations. The reasons for it may be debated. But the reality of it is incontestable.

We broaden our understanding of the FCC by two dimensions
if we add to the usual, single-actor view of the agency an understanding of it as an organization and as a political actor. We add
a fourth dimension if we look to all three models over time. History suggests how the antinomies of reform may be overcome
by a general process of institutional development.

Why do we usually dismiss improvement of this evolutionary
kind? In part because it occurs too slowly to meet the short-
run personal and political needs of reformers. In part because
it seems due more to the rising education, professionalism, and
affluence of the country as a whole than to specific reform actions. Far from causing us to think better of the FCC, or any
agency, such change merely elevates the standards by which we
criticize.

A more basic reason is that the theories underlying our analytic
models have difficulty perceiving and valuing institutional development. The theories tend to assume "economic man" or
self-interested behavior on the part of people. Institutional improvement, on the contrary, requires an increase in public mo-
rality. Improvement, for instance, requires that public interest
groups and the general public take an interest in regulatory issues in which they have no immediate economic interest --
something inexplicable to an economic analysis. [180] To perform better as citizens or officials in *any* collective endeavor,
people commonly need non-economic, even altruistic induce-
ments outside the realm of strictly selfish motivation. [181]

A yet more basic reason is that Americans do not view public
institutions in terms of development at all. The Federal government is perceived to have been created all at once with the writing
of the Constitution in 1787. It has not evolved; it was "founded"
once and for all. Even the numerous social institutions created
at the New Deal and since, such as Social Security, are perceived
to have been legislated, not developed over time. Static political
beliefs have provided consensus in a nation without many other
bases of agreement. In a society where all other structures per-
petually change and modernize, stability is provided by political

institutions that are perceived to change hardly at all. [182] Reforms will occur, but for immediate personal or political reasons, not in pursuit of a long-term developmental design.

The regulatory commissions have been viewed in especially static terms. Their original purpose was to take an area of policy "out of politics" and entrust it to the hands of "experts." Legitimized by powerful rationalist values, the FCC and the other commissions have not been allowed human failings. Their deficiencies have been recognized, but viewed more as scandals to be rooted out than as mandates for long-term change. Reformers have shown limited patience, and limited determination, in working with the Commission's problems over time. There is a tendency to decry evident abuses and propose urgent reforms, but not for long enough or in terms practical enough to have effect.

Reformers succumb to impatience when dealing with the FCC as an *institution.* Economists flirt with the idea of abolishing much communications regulation and trying to control the industry through market competition instead. [183] Public interest advocates would like to subject broadcasting to direct "community" control. The assumption is that these steps would end the need for a public agency with all its problems of competence and control. Consumers, or voters, would tell the industry what to do directly through the economic or political marketplace, without the need for the FCC as an intermediary. Control would be accomplished effortlessly through the "invisible hand," ending the need to wrestle with the public hand of government.

These proposals *are* utopian. The FCC and other public agencies are necessary for practical reasons. They must be wrestled with. Americans too readily turn away from low-performing organizations in search of alternatives. They need to stick with the institutions until reforms are pushed through. [184] What the FCC needs more than any single change is sustained political commitment from its constituency.

The impact of single reforms may be limited, but public caring sustains and builds civility, on which the more global process of institutional development depends. The critics and FCC officials may view each other as adversaries. But to the eye of history the Commission is carried forward by their mutual concern.

REFERENCES

[1]   Krasnow, Erwin G. and Lawrence D. Longley, *The Politics of Broadcast Regulation* (New York: St. Martin's Press, 1973), p. 73.

[2]   Cox, Kenneth A., "The Federal Communications Commission," *Boston College Industrial and Commercial Law Review*, Vol. 11, No. 4, May 1970, pp. 595-688.

[3]   Capron, William M. (ed.), *Technological Change in Regulated Industries*, Studies in the Regulation of Economic Activity (Washington, D.C.: The Brookings Institution, 1971), pp. 200, 223.

[4]   Allison, Graham T., *Essence of Decision: Explaining the Cuban Missile Crisis* (Boston: Little, Brown and Company, 1971).

[5]   *Ibid*, Chapter 1.

[6]   Weber, Max, *From Max Weber: Essays in Sociology*, ed. and trans. by H.H. Gerth and C. Wright Mills (New York: Oxford University Press, 1958), Chapters 4 and 8, and *The Theory of Social and Economic Organization*, trans. by A.M. Henderson and Talcott Parsons, ed. by Talcott Parsons (New York: Free Press, 1964), pp. 329-41.

[7]   *Communications Act of 1934*, 48 Stat. 10064 (1934), Title 47, U.S.C.; *Federal Radio Act of 1927*, 44 Stat. 1162 (1927); *Interstate Commerce Act of 1887*, 24 Stat. L 379 (1887).

[8]   *Communications Satellite Act of 1962*, 72 Stat. 419, 47 U.S.C. 701-744.

[9]   *Communications Act,* Sections 356, 214(a), and 4(k).

[10]   *Communications Act,* Section 326.

[11]   *FCC Report on Allocations from 44 to 108 Megacycles,*
       Docket No. 6651, June 27, 1945, reprinted in *Broad-
       casting,* July 2, 1945; see Krasnow and Longley, pp. 85-95.

[12]   *Sixth Report and Order,* Docket Nos. 8736, 8975, 9175
       and 8976, 41 FCC 148 (1952).

[13]   *First Report and Order and Second Notice of Inquiry,*
       Docket No. 18262, 35 FR 8644, June 4, 1970; *Second
       Report and Order,* Docket No. 18262, 46 FCC 2d 752
       (1974); *Memorandum Opinion and Order,* Docket No.
       18262, 51 FCC 2d 945 (1975).

[14]   *Specialized Common Carriers, First Report and Order,*
       Docket No. 18920, 29 FCC 2d 870 (1971).

[15]   Strassburg, Bernard, "Case Study: FCC's Specialized
       Common Carrier (SCC) Decision," in *Analysis of the
       Regulatory Process: A Comparative Study of the De-
       cision Making Process in the Federal Communications
       Commission and the Environmental Protection Agency,*
       Report prepared for the National Science Foundation
       under Grant No. APR 75-16718, unpublished (Washington,
       D.C.: The Urban Institute, November 30, 1977). pp. III
       29-40.

[16]   Weber (1964), pp. 341-58.

[17]   *Communications Act,* Section 309 (e).

[18]   Strassburg, pp. III 3-5.

[19]   *Home Box Office, Inc. v. FCC,* No. 76-1280 (D.C. Cir.
       Mar. 25, 1977), *cert. denied,* 46 U.S.L.W. 3216 (U.S.
       Oct. 3, 1977) (Dkt. Nos. 76-1841 and 1842); see *Poli-
       cies and Procedures Regarding Ex Parte Communica-
       tions During Informal Rulemaking Proceedings,* Order,
       Notice of Inquiry and Interim Policy, Gen. Docket 78-
       167, June 9, 1978.

[20]   Cox, pp. 671-74, and Strassburg, pp. III 29-34.

[21]   Geller, Henry, "Case Study: 1972 Cable TV Rules, "
        in *Analysis of the Regulatory Process: A Comparative
        Study of the Decision Making Process in the Federal
        Communications Commission and the Environmental
        Protection Agency,* Report prepared for the National
        Science Foundation under Grant No. APR 75-16718,
        unpublished (Washington, D.C.: The Urban Institute,
        November 30, 1977), pp. IV 11-17.

[22]   *Ibid,* pp. IV 13, 14.

[23]   Cole, Barry, and Mal Oettinger, *Reluctant Regulators:
        The FCC and the Broadcast Audience* (Reading, Mass.:
        Addison-Wesley Publishing Co., 1978), Part III, pp.
        131-241.

[24]   *KORD, Inc.,* 31 FCC 85 (1961); see also Geller, Henry,
        *A Modest Proposal to Reform the Federal Communi-
        cations Commission* (Washington, D.C.: The Rand Corp.,
        1974), p. 15.

[25]   *WHDH, Inc.,* 16 FCC 2d 1 (1969).

[26]   *The Evening Star Broadcasting Co.,* 27 FCC 2d 316
        (1971); *Chuck Stone et al. v. FCC,* 466 F.2d 316 (1972);
        see also Cole and Oettinger, p. 222.

[27]   *Black Broadcasting Coalition of Richmond v. FCC,* 556 F.
        2d 59(1977); see also Cole and Oettinger, pp. 224-25.

[28]   *Office of Communications of the United Church of
        Christ, et al. v. Federal Communications Commission,*
        359 F. 2 (1966); see also Krasnow and Longley, pp.
        36-41.

[29]   Cole and Oettinger, Part III.

[30]   *Policy Statement on Comparative Broadcast Hearings
        Involving Regular Renewal Applications,* 22 FCC 2d
        424 (1970); *Citizens Communications Center v. FCC,*
        447 F. 2d 1201 (1971); see also Krasnow and Longley,
        pp. 118-24.

[31]   Cole and Oettinger, pp. 138-42.

[32]   "FCC Directs Staff to Use Revised EEO Processing
       Criteria," News Release, Mimeo. No. 79238 (March
       10, 1977).

[33]   See Cole and Oettinger, p. 7; Krasnow and Longley,
       pp. 26-8; and Noll, Roger G., *Reforming Regulation:
       An Evaluation of the Ash Council Proposals.* Studies
       in the Regulation of Economic Activity (Washington,
       D.C.: The Brookings Institution, 1971), p. 43.

[34]   Weber (1964), pp. 358-63, 386-92.

[35]   Brown, Les, *Television: The Business Behind the Box*
       (New York: Harcourt Brace Jovanovich, Inc., 1971),
       pp. 167-69, 255-57.

[36]   Johnson, Nicholas, *How to Talk Back to Your Television
       Set* (Boston: Little, Brown and Company, 1967).

[37]   Krasnow and Longley, p. 31; see also Minow, Newton M.,
       *Equal Time: The Private Broadcaster and the Public
       Interest* (New York: Atheneum, 1964).

[38]   Brown, pp. 160-67.

[39]   *Children's Television Programs, Report and Policy
       Statement,* Docket No. 19142, 50 FCC 2d 1 (1974);
       Action For Children's Television v. FCC, 564 F.2d
       458 (1977).

[40]   *Citizen's Complaint Against Pacifica Foundation
       Station WBAI (FM), Memorandum Opinion and
       Order,* 56 FCC 2d 94 (1975).

[41]   Cole and Oettinger, pp. 113-15.

[42]   *Pacifica Foundation et al. v. FCC and U.S.A.,* 556 F. 2d 9
       (1977); see also *FCC v. Pacifica Foundation, et al.,* No. 77-
       528 (S.C., July 3, 1978).

[43]   *Writers Guild of America, West, Inc. et al. v. FCC,
       et al.,* 423 F. Supp. 1064 (1976).

[44]   Weber (1958), p. 247.

[45]   *Prime Time Access Rule, Report and Order,* 23 FCC 2d
       395 (1970); *Second Report and Order (PART III),* 50
       FCC 2d 829 (1974).

[46]   Brown, pp. 357-60.

[47]   *Cable Television Report and Order,* 36 FCC 2d 141
       (1972); see Geller (1977), pp. IV 5, 18, and *passim.*

[48]   Krasnow and Longley, pp. 29-31.

[49]   Allison, pp. 18-9, 22, 24, 247 and 251.

[50]   Rourke, Francis E., *Bureaucracy, Politics, and Public
       Policy* (Boston: Little, Brown and Company, 1969),
       pp. 1-8, 39-61; Weber (1964), pp. 330-31, 333-34,
       339; Weber (1958), pp. 90-1, 196, 198.

[51]   Weber (1958), pp. 82, 197, 221-24; Weber (1964),
       pp. 331-32, 334, 341-58.

[52]   Weber (1958), p. 214.

[53]   Simon, Herbert A., *Administrative Behavior: A Study
       of Decision-Making Processes in Administrative Organi-
       zation,* Second Edition (New York: The Free Press,
       1957), pp. 4-16.

[54]   Weber (1964), p. 340.

[55]   *Administrative Procedure and Judicial Review,* 80 Stat.
       381-388, 392-393; 5 U.S.C. 551-559, 701-706; and 81
       Stat. 54-56; 5 U.S.C. 552.

[56]   *Revision of Operating Rules for Class D Stations in the
       Citizens Radio Service,* Docket No. 21020; Notice of Pro-
       posed Rulemaking, 47 FCC 2d 1022 (1974); First Re-
       port and Order, 54 FCC 2d 841 (1975); Notice of
       Inquiry and Further Notice of Proposed Rulemaking,
       58 FCC 2d 928 (1976); Second Report and Order, 60
       FCC 2d 762 (1976); and Memorandum Opinion and
       Order, 62 FCC 2d 646 (1976).

[57]   Weber (1958), pp. 198, 215-16; Weber (1964), p. 330;
       Simon, pp. 103-8, 222-26, 240-41.

[58]   Noll, Chapter 3; President's Advisory Council on
       Executive Organization (Ash Council), *A New Reg-
       ulatory Framework: Report on Selected Independent
       Regulatory Agencies* (Washington, D.C.: Government
       Printing Office, 1971), pp. 13-27, 115-19; James H.

Landis, *Report on Regulatory Agencies to the President-Elect* (Washington, D.C.: Government Printing Office, 1960), pp. 53-4, and *passim.*

[59]   March, James G., and Herbert A. Simon, *Organizations* (New York: John Wiley & Sons, Inc., 1958), pp. 137-50; Simon, Chapter 5.

[60]   Leavitt, Harold J., William R. Dill and Henry B. Eyring, *The Organizational World* (New York: Harcourt Brace Jovanovich, Inc., 1973), Chapter 3.

[61]   Simon, Chapter 7; Tullock, Gordon, *The Politics of Bureaucracy* (Washington, D.C.: Public Affairs Press, 1965), Chapters 6, 11-20, 25; Downs, Anthony, *Inside Bureaucracy* (Boston: Little, Brown and Company, 1967), Chapter 11.

[62]   *First Report and Order,* Docket No. 21318, 67 FCC 2d (1978).

[63]   Cole and Oettinger, pp. 272-73.

[64]   Johnson, Nicholas and John Jay Dystel, "A Day in the Life: The Federal Communications Commission," *Yale Law Journal,* Vol. 82, No. 8., July 1973.

[65]   Simon, Chapter 8; Downs, Chapter 10.

[66]   Cole and Oettinger, pp. 11-12, 215-17; Krasnow and Longley, pp. 24-6.

[67]   *Report and Order,* 60 FCC 2d 591 (1976).

[68]   Geller (1977), pp. IV 17-8, 22, 25.

[69]   Strassburg, pp. III 36, 39 and *passim.*

[70]   Simon, Chapters 4-5; March and Simon, Chapters 6-7; Downs, Chapters 14-6.

[71]   Lindblom, Charles E., *The Intelligence of Democracy: Decision Making Through Mutual Adjustment* (New York: The Free Press, 1965), and "The Science of 'Muddling Through'," *Public Administration Review,* Vol. 19, No. 2, Spring 1959, pp. 79-88.

[72]   *First Report and Order and Second Notice of Inquiry,*
       Docket No. 18262, 35 FR 8644, June 4, 1970; see
       Docket No. 18861 and *FCC Rules and Regulations,*
       Section 2.106, footnote NG63.

[73]   *Report and Order,* Docket No. 20693, 63 FCC 2d 378
       (1976).

[74]   Simon, Chapter 5; March and Simon, pp. 141-50; Downs
       Chapter 16.

[75]   Cole and Oettinger, Part III.

[76]   Strassburg, p. III 29; Cox, pp. 671-74, 677.

[77]   Loevinger, Lee, "The Sociology of Bureaucracy," Speech
       to IEEE International Conference on Communications,
       Philadelphia, Pa., June 13, 1968.

[78]   Allison, Chapter 3.

[79]   Geller (1977), pp. IV 10-11, 17.

[80]   Cole and Oettinger, pp. 261-65.

[81]   Simon, Chapter 10; Downs, Chapters 18-19.

[82]   Downs, Chapters 2, 13.

[83]   Merton, Robert K., "Bureaucratic Structure and Person-
       ality," *Social Theory and Social Structure,* Enlarged
       Edition (New York: The Free Press, 1968), pp. 249-60.

[84]   Seidman, Harold, *Politics, Position, and Power: The
       Dynamics of Federal Organization,* Second Edition
       (New York: Oxford University Press, 1975), Chapter 5.

[85]   Cole and Oettinger, pp. 215-16.

[86]   Downs, pp. 61-3, 69-70.

[87]   Blau, Peter M., *The Dynamics of Bureaucracy: A Study
       of Interpersonal Relations in Two Government Agencies,*
       Revised Edition (Chicago: University of Chicago Press,
       1963), and *On the Nature of Organizations* (New
       York: John Wiley & Sons, 1974), Chapter 2; Peter
       M. Blau and W. Richard Scott, *Formal Organizations:
       A Comparative Approach* (San Francisco: Chandler
       Publishing Co., 1962).

[88]   March and Simon, Chapters 3-4.

[89]   Leavitt, pp. 30-44, 128-47, 167-8; Berkley, George E.,
        *The Administrative Revolution: Notes on the Passing
        of Organization Man* (Englewood Cliffs, N.J.: Prentice-
        Hall, Inc., 1969).

[90]   Tullock, p. 160.

[91]   Weber (1958), pp. 88, 108-10, 200-2, 242.

[92]   Savas, E.S., and Sigmund G. Ginsburg, "The Civil Service:
        A Meritless System?" *The Public Interest,* No. 32, Sum-
        mer 1973, pp. 7-85.

[93]   Stanley, David T., "What are Unions Doing to Merit
        Systems?" *Civil Service Journal,* Vol. 12, No. 13,
        January-March 1972, pp. 10-14.

[94]   Weber (1964), p. 384.

[95]   Merton, Robert K., "Role of the Intellectual in Public
        Bureaucracy," *Social Theory and Social Structure,* En-
        larged Edition (New York: The Free Press, 1968), pp. 261-
        78; Downs, p. 203; Berkley, pp. 66-90.

[96]   "FCC Issues Post-Facto Permit to FM," *Broadcasting,*
        May 30, 1977, pp. 39-40.

[97]   "Radiotelegraph Operator License Certificates to Bear
        Photographs," *Public Notice,* FCC 77-600, Mimeo No.
        83072, August 31, 1977.

[98]   Rourke, *passim.*

[99]   Weber (1958), pp. 90-1; Weber (1964), pp. 333-35.

[100]   Weber (1968), pp. 392-404.

[101]   Geller (1977), pp. IV 18-19.

[102]   Brown, pp. 253-54, 259-60.

[103]   Krasnow and Longley, pp. 112-120; *Citizens Communica-
        tions Center v. FCC,* 447 F. 2d 1201 (1971).

[104]   Downs, Chapter 17.

[105]   Allison, Chapter 5.

[106]   Cole and Oettinger, p. 264.

[107] *Program Review of the Citizens Radio Service,* FCC
       Staff Report, December 6, 1976.

[108] Cole and Oettinger, p. 264.

[109] Allison, p. 257.

[110] Bernstein, Marvin H., *Regulating Business by Inde-
       pendent Commission* (Princeton, N.J.: Princeton
       University Press, 1955), p. 219; Noll, pp. 39-45;
       Olson, Mancur, *The Logic of Collective Action:
       Public Goods and the Theory of Groups* (Cambridge,
       Mass.: Harvard University Press, 1971).

[111] Cole and Oettinger, Chapter 3; Krasnow and Longley,
       pp. 31-5.

[112] Cole and Oettinger, pp. 8-10, 30-34.

[113] Noll, Roger G., Merton J. Peck and John J. McGowan,
       Economic Aspects of Television Regulation. Studies in
       the Regulation of Economic Activity (Washington, D.C.:
       The Brookings Institution, 1973), p. 123.

[114] Cole and Oettinger, pp. 44-8; see *Policies and Procedures
       Regarding Ex Parte Communications During Informal Rule-
       making Proceedings,* Order, Notice of Inquiry and Interim
       Policy, Gen. Docket No. 78-167, June 9, 1978.

[115] U.S., Congress, House, Committee on Interstate and
       Foreign Commerce, Subcommittee on Oversight and
       Investigations, *Federal Regulation and Regulatory Reform,*
       Subcommittee Print, 94th Cong., 2d sess., 1976, pp.
       251-53.

[116] Krasnow and Longley, Chapter 5.

[117] Noll (1971), pp. 23-7.

[118] Cole and Oettinger, Part IV; *Action For Children's Television v.
       FCC,* 564 F. 2d 458 (1977).

[119] Crotts, G. Gail, "The Public Information Function of
       the Federal Communications Commission," Ph.D. dis-
       sertation, University of Illinois at Urbana-Champaign,
       1974, pp. 253-58.

[120] Bernstein, pp. 156, 285.

[121]   Krasnow and Longley, Chapter 3.

[122]   *Ibid,* pp. 55-6.

[123]   Geller (1977), pp. IV 19-21.

[124]   See U.S., Congress, House, Committee on Interstate
        and Foreign Commerce, Subcommittee on Communi-
        cations, *Option Papers,* Subcommittee Print, 95th Cong.,
        1st sess., May 1977; *The Communications Act of 1978,*
        H.R. 13015, 95th Cong., 2d sess., June 7, 1978.

[125]   Krasnow and Longley, Chapter 6; *All Channel Televi-
        sion Receiver Act,* P.L. 87-529, 87th Cong., July 10,
        1962.

[126]   Krasnow and Longley, Chapter 7.

[127]   *Ibid,* Chapter 8.

[128]   Strassburg, pp. III 24-30.

[129]   *Consumer Communications Reform Act of 1976, S.*
        530, 95th Cong., 1st sess., 1976.

[130]   Krasnow and Longley, pp. 44-7.

[131]   *Ibid,* pp. 48-50.

[132]   *Ibid,* pp. 47-8; see *FCC v. National Citizens Committee For
        Broadcasting et al.,* No. 76-1471 (S.C. June 12, 1978).

[133]   Strassburg, pp. III 25-6.

[134]   *Writers Guild of America, West., Inc. et al. v. FCC et al.,*
        423 F. Supp. 1064 (1976).

[135]   *Action For Children's Television v. FCC,* 564 F. 2d
        458 (1977).

[136]   *United States v. Southwestern Cable Co.,* 392 U.S. 157,
        178 (1968); see also *Midwest Video Corp. v. FCC,*
        No. 76-1496 (8th Cir. Feb. 27, 1978).

[137]   Lowi, Theodore J., *The End of Liberalism: Ideology,
        Policy, and the Crisis of Public Authority* (New York:
        W. W. Norton & Co., 1969).

[138] Chayes, Abram, "The Role of the Judge in Public Law
       Litigation," *Harvard Law Review,* Vol. 89, No. 7, May
       1976, pp. 1281-1316; Stewart, Richard B., "The Re-
       formation of American Administrative Law," *Harvard
       Law Review,* Vol. 88, No. 8, June 1975, pp. 1669-1813.

[139] *United States v. Midwest Video Corp.,* 406 U.S. 649,
       667-69 (1972).

[140] *Home Box Office, Inc. v. FCC,* No. 76-1280 (D.C.
       Cir. March 25, 1977), *cert. denied,* 46 U.S.L.W.
       3216 (U.S. Oct. 3, 1977) (Dkt. Nos. 76-1841 and
       1842).

[141] *Midwest Video Corp. v. FCC,* No. 76-1496 (8th Cir.
       Feb. 27, 1978); FCC has filed an appeal.

[142] *Brookhaven Cable et al. v. Robert F. Kelly et al.,*
       Docket Nos. 77-6156, 77-6157 (2nd Cir. March 29,
       1978).

[143] Noll (1971), pp. 34-8, 101-02; Krasnow and Longley,
       pp. 16-17, 57.

[144] Seidman, Chapters 1 and 10.

[145] President's Advisory Council on Executive Organization
       (Ash Council), p. 13.

[146] Madison, James, *The Federalist Papers #10;* Dahl,
       Robert A., *Democracy in the United States: Promise
       and Performance,* Second Edition (Chicago, Rand
       McNally & Co., 1972); Dahl, Robert A., and Charles
       E. Lindblom, *Politics, Economics and Welfare* (New York:
       Harper & Row, 1963); Leavitt, Chapter 17.

[147] Parks, Rolla Edward (ed.), *The Role of Analysis in
       Regulatory Decision Making: The Case of Cable Tele-
       vision* (Lexington, Mass.: D.C. Heath and Company,
       Lexington Books, 1973).

[148] Kaufman, Herbert, *Administrative Feedback: Monitoring
       Subordinates' Behavior* (Washington, D.C.: The Brookings
       Institution, 1973), Chapter 9; Hargrove, Erwin C.,

The Missing Link: The Study of the Implementation of Social Policy, Urban Institute Paper 797-1 (Washington, D.C.: The Urban Institute, July 1975), pp. 110-17.

[149]   Rourke, pp. 11-37, 63-86.

[150]   U.S., Congress, House, Committee on Interstate and Foreign Commerce, Subcommittee on Oversight and Investigations, *Federal Regulation and Regulatory Reform,* Subcommittee Print, 94th Cong., 2d sess., October 1976, pp. 245-77.

[151]   Noll (1971), Chapter 3.

[152]   Noll (1971), pp. 93-4, 108; Rourke, pp. 122-28.

[153]   *Administrative Procedure Act,* P.L. 404, 79th Cong., 2d Sess., June 11, 1946, 60 Stat. 237, 5 U.S.C. 1001-1011.

[154]   Noll, pp. 103-5.

[155]   President's Advisory Council on Executive Organization (Ash Council), pp. 21-2, 53-5.

[156]   Geller (1974), pp. 48-9; *The Communications Act of 1978,* Sec. 236 (a)(1).

[157]   Noll (1971), pp. 94-5.

[158]   Weber (1964), p. 324.

[159]   Noll (1971), pp. 2, 4-5, 13-14, 97-8.

[160]   Leibenstein, Harvey, "Allocative Efficiency vs. 'X-Efficiency'," *American Economic Review,* Vol. 56, No. 3, June 1966, pp. 392-415.

[161]   Noll (1971), pp. 1-2, 79-80; President's Advisory Council on Executive Organization (Ash Council), pp. 20-2, 36-40, 48-52.

[162]   Weber (1958), Chapter 11.

[163]   Johnson and Dystel, *passim.*

[164]   *Ibid,* pp. 1633-34.

[165]   Landis, *passim.*

[166] The President's Committe on Administrative Management
       (Brownlow Committee), *Administrative Management in
       the Government of the United States* (Washington, D.C.:
       U.S. Government Printing Office, January 1937), Senate
       Document No. 8, 75th Cong. 1st sess.; President's
       Advisory Council on Executive Organization (Ash
       Council).

[167] Commission on Organization of the Executive Branch
       of the Government (Hoover Commission), *Task Force
       Report on Regulatory Commissions,* Appendix N
       (Washington, D.C.: U.S. Government Printing Office, Jan-
       uary 1949); Pritchett, C. Herman, "The Regulatory
       Commissions Revisited," *American Political Science
       Review,* Vol. 43, October 1949, pp. 978-89.

[168] *Federal Communications Commission Annual Report
       to Congress* (Washington, D.C.: Government Printing
       Office, 1950).

[169] Minow, pp. 290-306.

[170] Downs, pp. 118-27; Tullock, pp. 217-20.

[171] Wilson, James Q., "The Bureaucracy Problem," *The
       Public Interest,* No. 6, Winter 1967, pp. 3-9.

[172] Levine, Robert A., *Public Planning: Failure and
       Redirection* (New York: Basic Books, Inc., 1972),
       pp. 82-102, 174-75; Murphy, Jerome T., *State Edu-
       cation Agencies and Discretionary Funds: Grease the
       Squeaky Wheel* (Lexington, Mass.: D.C. Heath and Co.,
       Lexington Books, 1974), pp. 149-56; Kaufman, pp. 74-9.

[173] Johnson, *passim.*

[174] *Reimbursement of Expenses For Participation in FCC Pro-
       ceedings,* Notice of Inquiry, Gen. Docket No. 78-205, FCC
       No. 78-478 (June 30, 1978).

[175] Noll (1971), pp. 103-5; Johnson and Dystel, p. 1634.

[176] Lowi, Theodore J., *The Politics of Disorder* (New York:
       Basic Books, 1971), Chapter 2.

[177] Geller (1974), *passim.*

[178]   Lowi (1969), *passim.*

[179]   Lowi (1971), *passim.*

[180]   Noll (1971), p. 46.

[181]   Olson, *passim.*

[182]   Hartz, Louis, *The Founding of New Societies* (New York: Harcourt, Brace & World, 1964), Chapters 1-4.

[183]   Noll (1971), *passim.*

[184]   Hirschman, Albert O., *Exit, Voice, and Loyalty: Responses to Decline in Firms, Organizations, and States* (Cambridge, Mass.: Harvard University Press, 1970).

Chapter 3

FCC REGULATION OF LAND MOBILE RADIO —
A CASE HISTORY*

Dale N. Hatfield
U.S. Department of Commerce
Boulder, Colorado

*The views expressed in this Chapter are the private ones of the author and do not necessarily represent the position of the U.S. Department of Commerce.

3.0 INTRODUCTION

Some form of radio is the only practical means of communicating
with vehicles that move about widely on the land, on the sea, in the
air, or through outer space. Communication with ships at sea was
one of the earliest applications of radio, and lives were saved
through its use when the art was still in its infancy. Because there
are always alternatives for communicating between two fixed
points on land (e.g. by wirelines or underseas cable), land mobile,
aeronautical mobile, maritime mobile, and, more recently, space
communications have always been regarded as being among the
highest priority uses of the radio spectrum. This chapter will concentrate solely on *land* mobile radio in the United States and its
regulation by the Federal Communications Commission (FCC).
Land mobile radio includes communication through portable radio
units as well as through mobile units permanently installed in vehicles. Only business and personal type communications will be
considered, not communications for entertainment or as a hobby.

FCC regulation of land mobile radio is a particularly interesting
case history for a number of reasons. First, it has involved a series
of extremely difficult and far-reaching decisions in a fundamental
area of FCC responsibility — the allocation of the scarce radio
spectrum among competing users. Second, it provides important
examples of the profound influence of FCC actions on communications industry structure and the types, quality, quantity, and
costs of services ultimately provided to the public. Third, and
finally, the delays in the introduction of new technologies and the
uncertainties produced by FCC involvement in land mobile radio
are often cited as reasons for proposing basic changes in the present regulatory scheme.

3.1 DEFINITIONS

There are currently three basic types of land mobile radio services:
one-way signaling (paging), dispatch, and mobile telephone.

The first, one-way signaling or paging, uses a radio signal to merely
alert or instruct the user to do something. The user (an office
machine repairman or doctor, for example) carries a small receiver
which is automatically actuated when a message is directed to it.
There are two types of paging systems: tone-only and tone-voice.

In the former, the receiver emits a tone and the user takes some predetermined action such as calling his or her office or telephone answering service. In the latter, the tone is followed by a short voice message which allows alternative instructions; e.g., call a specific number. Tone-only is cheaper because less "air-time" is used and the equipment can be somewhat less complex.

Dispatch communications allows two-way communications between a base station and mobile units or between mobile units without access to the regular, switched public message telephone system. Dispatch communications are used to coordinate and control a fleet of mobiles; e.g., police cars, taxis, or cement trucks. The mobile user can normally talk only to the dispatcher or another mobile; they cannot dial a telephone in the regular landline network. Messages on a dispatch system are typically short, usually one minute or less.

Mobile telephone services allow the mobile user to receive or place regular calls through the exchange (local) and message toll (long distance) facilities of the regular landline telephone system. The service can be exactly equivalent to regular landline service except that the telephone can be mounted in a vehicle or perhaps carried in a briefcase. Messages on a mobile telephone system are typically of longer duration.

These three types of services can be provided on a private or common carrier basis and there are also shared systems. In a private system, the user — a taxicab company for example — is independently licensed and owns (or leases) the base station and mobile equipment for his own exclusive use. In a shared system, several users may band together and share the use and cost of a system. In a common carrier system, the service is provided for hire by the common carrier who owns the base station equipment and other facilities. The subscriber can own his own mobile equipment or lease it from the carrier.

3.2 EARLY TECHNOLOGICAL DEVELOPMENTS

In a number of articles [1, 2, 3, 4], the history of the technological development of land mobile radio has been traced and hence it will only be summarized here. One of the earliest uses of land mobile radio was by the Detroit Police Department who began ex-

perimenting with a one-way (i.e. base-to-car) system in early
1921. By the late 1920's they had achieved widely publicized success, and by 1930 there were police radio stations in 29 cities. In
1932 the first license for a mobile transmitter in a vehicle was issued. This permitted two-way communications. Growth from this
point was rapid, and by the onset of World War II several other
services besides police radio were in existence. These were private
dispatch systems as described above; that is, communications occurred only between the dispatcher and the associated vehicles and
no interconnection with the conventional telephone network was
provided.

In the late 1930's and early 1940's, the first Very High Frequency
(VHF) band came into use and Frequency Modulation (FM) was
introduced. These specific advances and the extensive research
and development to support military requirements for mobile
communications during the war years brought about considerable
improvements in performance and greatly stimulated further
growth of land mobile radio.

3.3 REGULATORY AND FURTHER TECHNOLOGICAL
DEVELOPMENTS

Problems of channel congestion caused by the rapid growth of
land mobile communications plagued the industry and the regulators almost from the beginning. In the early 1930's, the Federal
Radio Commission, the predecessor of the FCC, allocated frequencies just above the broadcast band for police use and the FCC allocated the VHF channels, referred to above, soon after it was
formed in 1934. Because of the rapid growth, the FCC formally
recognized several categories of mobile radio services in the general
frequency allocation proceeding (Docket 6651) of 1945. Late in
this proceeding, the Bell System requested the allocation of exclusive channels for common carrier mobile telephone service. The
FCC then initiated a rule-making proceeding to address the entire
question of frequency allocations for land mobile radio. This was
entitled the General Mobile Radio Proceeding (Docket 8658) and
the Commission reached its decision in 1949. During the period

from 1945 to 1949 the Bell System and the independent telephone companies started plans to provide mobile telephone service throughout the nation. The first urban mobile telephone system was established in 1946 in St. Louis, Missouri, on an experimental basis.

The growth in private specialized uses was dramatic during this period, and, in the decision in Docket 8658, the Commission recognized a number of specific subcategories of service including Police Radio, Land Transportation, Automobile Emergency, and several others. At the same time, the Commission allocated channels for mobile telephone service. Significantly, it provided separate sets of frequencies for the landline (wireline) common carriers (WCCs) and miscellaneous (radio) common carriers (RCCs), thus consciously providing for competition in the provision of public mobile telephone service.

The next major regulatory decision affecting land mobile radio came in 1958 under FCC Docket 11991. This proceeding established the business radio service, a category that yielded a substantially greater number of eligible licensees. This service rapidly outgrew the others in terms of numbers of licensed transmitters in service. Another major milestone for the RCCs occurred in 1961 when they were able to negotiate an interconnection agreement with the Bell System. Up until that time they were unable to effectively compete with the WCCs because they could not interconnect their base station facilities with the public message landline network. This meant that an operator had to relay each message between a mobile and a wireline telephone. In the Bell mobile telephone system of the era, each call was handled manually by an operator, but once connected, the mobile could carry out essentially a normal telephone conversation with any landline telephone. Dial mobile operation was introduced much earlier than this, but did not come into wide use until the advent of the Improved Mobile Telephone System (IMTS) in the mid-1960's.

Another major growth area was paging. This type of one-way service to vehicles was offered by the Bell System on an experimental basis as early as 1946. In 1960 small portable receivers were developed and extension of the service to these convenient types of units added greatly to the appeal. In Docket 16776, the

FCC reallocated channels for private paging systems in the Industrial Radio Service. The RCCs in particular exploited the common carrier supplied paging market and, recognizing the greatly increased demand, the FCC in 1968 (Docket 16778) allocated two new channels to the RCCs and two new channels to the WCCs for exclusive use in paging systems.

By the mid-1960's, the continuing rapid growth of land mobile again produced congestion on the channels. This crowding was especially severe in large urban areas. The problem was addressed in reports by a number of groups including the President's Task Force on Communications Policy [5], the Advisory Committee for the Land Mobile Radio Services (ACLMRS) [6], and the Joint Technical Advisory Committee [7]. In 1967 the Commission itself established a Land Mobile Relief Committee to study various options. As a result of these studies, the Commission undertook two steps in the late 1960's to alleviate the situation. On the one hand, they established a Spectrum Management Task Force to design a system to more effectively manage the existing land mobile frequency resources (Docket 19150) [8]. On the other hand, the Commission moved to allocate additional frequencies for land mobile use in Dockets 18261 and 18262.

In Docket 18261, the Commission proposed in 1968 to provide for almost immediate relief by allowing the land mobile services to share the lower seven UHF television channels (i.e. channels 14-20) in the largest urban areas. These seven channels, which lie in the frequency range of 470 to 512 MHz, are immediately adjacent to an existing land mobile radio allocation (450 to 470 MHz). This proximity to an existing allocation greatly facilitated equipment availability.

Broadcasters strongly opposed the proposal and sought to thwart it entirely or to sharply limit its scope. This confrontation between broadcasters and land mobile radio interests has been a continual one and persists to this day. It has been compounded by the rapid growth of land mobile radio from the early police systems serving a few hundred mobiles on a one-way basis to a market where several million mobiles are being served. Faced with this growth and acknowledging both the importance of the land mobile service and that further technological improvements could do little to improve the utilization of existing allocations, the Commission finally voted to provide short-term relief by allowing sharing on two of the lower seven UHF television channels in each of ten major urban areas (Boston, Chicago, Cleveland, Detroit, Los Angeles, New York, Philadelphia, Pittsburgh, San Francisco, and Washington, D.C.). This decision was reached in June of 1970.

At the same time it proposed the sharing in Docket 18261, the Commission also proposed in another proceeding to reallocate the uppermost fourteen UHF television channels (i.e. channels 70-83) to the land mobile service. Because of the importance of this proceeding (Docket 18262) to the future of land mobile radio, it will be described in more detail in the following section. Before turning to that description, however, it may be useful to briefly review the frequency allocations for land mobile radio as they stood prior to the reallocations of additional spectrum in Dockets 18261 and 18262.

Private land mobile radio systems, regulated under Parts 89, 91, and 93 of the FCC Rules and Regulations, fall under the general classification of Safety and Special Radio Services. Within this general classification, there are three broad categories of private land mobile radio services defined in the rules: Public Safety, Industrial, and Land Transportation. These categories are, in turn, divided into subcategories as shown in Table 3-1.

Table 3-1 Safety and Special Radio Services

Public Safety Services
Fire
Forestry Conservation
Highway Maintenance
Local Government
Police
Special Emergency
State Guard

Industrial Services
Business
Forest Products
Industrial Radiolocation
Manufacturers
Motion Picture
Petroleum
Power
Relay Press
Special Industrial
Telephone Maintenance

Land Transportation Services
Automobile Emergency
Interurban Passenger
Interurban Property
Railroad
Taxicab
Urban Passenger
Urban Property

Before the reallocations of Dockets 18261 and 18262, these three broad categories of private land mobile services had access to radio channels in the 30-50 MHz, 150 MHz, and 450 MHz regions of the spectrum. These regions are referred to in the land mobile community as the low band, high band, and UHF band, respectively. The actual amount of spectrum allocated to the private land mobile services in these three bands is summarized in Table 3-2 [9].

Table 3-2 Pre-Docket 18261/18262 Spectrum Allocations to
*   Private LMR Services*

Spectrum in Megahertz (MHz)

Band	Public Safety	Industrial	Transportation	Total
Low Band	7.160	5.880	1.420	14.460
High Band	4.110	2.295	2.610	9.015
UHF Band	3.375	10.475	1.750	15.600
	14.645	18.650	5.780	39.075

In addition to the approximately 39 MHz of spectrum allocated to
private systems, another 3 MHz of spectrum spread among the
same bands was (and is) allocated to the WCCs and RCCs for the
provision of radio telephone and radio paging service. The spectrum allocated to this service, which is formally known as the
Domestic Public Land Mobile Radio Service, is divided evenly between the two types of common carriers.

In the early 1970's, not long after the initiation of Docket 18262,
Motorola estimated the total land mobile radio market as shown in
the following table [10] :

Table 3-3 Relative Shares of the LMR Market (Based on Figures
*   from the FCC Annual Report for 1971)*

	Mobiles in Service	Percent
Private Dispatch	2,537,996	94.2
Common Carrier —		
Bell System	62,372	2.2
Non-Bell WCC	11,130	.4
RCCs	84,025	3.1
Total	2,695,523	100.0

Growth of the land mobile radio market is estimated to be about
10 percent per year except for radio paging which is somewhat
higher. Today (1978), it is estimated that there are over four million mobile units in service. The extent to which this growth has

been limited by lack of available channels cannot be accurately estimated. However, in the case of the Domestic Public Land Mobile Radio Service, the congestion in major metropolitan areas was so severe by 1957 that priorities had to be established for determining which types of subscribers would be afforded service.

3.4 DOCKET 18262: REALLOCATION OF SPECTRUM SPACE NEAR 900 MHz FOR LAND MOBILE SERVICES

3.4.1. Background

As noted in Section 3.3, Docket 18262 stemmed from a number of studies of land mobile radio conducted in the mid-1960's. The most direct impetus, however, came from the release on March 20, 1968, of the previously referenced report of the Land Mobile Frequency Relief Committee. The report consisted of three FCC staff studies, the last of which addressed the problems and costs associated with reallocating all or a part of the 806-960 MHz band to the land mobile radio services. This staff study was the genesis of Docket 18262. The remainder of this section is devoted to (1) a description of the issues that were raised as a result of the subsequent reallocation, and (2) a review, in chronological order, of the Commission's actions in resolving these issues.

3.4.2. Major Issues

Land Mobile Radio Versus Competing Uses. The threshold question in Docket 18262 was whether or not spectrum in the vicinity of 900 MHz should be reallocated to land mobile radio. The radio spectrum is a scarce natural resource which has the rather interesting property that, unlike most other natural resources such as oil or coal, it is not depleted by use. However, if two or more users attempt to use the same frequency at the same time and in the same vicinity, harmful interference can result. Since the radio spectrum is limited in extent and since various portions have more valuable characteristics than others in certain applications, society must apportion spectrum among competing users in some coordinated fashion to avoid chaotic interference. Under the Communications Act of 1934, the FCC has the responsibility for allocating spectrum for all non-federal government users. This responsibility is described in general in Chapter 2 of this book.

In Docket 18262, the competing service was predominantly television broadcasting since the proposal eventually called for a direct reallocation of UHF television channels 70 through 83, a total of 84 MHz of spectrum. Thus the Commission was faced with the fundamental issue of whether the public interest would be better served by reallocating the spectrum to land mobile radio or preserving it for UHF television.

The competition between land mobile radio and broadcast interests over frequency space in the UHF portion of the spectrum began before the initiation of Docket 18262. As early as 1949 AT&T proposed a high capacity, common carrier mobile radio system for operation in the 470-500 MHz band. In a July 11, 1951 decision in Docket 8976, the Commission allocated the spectrum to television broadcasting instead. In deciding the needs of television were paramount in the 470-500 MHz band, the Commission stated that:

> "In arriving at this conclusion we are forced to resolve
> a conflict between two socially valuable services for
> the precious spectrum space involved. We find that
> needs of each of the two services are compelling."[11]

The conflict appeared again in Docket 11997, a proceeding initiated by the Commission in April of 1957. This proceeding dealt with the broad issues of non-federal government spectrum allocations in the entire 25 to 890 MHz band, and AT&T once again indicated the need for a substantial amount of spectrum for an efficient, high capacity domestic public land mobile service. Although the Commission acknowledged the importance of land mobile service, in 1964 it denied the request and reaffirmed its stance that all 70 UHF television channels (i.e. channels 14 to 83) were necessary to insure a competitive and effective system of television broadcasting.

As an alternative to encroaching on the UHF channels, the Commission sought technical solutions short of outright reallocation. These proved insufficient, and the Commission was caught between a burgeoning demand for land mobile services on the one hand and a disappointing growth of UHF television broadcasting on the other.

The disappointing growth in UHF television was originally attributed to a number of factors including (a) the Commission's decision to allow both VHF and UHF stations in the same market, (b) the lack in many households of television sets with UHF reception capability, (c) significant technical limitations of UHF relative to VHF, and (d) the frequent lack of network affiliation or other access to attractive programming by UHF stations. The Commission tried a policy of "selective deintermixture" and the All Channel Receiver Act was enacted. The latter required that all television sets manufactured after 1964 be equipped to receive both VHF and UHF channels. Even with these and other policies favoring UHF development, growth continued to be slow and a large percentage of UHF channels remained unused. This lack of use, coupled with the large numbers of UHF channels that were unused because of television receiver "taboos," presented an inviting target for the advocates of additional spectrum for land mobile radio. At the time Docket 18262 was initiated, UHF channels 70-83 were used almost exclusively for unattended translators* rather than by regular broadcast stations.

Groups associated with land mobile radio interests pointed to this apparent underutilization and advocated reallocation. Broadcasting groups, including educational broadcasters, argued that the congestion on existing land mobile channels was the result of poor frequency management techniques rather than lack of spectrum, and that reallocations could be avoided by improvements in these methods. They argued that the Commission's policies toward UHF broadcasting were beginning to bear fruit and that eventually all 70 channels would be needed to provide a sufficient number of outlets in urban areas and to provide for educational broadcasting needs.

*Translators are used to extend the coverage area of a broadcast station. A translator station receives a signal from a distant television station and retransmits it on another channel. When the translator is located at an advantageous place (on a hilltop, for example), viewers who cannot pick up the station directly can receive it via the translator.

Additional spectrum in the vicinity of 900 MHz for land mobile use was also contemplated through reducing the amount
of spectrum allocated to the Industrial, Scientific, and Medical
(ISM) Service. This band near 900 MHz was used primarily for
microwave ovens. Reducing the size of the ISM band increases
the costs of the microwave ovens. The Commission also proposed
reducing the amount of spectrum allocated to broadcasters for
studio-to-transmitter links (STLs) in the vicinity of 900 MHz.

In summary, then, the threshold question in Docket 18262 was
dominated by the UHF television broadcasting versus land mobile spectrum allocation issue. Lesser issues centered around
the STL and ISM band reallocations.

Types and Sizes of Suballocations.   Following the resolution
of the threshold reallocation issue just described, the next major issue centered around how the new allocation should be
subdivided among the various land mobile services: domestic
public (common carrier), public safety, industrial, and land
transportation. The Commission simplified the debate by
lumping the latter three categories under the general heading
of private systems. Essentially the competition for spectrum
within the new allocation was between the domestic public
land mobile radio service offered by the WCCs, such as AT&T,
and the private land mobile systems. In addition, the RCCs
also sought spectrum for public offerings. The resulting intra-
LMR service rivalry was in many ways as intense as the inter-
service rivalry between UHF television broadcasting and land
mobile radio. This rivalry was further intensified when the Commission provided for the licensing of "specialized mobile radio
(SMR) systems", a concept which will be described in a later
section.

The Bell System argued that their high capacity mobile telecommunication system (HCMTS) required substantial amounts
of spectrum to achieve economic and spectral efficiency and to
accommodate future growth. Proponents of private systems
argued that the common carrier growth forecasts were overly
optimistic, that growth had been and would continue to be in

private conventional and shared dispatch systems, and that Bell's HCMTS was not necessarily the most efficient way of handling dispatch communications.

The Commission changed the proposed suballocations from time to time during the course of the proceeding. These changes will also be described in a later section.

Technology Choices. In addition to the private versus common carrier service dichotomy and the disputes over the amount of spectrum to be allocated to them, the Commission was also faced with a choice among competing technologies for providing LMR services at 900 MHz. Three such classes of systems emerged during the proceeding: conventional, trunked, and cellular.

A private, single-channel system used for dispatch service is perhaps the simplest conventional system to visualize. It consists of a base station transmitter-receiver combination with minor accessories, an antenna tower and antenna, and the individual mobile transceivers. The base station equipment is frequently located on the premises of the business or public agency, but it may be located remotely at an advantageous location for good coverage. The channel may be shared on an informal basis with a number of similar systems in the same service.

Increased geographic coverage, especially for mobile-to-mobile or portable-to-portable communication, can be accomplished through the use of a repeater located on a very high tower, building, or mountain top. A repeater receives a mobile (or base station) signal on one frequency and retransmits it on a second frequency. Because of its advantageous location, the repeater can normally receive and transmit over a large area. Thus, two low-powered mobiles or portables which are only a few miles apart may not be able to communicate directly with each other, but, because they are both within line-of-sight of the repeater, they are able to communicate through it. It also has a cost advantage if several systems are sharing a single-channel pair because, for equivalent coverage, each system would need an expensive base station on a high building or mountain top. With a repeater, the individual base stations can use very simple low-power equipment.

Conventional systems can provide access to more than one channel. They can be distinguished from multi-channel trunked systems described below by the fact that the access is provided manually rather than under computer control.

A conventional common carrier mobile telephone system is similar in principle to the non-repeater system described above. Added base station equipment would be required, principally in terms of added control equipment and facilities for interconnecting with the landline network. The mobile units cost significantly more due to added control, dialing, and duty cycle requirements.

The second class of system considered by the Commission was the multi-channel trunked system. Both Motorola and General Electric proposed such systems for use at 900 MHz.

To fully appreciate the advantages of these systems, it is instructive to consider one of the technical objections to conventional single channel systems: low spectral efficiency. This inefficiency can be visualized by considering a collection of channels each occupied by one or more conventional single-channel systems in the same or different service. Spectral inefficiency can result when some frequencies are heavily used and others are only lightly used in a particular area. This can occur, for example, in a major urban area where forestry frequencies are lightly used while taxicab frequencies are overcrowded. In a long-term sense, this can be relieved to a certain extent by managing frequencies on a regional rather than a national basis. Even then, during a given hour or shorter period of time, there may be certain channels that are being heavily used and others which are only being lightly used or perhaps not being used at all. It is intuitively obvious that pooling a group of channels together and giving the users access to all channels will give them better service, or, for the same quality of service, it will enable a greater volume of traffic to be handled. This is known as trunking and the resulting improvements can be easily predicted.

In the systems proposed for the new band, access to a large number of channels (say 6 to 20) is controlled by a computer which gives the user a channel for the duration of his message, or, if all channels are busy, it places him in a queue stored in the machine. Such systems are essentially precluded from existing bands because of assignment regulations and lack of sufficient channels to fully exploit trunking. There is fairly extensive use of trunking in the common carrier supplied mobile telephone service, but only a few channels are available.

The cellular system is the third class of system considered by the Commission. It was proposed by AT&T and, subsequently, by others. The availability of a wide block of spectrum and the propagation characteristics at 900 MHz made cellular systems ideal for the new band. The cellular system gets its name from the concept of dividing a geographic area into a series of hexagonal cells. The hexagonal cell was chosen because its shape roughly approximates the circular coverage of a base station and because when they are fitted together they completely cover the area. Base station transmitters and receivers are placed in each cell and connected by wirelines to the central switching computer and from there into the regular telephone network. The base station transceivers provide coverage for the particular cell, but naturally a signal strong enough to provide good service at the edge of the cell will also enter the immediately adjacent cells. This means that the frequencies used in one cell cannot be reused in these adjacent cells, but they can be used again in more distant cells. AT&T proposed a seven-cell repeating pattern.

The key to the cellular concept is that in an initial system covering a market area the cells can be quite large, and as demand develops these large cells can be subdivided into smaller cells. When the cells are large, a particular channel might only be used a few times in the area; while in the final system, it might be used 10 or more times. Thus, the system becomes more and more spectrally efficient; i.e., more and more users are accommodated in the same amount of spectrum. In the 900 MHz band, there is a sufficient number of channels available that each cell will have an adequate number to achieve good trunking efficiency. The overall spectral efficiency of the cellular system, then, comes from both trunking and extensive geographic reuse of channels. In a conventional system, reuse of a channel may be precluded over an entire metropolitan area and is spectrally inefficient in that sense.

A major technical problem in the cellular systems is locating a particular mobile and transferring a call from cell to cell as the mobile moves about. For example, when a call is placed to a mobile from the landline network, the mobile must be located in any one of perhaps 50 to 100 cells in the area. Once the conversation is initiated the mobile may move from one cell to another, and this requires that the system have the capability of tracking the mobile. The smallest cell proposed by AT&T has a radius on the order of one mile, hence during a normal length telephone conversation at freeway speeds, the mobile may transit several cells.

Industry Structure. It is clear that the Commission's decisions regarding the suballocations of the spectrum in the 900 MHz band among the different services and among different types of service providers would have a significant impact on the resulting industry structure. In the extreme, for example, the Commission might have excluded either common carrier or private systems entirely. In fact, at one point in the proceeding, the Commission concluded that only wireline common carriers should be licensed to operate high capacity cellular systems. Similarly, its decisions regarding the number of SMR systems to be allowed in an area and the ease of entry of such systems would have a tremendous effect on the degree of competition among SMR systems and between SMR and private and common carrier systems.

In addition to these rather obvious decisions, the Commission was also faced with other issues concerning industry structure. Some parties in the proceeding feared that the wireline common carriers might use revenues from their monopoly telephone services to cross-subsidize their mobile services. They argued that such cross-subsidies could give wireline carriers an unfair advantage and could destroy desirable competition in the supply of dispatch services and equipment. To paritally combat this potential problem, various parties urged the Commission to require the wireline common carriers to set up a separate corporate subsidiary to offer mobile services. The Commission was also urged not to allow the wireline common carriers to manufacture mobile (as opposed to base station) transmitting and receiving equipment. Again the fear was that the wireline common carriers might destroy competition in the provision and maintenance of LMR equipment.

The fear was also expressed that mobile equipment manufacturers (e.g. Motorola, GE, and RCA) might dominate the market for trunked type SMR system offerings, and the Commission was urged to limit the number of such systems each manufacturer could own.

3.4.3   Chronological Review of Commission Actions

Docket 18262 was formally initiated by the Commission on July 17, 1968, with the adoption of a Notice of Inquiry and Notice of Proposed Rulemaking (NOI/NPRM) [12]. The 1968 NOI/NPRM was addressed to the entire spectrum range from 806 to 960 MHz, but the Commission proposed changes in only the 806-947 MHz portion of the band. Based primarily on the work of its Land Mobile Frequency Relief Committee, the Commission proposed to reallocate a total of 115 MHz of spectrum — 40 MHz for private systems and 75 MHz for common carrier systems. This 115 MHz of spectrum was to be provided by reallocating 84 MHz from television broadcasting (UHF channels 70-83), 26 MHz by a reduction in the size of an ISM band, and 5 MHz by a 50 percent reduction in the spectrum allocated for broadcast studio to transmitter links.

The actual proposal called for the reallocation of UHF television channels 70-83 for use in the 25 largest urban areas on a co-equal basis with broadcast translators. In other words, land mobile use of the spectrum would have been confined to designated large cities.

The next Commission action in the proceeding came on May 20, 1970, when it adopted its First Report and Order and Second Notice of Inquiry [13]. In the action, the Commission firmed up its proposed reallocations and made an adjustment in the placement of the ISM band. It concluded that the LMR-translator sharing was not practical; it proposed, instead, to make the allocation exclusively for the former. It proposed to accommodate displaced translators on UHF channel 69 and below. The resulting allocations looked like the following:

806-881 MHz	Common Carrier LMR
881-902 MHz	Private LMR
902-928 MHz	ISM
928-947 MHz	Private LMR

The Commission also concluded that development of the common carrier band should be limited to wireline common carriers, and it said that the needs of the radio common carriers would be accommodated in the 470-512 MHz band being treated in Docket 18261.

Most significantly, the Commission called on AT&T and others to undertake a "comprehensive study of market potentials, optimum system configurations, and equipment design looking toward the development and implementation of an effective, high capacity common carrier service in the band 806-881 MHz" [14].

In the Second Memorandum Opinion and Order [15] adopted on July 28, 1971, the Commission responded to various petitions for reconsideration of parts of its May 20, 1970, order. In the relatively brief action, it deleted the restriction limiting the development of the 806-881 MHz band to wireline telephone companies. In addition, it clarified the earlier order by stating that it also welcomed studies and proposals for spectrally efficient, high capacity systems by private land mobile interests.

The next major Commission action was the Second Report and Order [16] adopted on May 1, 1974. In the interim between the July 1971 order and this action, the Commission received the results of a number of studies by parties responding to

their earlier request. These included extensive technical and
marketing studies for common carrier cellular systems submitted
in December 1971 by AT&T and Motorola. The latter also filed
information on a computer controlled trunked system for private systems and, in April 1973, it submitted a further technical
report on a portable telephone system. The AT&T cellular system proposal called for the use of 64 MHz of spectrum for
the offering of domestic public land mobile service and for
11 MHz of spectrum for a public air-to-ground service.

The Second Report and Order of May 1, 1974, was the Commission's major decision in Docket 18262. It rejected outright
AT&T's proposal to use 11 MHz of spectrum for public air-to-
ground service. Furthermore, it found that the remaining 64 MHz
of spectrum for common carrier cellular systems was excessive
and it reduced it to 40 MHz. It also reinstated the wireline common carrier restriction of the development of cellular systems in
the 40 MHz suballocation. In an important aspect of the decision
and one that generated perhaps the most controversy, the Commission created the new class of service providers which eventually became known as Specialized Mobile Radio (SMR) systems. At the same time, the Commission reduced from 40 to
30 MHz the amount of spectrum available to conventional and
trunked systems operated on a private or shared basis. It also
made the same spectrum available to the new SMR class of service providers. Thus dispatch users eligible for licensing under
Parts 89,91, and 93 of the Commission's rules (i.e. the public
safety, industrial, and land transportation services) would be
able to receive dispatch service on a private basis, on a not-for-
profit, cost-shared basis with other eligible users, or from an
SMR service provider—*a commercial firm offering common-
user service on a for-profit basis.* In the regulatory scheme developed by the Commission, the SMR systems were to operate
on a competitive, open entry basis with a minimum amount of
regulation. The Commission found that the SMR systems would
not be common carriers for regulatory purposes, and to prevent
individual states from taking regulatory action which might
thwart its rules and policies, the Commission asserted Federal
primacy in the area.

Because of the uncertainty with regard to the market demand for the various types of services and in recognition of the possibility of future technological developments, the Commission set aside 45 MHz to be held in reserve.

With regard to other industry structure questions, the Commission decided to require the wireline common carriers to establish separate subsidiaries to offer cellular mobile radio service. It decided against restricting the wireline carriers from offering dispatch service, but it did preclude them from offering a "fleet-call" capability. It also precluded the wireline common carriers from manufacturing, providing, or maintaining mobile equipment. To prevent equipment manufacturers from dominating the SMR service market, the Commission decided to limit them to owning one SMR system per market and a total of five nationwide.

On March 19, 1975, the Commission adopted a Memorandum Opinion and Order [17] dealing with certain petitions for reconsideration of portions of the decision. The Commission generally denied the petitions except that it removed its restriction that only wireline common carriers would be allowed to develop cellular systems. It also discussed at some length the rationale for creation of the SMR license category.

The Commission terminated Docket 18262 with a final Memorandum Opinion and Order [18] dated July 16, 1975. In the brief order, the Commission disposed of further petitions for reconsideration and clarified its position on the licensing of repeater systems. It said that repeaters could be shared at 900 MHz under existing "multiple licensing" practices used in other bands.

3.5 POSTSCRIPT

In Docket 18262, the Commission nearly quadrupled the amount of spectrum available to land mobile radio services. This vast reallocation coupled with the policies and rules adopted during the course of the proceeding will largely determine the LMR industry structure between now and the end of the century. This industry structure and further technological developments will, in

turn, determine the cost, availability, and quality of the land mobile communication services for the public. The true efficacy of
the Commission's decisions in Docket 18262 cannot be finally
evaluated for years or, perhaps, even decades. A comprehensive
assessment of the long-run impact of land mobile radio technology on society was recently published by a group of researchers
at Cornell University [19]. It includes an evaluation of the FCC's
actions.

The most obvious criticism of the proceeding was the long delay
in completing it — a period of almost exactly seven years from
the original Notice in 1968 to the action terminating it in 1975.
The period is even longer when it is measured from the time of
AT&T's early proposal concerning a high capacity LMR system.
The costs of such delays are difficult to measure, but, with the
current pace of engineering development, delays of this magnitude can easily equal a technological generation.

The major criticism of the substantive parts of the decision has
come from the National Association of Radiotelephone Systems*
(NARS), a trade association of radio common carriers, and the
National Association of Regulatory Utility Commissioners
(NARUC), a quasi-governmental organization made up of members of the regulatory bodies of each state. NARS was critical
"of the FCC's historic allocation policies favoring private radio
services" [20] and the lack of their own spectrum allocation
at 900 MHz. They also questioned the legality and fairness of
unregulated competition by SMR systems with common carrier systems. Both NARS and NARUC were strongly opposed
to the Commission's preemption of state jurisdiction over the
SMR system licensees. These and other parts of the decision
were appealed but the courts upheld the Commission [21].

The criticism has also been made that by allowing shared repeaters and by not imposing strict enough standards on the
number of units on each channel, the Commission has removed
the incentive for the more spectrally efficient trunked systems.
This lack of incentive and the threat of common carrier regulation under certain circumstances may account for the slow
development of such systems.

*NARS is now known as the Telocator Network of America.

On a more positive note, there has been a significant amount of
activity in developing private 900 MHz systems in urban areas,
and the Commission has granted developmental authority to
the Illinois Bell Telephone Company and the American Radio
Telephone Service, Inc. to construct cellular mobile telephone
systems in the Chicago and the Washington D.C.–Baltimore
areas, respectively.

REFERENCES

[1] "The Origin and Development of Land Mobile Radio,"
     Appendix A of the Reply Comments of Motorola, Inc.
     in FCC Docket 18262, dated July 20, 1972.

[2] Noble, Daniel E., "The History of Land Mobile Radio
     Communications," *Proceedings of the IRE,* May 1962.

[3] Tally, David, "A Prognosis of Mobile Telephone Com-
     munications," *IRE Transactions on Vehicular Communi-
     cations,* August 1962.

[4] Kargman, Heidi, " Land Mobile Communications: The
     Historical Roots," Contained in a Report by Raymond
     Bowers *et al., Communications for a Mobile Society,*
     Cornell University, 1977.

[5] *Final Report,* President's Task Force on Communications
     Policy, U.S. Government Printing Office, Washington,
     D.C., 1968.

[6] *Report of the Advisory Committee for the Land Mobile
     Radio Services,* U.S. Government Printing Office, Wash-
     ington, D.C., 1968.

[7] Joint Technical Advisory Committee, *Spectrum Engineer-
     ing -- The Key to Progress,* The Institute of Electrical
     and Electronics Engineers, Inc., New York, 1968.

[8] King, Diane, "Chronology of the National Spectrum Man-
     agement Program," Federal Communications Commission,
     Office of the Chief Engineer, Report No. SMTF 76-01,
     August 1976.

[9] Agy, Vaughn L., "A Review of Land Mobile Radio," Technical Memorandum 75-200, Office of Telecommunications, U.S. Department of Commerce, 1975.

[10] Reply Comments of Motorola, Inc., in FCC Docket 18262, July 20, 1972.

[11] Fourth Report of Commission and Order, Docket 8976, July 11, 1951, Quoted in Notice of Inquiry and Notice of Proposed Rulemaking, Docket 18262, July 17, 1968 (14 FCC 2d 311)

[12] Notice of Inquiry and Notice of Proposed Rulemaking, Docket 18262, July 17, 1968 (14 FCC 2d 311).

[13] First Report and Order and Second Notice of Inquiry, May 20, 1970 (19 RR2d 1663).

[14] *Ibid.,* p. 1677.

[15] Second Memorandum Opinion and Order, Docket 18262, July 28, 1971 (31 FCC2d 50).

[16] Second Report and Order, Docket 18262, May 1, 1974 (46 FCC 2d 751)

[17] Memorandum Opinion and Order, Docket 18262, March 19, 1975 (51 FCC 2d 945)

[18] Memorandum Opinion and Order, Docket 18262, July 16, 1975 (55 FCC 2d 771)

[19] Bowers, Raymond, Alfred M. Lee, and Cary Hershey, *Communications for a Mobile Society — An Assessment of New Technology,* Cornell University, Ithaca, NY, 1977.

[20] Excerpts of Testimony by George Perrin and Kenneth Hardman, For Independent Mobile Radio — 'Preferred Station' Classification, *Telocator,* December 1977.

[21] National Association of Regulatory Commissioners v. FCC, 173 U.S. App. D.C. 413, 525 F.2d 630, cert. denied, 425 U.S. 992 (1976).

Chapter 4

BROADCAST CONTROL AND REGULATION

Harold E. Hill
Department of Communication
University of Colorado
Boulder, Colorado

4.0 Introduction

4.1 Distinction between Governmental and Non-Governmental Controls

4.2 Governmental Controls
  4.2.1 Extent and Desirability
  4.2.2 The Federal Communications Commission
  4.2.3 The Federal Trade Commission
  4.2.4 Other Governmental Agencies
  4.2.5 Special Problems with Cable Television
  4.2.6 General Considerations

4.3 Control by Ownership

4.4 Economic Controls
  4.4.1 Advertisers
  4.4.2 Networking and Syndication
  4.4.3 "Profession" vs "Business"
  4.4.4 General Considerations

4.0   INTRODUCTION

Broadcasting as an "institution" in American life and the regulation of broadcasting have developed, and continue to do so, by a process of interaction between the broadcast industry and society. Broadcasting both reflects and molds the mores of society, thus each has a definite effect upon the other, shaping and developing attitudes and values. Therefore, it is important to examine and try to understand the various pressures and influences that are strong determinants as to what we see and hear via the broadcast media. It is the purpose of this chapter to provide a preliminary and abbreviated overview of many of these "controls" in the hope that the reader will be encouraged, even challenged, to seek further, more detailed information.

4.1   DISTINCTION BETWEEN GOVERNMENTAL AND NON-GOVERNMENTAL CONTROLS

It is necessary that we understand the distinction between governmental and non-governmental controls. The first are a result of legislative or regulatory action, often supported by judicial decision. The latter are those imposed by the various elements of society: broadcast owners, economics, audiences, pressure groups, sociological and psychological factors, technology, and numerous other factors. Contrary to what one might believe, governmental controls are rather insignificant when compared to all of the "others." Interestingly enough, some governmental controls are the result of changing societal interests and needs and thus reflect in a sense, what may have been formerly considered a non-governmental control or influence.

During the discussions of controls in this chapter, there will be opinions expressed and criticisms raised, but no attempt will be made to "dissect" the pros and cons of the various forces involved, nor of the influence they have on broadcasting content. The primary purpose is to provide some basic information to enable the reader to make judgemental decisions without too much undue influence.

4.2 GOVERNMENTAL CONTROLS

The Federal government has exercised varying degrees of control over broadcasting since the Wireless Ship Act of 1910, which required larger ships to have wireless radio for safety reasons. This was followed in 1912 by the Radio Act which simply stipulated that no one could operate a radio station unless a license was obtained from the Secretary of Commerce and Labor. However, no standards were established for the establishment or operation of any station, nor were guidelines provided to assist the Secretary of Commerce and Labor in making a decision. In fact, there was little need for any decisions because there was no legislative provision for denying a license.

Naturally, the proliferation of stations was so rapid under these "non-rules" that the airwaves soon became intolerably crowded and interference was rampant. Subsequently then-Secretary of Commerce and Labor Herbert Hoover began to make specific frequency assignments. At the same time (in the early 1920's), Hoover was attempting to convince Congress that regulating legislation was needed in the rapidly expanding field of radio broadcasting.

Finally, with some 800 stations on the air and no control over frequency use, Congress passed the Radio Act of 1927 which established the Federal Radio Commission to license broadcasting stations and generally to oversee the industry. Seven years later when Congress determined that it was necessary to have coordinated Government regulation of all facets of communication, the old Radio Act of 1927 became, practically word-for-word, the broadcast section of the Communications Act of 1934. Therefore, although there have been numerous amendments to

the act of 1934, none of them have been substantive so far as
program control is concerned, and the broadcast industry is operating under Federal legislation passed more than 50 years ago!

4.2.1   Extent and Desirability

The Communications Act of 1934, as was true of the Radio Act
of 1927, deals very little with program content while dealing extensively with the technical aspects of broadcasting. Congress
gave the Federal Communications Commission (FCC), which was
established by the Act, broad authority to license stations "in the
public interest, convenience, and necessity." Such general language
has given the FCC, as well as the broadcasters, considerable leeway,
and decisions have generally tended to follow the mores of society
at the time, although lagging by a period of months or even years.

The only specific program restrictions in the original law deal with
political broadcasts, lotteries, and obscenity. In brief, the law stipu-
lates that if a station provides time for a political candidate from
one party, equal opportunity must be provided for all other qualified candidates for the same office. Note that no station is required
to provide time for candidates in the first place. "Equal opportunity" applies to time and day, and simply means that the broadcaster
cannot put his friend, Candidate A, on during prime time, and then
offer 6:30 Sunday morning to Candidate B. The law does not attempt to define "qualified candidate" because state statutes provide the information regarding what conditions must be met for
a group to qualify as a "legally qualified political party." Finally,
note that this section of the law applies only to candidates and not
to anyone who might speak in their behalf.

Congress subsequently amended the original act to exempt interviews, bonafide news events, newscasts, and certain types of documentary programs from the equal time provision. However, there
are still some problems which must be faced. For example, there is
an obvious advantage to an incumbent who appears at press conferences, whereas his opponent(s) does not have such an opportunity.
This is especially true so far as the Presidency is concerned. In order
to provide a modicum of equality, the networks, after covering a
major Presidential address or press conference, will often make time
for rebuttal available to leading spokesmen for the opposition party.

A related problem stems from the fact that we no longer have just two parties with candidates for office, even for the Presidency, and it is virtually impossible, therefore, to broadcast such important events as Presidential debates. We have had only two exceptions: in 1960 Congress enacted special legislation to exempt the Kennedy–Nixon debates from the provisions of the law, and, in 1976, the Ford–Carter debates were "staged" as a bona fide news event in a maneuver designed to circumvent the true intent of the law.

In addition to the provisions of the law relating to programing, the FCC has the authority to promulgate regulations designed to facilitate their task of regulation. One of the most far-reaching has been the Fairness Doctrine. Quite briefly the Doctrine evolved from the FCC's so-called "Mayflower Case" in 1941 when the Commission ruled that stations could not editorialize. Naturally, this created considerable concern and, after extensive hearings, the FCC reversed itself in 1948 and ruled that stations could editorialize so long as they maintained a fair overall balance in their presentation of views. Subsequently, over the intervening years, the FCC has determined that stations may present either editorials or statements by outsiders on matters of public concern, provided time is made available for spokesmen for opposing views. This was followed by a directive indicating that making time available for opposing views was not sufficient, but that stations must seek out such views. The directive was later amended to require that stations may not avoid the whole matter by simply not presenting either side, but that stations must seek out and air varying points of view about issues of concern to their communities.

So far as lotteries are concerned, the Communications Act of 1934 explicitly forbids broadcasting information about them. The courts have held that three elements must be present to constitute a lottery: prize, consideration, and chance. It is because all three of these elements are not present that stations are able to broadcast the various contests and game-shows that they do.

Obscenity, also forbidden by the law, is a more complicated matter. This is one of the examples of broadcasting regulation tending to "relax" as the mores of the country become more lenient regarding obscenity, sex, etc. A number of years ago, broadcast stations would not have dared broadcast programs that contained even such words as "damn" and "hell." Now these words are relatively mild compared to some of the language we hear. During the summer of 1978, the United States Supreme Court upheld the right of the FCC to reprimand WBAI-FM, a public radio station in New York City, for airing comedian George Carlin's "Seven Dirty Words" monologue. This resulted in an immediate response from spokesmen of the broadcast industry who claimed this was tantamount to allowing the FCC to censor programing material, which is forbidden by the Communications Act of 1934.

In reply to this reaction, FCC Chairman Charles Ferris said that the Commission did not consider the Supreme Court ruling a carte blanche permission to rigidly control broadcast programing, but that the Commission would continue to handle obscenity cases on a case-by-case basis. And, in fact, the FCC soon thereafter declined to censure public television station KQED, San Francisco, for an alleged obscenity infringement.

Thus, we see that the extent of FCC control over programing content is generally limited to the specific areas indicated in the foregoing, with the exception of that rather nebulous area, "the public interest, convenience, and necessity." However, it is questionable whether it is desirable to have *any* Federal control over program content (this does not apply to commercial content which will be discussed in section 4.2.3).

Since commercial broadcasting (which constitutes by far the largest portion of broadcasting in the United States) is dependent for survival upon the selling of time to sponsors, and those sponsors will buy time only on stations which appeal to a fairly sizeable audience, isn't it reasonable to assume that stations will program materials which will appeal to at least a large enough segment of the audience to attract sponsors? Just as there are "customers" for XXX-rated motion pictures and "girlie" magazines, there will always be a segment of the audience who would like to have similar materials available on radio and television. If, in a given community, there is a sufficient number of viewers and listeners to that type of programing to attract sufficient numbers of sponsors to make the programing economically viable, why should it not be broadcast? For those who claim that there is danger of "contaminating" children, the only response is, "There is always the 'off button'."

Basically, the question is, does the government have the right, either legally or ethically, to dictate what will be available to the broadcast audience? There may be just as many people opposed to loud rock and roll as there are opposed to so-called "dirty" broadcast material. Does that mean that we should ban rock and roll music from the airwaves? Although these may appear to be frivolous questions on the surface, they actually go to the very heart of the question of what should be the extent of the government's control over broadcast program content. Furthermore, as will be discussed later is this chapter, other forms of influence and control are perhaps much more insidious and potentially harmful.

4.2.2   The Federal Communications Commission

Although the FCC has been discussed to some extent in the previous section, it is important to understand a little more about this body which has so very much to say about the broadcasting industry in this country. The Commissioners are appointed by the President, and they must be approved by the Senate. Their terms are for seven years and not more than four may be from the same political party. This commendable effort by Congress to keep politics out of the FCC has been only partially successful, because, generally, the Commissioners of the same party as the President tend to echo his philosophy. However, a larger problem is the fact that the Commission does not have sufficient staff nor budget to perform its functions properly. The FCC, in 1978, has a staff of approximately 2,000 and a budget of approximately $40,000,000 with which to regulate some 8,000 broadcast stations, plus consider applications for new stations, respond to challenges by would-be licensees or by the general public who are not satisfied with the job being done by the present licensee, and defend many of their actions in court.

In addition to the problems enumerated above, because the Commission is few in number, much of the work must be done by either "subcommittees" of one or more Commissioners, by Hearing Examiners, who then make recommendations to the Commission, or by senior staff members. Therefore, although it is fair to say that the Commissioners are generally honest men and women attempting to do their job well, the mere enormity of their task plus the paucity of resources, makes it extremely difficult. The result often is decisions which might not be in the best interests of the public, and actions which arouse the ire, correctly or not, of the broadcasting industry. If, for example, the FCC needed to concern itself only with the electronic aspects of control, it might be able to function much more efficiently, and the public might be better served in the long run.

Considerations such as those indicated in the foregoing several pages are part of the reason that the House of Representatives Subcommittee on Communications was considering, in 1978, new legislation which would considerably alter the role of the FCC. One of the recommendations receiving favorable reaction would result in de-regulating radio, except for technical aspects, while continuing television regulation much as it is now. Certainly, if this proposed legislation wins approval in the Congress, is approved by the President, and becomes law, we will have an opportunity to evaluate the pros and cons of governmental program control.

4.2.3   The Federal Trade Commission

Earlier a distinction was made between the control of program content and the control of commercial content. The Federal Trade Commission (FTC) is charged by law with the responsibility for, among other things, making sure that the consumer is protected from misleading and false advertising.

It is only in recent years that the FTC has concerned itself to any great extent with broadcast advertising. The change may be due to the general "toughening" of the FTC regarding all advertising, labeling, etc., or it may be the result of more blatant claims on the part of the sponsors. At any rate, during the past several years, the FTC has investigated numerous advertising campaigns which it deemed not to be in the best interests of the public.

As a result, several consent decrees (wherein the advertiser does not admit any wrongdoing but agrees to cease and desist from further tactics of the same nature) have been entered into between the advertiser and the FTC in which the advertiser agrees not only to stop what is being done but to admit on the air that misstatements had been made in earlier commercials. In some cases, a certain percentage of the advertising budget for the forthcoming year must be pledged to such corrective actions.

Although there is little direct evidence of the effect of such FTC action on the consumer, it would appear logical to assume that at least a portion of the viewing and listening audience would be interested in the "truth" and would possible alter their buying habits as a result. If this result does occur, this seems like adequate reason for continuing, perhaps even strengthening, the government's role in regulating commercial content.

4.2.4   Other Governmental Agencies

Other governmental agencies play a minor role in broadcasting control, but since these are concerned with the technical aspects they only need to be mentioned. The Federal Aviation Agency establishes limits on the heights of antennas, generally, and any proposed tower construction in the vicinity of airports and/or flight patterns must be approved by the FAA. Wage and hour laws, at both the Federal and state levels, as well as other Department of Labor and HEW rules and regulations have a modest influence on broadcasting, as do certain state laws, such as those dealing with broadcast coverage of trials, the so-called "blue laws," and zoning regulations. However, none of these, with the possible exception of the FAA tower restrictions. place any real burden on the broadcaster, and certainly no greater burden than is placed on any other business enterprise by the same or similar laws and regulations.

4.2.5   Special Problems with Cable Television

Because of the constantly changing posture of the FCC toward cable television, it is not safe to go into detail about specific regulations. Suffice it to say that, concerns of broadcasters to the contrary, there appears to be a potential for greater harm than benefit if the FCC continues or institutes regulations that are too stringent upon the cable television industry, because the most important consideration is to provide the best possible service to the general public with less regard for the more parochial interests of the broadcasters and the cable television operators. Naturally, due consideration needs to be given to the financial viability of these two groups, but this must be balanced against the need and desirability of providing the most extensive program content and choice to the greater number of people.

The FCC must address itself to several important questions: Is it in the public interest to restrict the availability of "distant signals?" Is the public best served by requiring, or not requiring, "public access?" Should the regulation of CATV be left entirely to the states (which might regulate the industry as a public utility)? Should competition be encouraged or discouraged, in the public interest?

These are only some of the questions that have been debated over the years with very few positive results. Should the FCC, or any other federal agency, have control over the cable industry, or should this be left to the states, or local governments, who might have a better realization of what the people in their particular areas need and want? Of course, some of the local governments have enacted franchise ordinances which have been so restrictive as to prevent the successful operation of the franchise. In spite of this, it seems on the surface that the local governments would have a better feel for the "pulse" of their communities, and therefore better public service would result. But again, this and the other questions raised above cannot be answered with any great degree of validity because the confusion at the national level has been deleterious to realistic research and study. This type of control appears to be detrimental to the best interests of the viewing and listening public.

4.2.6   General Considerations

In summary, with some admitted prejudice, it appears that there may well be too much Federal regulation of what we see and hear on television and radio. As will be explored in more detail near the end of this chapter, the public might be better served if serious attention were given to alternative methods. Again it should be emphasized that some control of commercial content seems advisable, but that program content control may well be ill-advised, or at least should be re-examined carefully. It would seem that the primary goal should be to serve the interests of the people who are the recipients of the vast amount of program material. These interests may well be best served with a decentralization of control. At the same time we concern ourselves with the governmental control, we must concomitantly address ourselves to all of the other current and potential influences that help determine what is broadcast. These latter concerns, in a variety of forms, will be addressed in the following section, and it is important to keep in mind government's actual or potential control as these other factors are considered.

4.3 CONTROL BY OWNERSHIP

It would be naive to assume that ownership interests do not have a considerable influence on broadcast content, whether it be at the station level — perhaps with just a one-family operation — or a national network — with a myriad of stockholders and interlocking directorates and management. Obviously, these various owners have a responsibility to make a profit, just as would be true in any business venture, but is it possible that too much attention is paid to profit to the potential detriment of programing?

". . . the greatest control and restricting force on the broadcasting industry (is) GREED. Because broadcasters are a greedy brood, they seek to maximize their profits by exploiting the underlying foundation of the commercial broadcasting business. In other words, they are constantly trying to accelerate the monetary intake they receive from the sale of their air time to advertisers."[1]

Harsh words? Perhaps, but listen to what Fred Friendly, well-known journalist, director, and producer had to say after he resigned as head of CBS News:

"Whatever bitterness I feel over my departure is toward the system that keeps such unremitting pressure on men like Paley and Stanton that they must react more to financial pressure than to their own tastes and responsibility. Possibly if I were in their jobs, I

would have behaved as they did. I would like to believe otherwise, but I must confess that in my almost
two years as head of CBS News, I tempered my news
judgement and tailored my conscience more than once. . .
*The fact that I am not sure what I would have done
in these circumstances, had I been chairman or president of CBS, perhaps tells more clearly than anything
else what is so disastrous about the mercantile advertising system that controls television, and why it must
be changed".* (emphasis added) [2]

As we pointed out earlier, there is absolutely nothing wrong
with the "profit motive." According to leading economists,
this is one of the factors that has made this nation great. But
the almost religious dedication to profit to the exclusion of
consideration of the public good has spread throughout all
aspects of our society (automobile manufacturers are alleged
to be more concerned for profit margin than they are for passenger safety), and examples can be cited in all areas of business. Therefore, does it seem unreasonable to believe that those
who control broadcast properties, the owners, will forsake the
"profit motive" in order to bring the public more meaningful
programing at a fiscal loss? Thus, the owner, or his agent who
makes program decisions states that he is "giving the public
what it wants," a statement based not on scientific research
but on the "ratings," which at best indicate what members of
the audience have selected from among the offerings available
at that time. There is no attempt at evaluation of the "worth"
of the program to the audience in a real sense. "The Beverly
Hillbillies" might well outdraw the competition if the competition has even less appeal to the audience than "Hillbillies."

The owner, or his agent, therefore makes judgements based on
very little, and perhaps false, information about what the audience
likes or dislikes, wants or doesn't want, or even believes or disbe-
lieves. And this relates back to the matter of financial control and
what type of programing will result in the greatest profit. Professors Gurevitch and Elliott report that in commercial broadcasting

the shape of the beliefs and understandings of audience tastes and tolerances held by those in financial control of the industry play the largest part in setting the situation within which the program maker works. [3]

Another issue which must be considered is the matter of cross-ownership of various media outlets. For example, a few years ago, in a relatively small city in the Rocky Mountain-Great Plains region, a single family controlled the only television station in town, the only newspaper, the only cable system, and the leading AM and FM radio stations. In such a case (and there are others with similar implications if not quite so startling in impact) it seems reasonable to ask whether, under such circumstances the public in that community might expect to receive diversified media output, especially so far as news, public affairs, and comment are concerned. The FCC has looked into cross media ownership at various times, both in regard to owning broadcast properties in the same community and owning such properties as well as other forms of media. At the present time (1978), this matter is in somewhat of a state of flux so far as the Commission is concerned.

Of course, in fairness it should be pointed out that cross media ownership might be the only way some small communities could gain the benefits of both broadcast and print media, but each case needs to be examined with care to determine whether or not such operation is in the public interest. Again, it is important to avoid the danger of profit motivation overriding service to the community.

4.4 ECONOMIC CONTROLS

While the foregoing section deals with one aspect of economic controls, it is important to examine other factors of economic influence if we are to have a fairly comprehensive understanding of the financial pressures exerted upon and by the broadcasting industry, and, again, what role they play in determining what the public sees and hears.

4.4.1 Advertisers

In the Introduction, it was pointed out that "broadcasting both reflects and molds the mores of society, thus each has a definite effect upon the other, shaping and developing attitudes and values."

The statement is even more applicable to the commercials we see and hear daily. Patrick Walsh says the commercial is the *avant-garde* of change and both reflects and accelerates the societal trends and mores far more than does the rest of television programing.[4] We are seemingly addicted to attempting to emulate attitudes, actions, and buying habits of those we see in television commercials. For example, when cigarette commercials were still allowed on the air, most of them showed the smoker as being romantic, manly, good looking, or whatever, in an effort to convince the would-be smoker that taking up the habit would result in the new smoker having the same desirable attributes as displayed by the person in the commercial.

We seem to have the concept that there is an average, mass American, and it appears that it is this audience that the advertisers (and, indeed, the broadcasters) attempt to reach and appeal to. This has allowed the creation of certain images by advertisers and their agencies.

> ". . . there are probably five categories of easily recognizable females in today's media advertisements. We start with the bouncy-haired, acne-faced, pre-puberty adolescent, or a Pepsi generation member. We then proceed to the young, single woman who is hell-bent on marriage and trapping an adequate male. She is the consumer who is dousing herself with enticing perfumes, gargling with minty mouthwashes, brushing with sexy toothpastes, and (pardon me) getting all the confidence she can from her feminine hygiene spray. Next is the young mother who espouses the benefits for certain baby-soft shampoos and dovey detergents. She miraculously transforms into the older housewife, whose voice is reminiscent of a power saw, and whose nose is either stuck in an oven or someone else's business. Finally, we cannot forget the constipated grandmother who relates tales of laxatives and denture creams".[5]

Seriously, there is danger that we may cause "warping" of viewers' outlook, not only on life and society as a whole, but on one's own self, on ones's self-esteem and value. Is it not

possible for the following scenario to be played out in real
life (in fact, it is probable) and result in self-shattering trauma?
A rather plain, not too bright young man or woman, almost
completely devoid of any likeable personality traits, is generally
ignored by his/her peers and feels generally miserable about
his/her apparent failure as a human being. But, wait, there
is hope! Television commercials indicate that there is no need
for one to be plain, or lonely, or inarticulate, or whatever. All
one need do is use a particular deodorant, toothpaste, aftershave
lotion, cologne, and shampoo, while wearing certain clothing and
driving a certain car, and all troubles will vanish overnight and
popularity will zoom and friends will come running. So our "victim" follows all of the tips given by the commercials, but nothing
happens. It certainly can't be the fault of the products, because
the commercials have told us they will work wonders; therefore,
the saddened party must be at fault and impossible to improve.
What does that do for the ego?

Although the foregoing may seem to be a fantasy, it is not; there
is mounting evidence of the influence that broadcast commercials
have on our actions. Witness the current controversy over advertisements during children's viewing hours for sugar coated cereals,
to cite but one example. Manufacturers (the advertisers) on the
one hand claim the product creates no problem, yet health groups,
Action for Children's Television, parents, and others refute such
statements, indicating that the product is harmful for children.
There are many such examples of the effect of advertising on consumer behavior, and not just on children.

If broadcast commercials do, as we have indicated, affect our
general habits, values, and actions, as well as convince us to buy
a certain product, then it behooves the advertiser to reach as
large an audience as he can. How does he do this? By buying
into the most popular shows. Such "buying in" becomes very
competitive, with various potential sponsors attempting to out-
bid one another for the privilege of having their commercials
appear on the most popular program(s). Commercial time for
such programs thus becomes extremely expensive. With such a
significant investment, it follows logically that the sponsor wants

to make certain that the program retains its popularity, so he assumes as much control as possible over program content. Therefore, the sponsor not only controls what we see and hear during the commercials, but he also has considerable "clout" insofar as programing decisions are concerned, because the network or station does not want to give up the income derived from that sponsorship.

Sponsor influence, or even control, over program content is generally exercised through the advertising agencies. This control is not something new; it has existed from nearly the beginning of sponsored broadcasting, and became more apparent during the post-World War II boom in broadcasting, as evidenced in a column by *New York Times* television critic Jack Gould, reporting on the 1959 FCC hearings into practices and policies of TV networks: "In the case of shows in which they were active, for instance, the agencies said that they review all scripts in advance, scrutinize dialogue and story lines, and have their 'program representatives' on hand to check each day's production work."[6]

4.4.2   Networking and Syndication

Television stations generally, because of the tremendous expense involved, do not do a great deal of local production. Therefore, to fill the hours they program each day, they must turn to the networks or those companies which syndicate programs. (Radio stations depend to a very modest degree on such sources because the greatest share of the programs are locally originated.) Program content control, therefore, is in the hands of the program producer, influenced of course by the other forces discussed in this chapter, rather than by station ownership or management.

"Stations have the right to refuse to carry network programs they find objectionable, and, of course, they do not need to purchase those syndicated programs they do not like." Such is the response one gets when the question of program control by networks and syndication distributors is raised. Of course, such a response dodges the real issue of how much program control is relinquished by the stations. Except for local news programs and an occasional production regarding some matter of local concern, the television programing you see is determined by "decision-makers" at networks or in syndication companies.

Such organizations are "big business," with responsibilities to their stockholders which far outweigh any responsibility they might feel toward the general public, the broadcast audience. Because the major focus of the businesses controlling the programing agencies (networks and syndicators) is to show a profit, profit must be the overriding factor in programing decisions. Thus, programs must appeal to the widest possible audience, to a mass audience, not to individuals nor to minorities (and this includes all types of minorities, not just ethnic minorities). Therefore, the greatest number of programs (there are exceptions, of course, like CBS's *Sixty Minutes*) are geared to the lowest common denominator; they are purely entertainment programs with little, if any, attention to public affairs. This is due to the low cost-efficiency of public affairs programs. Neilson ratings are readily translated into numbers of viewers so that we arrive at so many viewers per rating point. In general (again, excepting those rarities like *Sixty Minutes*), network public affairs programing costs three to four times as much per rating point as any other form of program.

The large number of prime time programs on the air geared to entertainment, compared to the low number of public affairs programs gives rather strong indication that the networks have made the choice for economic viability over public service programs. Less profitable and less popular with the audience, they are not favored by advertisers if they introduce any kind of controversial subject.

For the same reasons, syndicators must sell those programs with the greatest mass appeal so that more stations will buy them. Thus we get re-runs of such old network programs as *Gilligan's Island, I Love Lucy,* and *Hogan's Heroes.* Not that there is anything wrong with these programs, *per se*, but some viewers might eventually tire of them and prefer something that stimulated their minds a little more.

In recent years, there has been a trend in syndicated programs toward "nature" programs, such as *Wild Kingdom* and *Last of the Wild*. These programs offer an attractive alternative to the re-runs mentioned earlier. They are educational as well as informative and interesting, and the producers are certainly to be complimented on making this type of program available. In fact, these indications of a greater public awareness on the part of the syndicators, plus the growing number (though still terribly small by comparison) of network public affairs programs is encouraging. It is hoped that this movement will continue and networks and syndicators will realize even more than now that there are, after all, large audiences for such stimulating programing, although it must be admitted that those audiences still don't approach the size of those for the "lighter" offerings of the networks and syndicators.

More important than the content itself, however, is the simple fact that local station management, which by law is supposed to be responsible for program content, does *not* have that control, and decisions as to what we see and hear are made by persons far removed from our local scene and our local concerns and interests.

4.4.3 "Profession" vs "Business"

Is it possible for a group of persons, broadcast programers, who consider themselves members of a "profession," to reconcile that posture with a constant drive for large audiences in order to bring in more money and make more profit? It can be said that doctors, dentists, and lawyers are also "out to make a buck," but they appear to have a greater dedication to serving the public, and their profession has a licensing system to prevent pure charlatans from practicing. No such professional licensing exists for individuals engaged in broadcasting, except for the technical licenses for which broadcast engineers must qualify, and those have nothing to do with program content.

Various proposals have been made over the years regarding systems for licensing, but so far none has appeared very workable nor have any been received with any degree of enthusiasm by broadcasters. Here again, unfortunately, it appears that "economics" takes precedence, just as was the case regarding the public good. However, it might be appropriate to look at what are generally accepted criteria for a professional so that some future thought might be given to how broadcasters "stack up."

According to Yoder, distinctive attributes of a professional are:
1. Acquisition of a specialized technique supported by a body of theory.
2. Development of a career supported by an association of colleagues which may have developed a code of ethics, with suitable disciplinary action imposed by colleagues.
3. Establishment of community recognition of professional status.
4. May have specific requirements for formal educational preparation.
5. May have specific requirements for admission to practice, usually established by law. [7]

4.4.4 General Considerations

The foregoing discussion of the various economic controls affecting broadcasting is by no means exhaustive. It is a brief overview of *some* of the present and potential problems which, it is hoped, will serve as a basis for discussion and further study. However, as a final word, it should be pointed out that there is really no justification for the assumption that broadcasting is "free" in this country, and that commercially based broadcasting is the only system which will work.

The public has three-to-four times as much invested (in receivers and antennas) in broadcasting equipment as does the entire broadcasting industry. We spend over one billion dollars a year just in repairs and parts for our receivers, to say nothing of the extra dollars we spend for products made more expensive because of the dollars the manufacturer has spent on advertising through the broadcast media.

The nations of Western Europe, as well as those of most of the rest of the free world, have provided sufficient evidence that forms of broadcasting, other than commercially oriented ones, not only can exist but can prosper and provide a greater degree of actual public service. In many cases, these long-existing public broadcasting systems have been joined, in recent years, by commercial systems. However, these latter systems generally are very closely supervised by the government so that there is absolutely no connection between sponsors and program content. The still struggling public broadcasting system in this country may provide, eventually, an alternate system, dedicated to enlightening and informing the public.

4.5 AUDIENCE INFLUENCE

4.5.1 "Giving the Public 'What it Wants' "

We often hear spokesmen for the broadcasting industry, responding to critics, say that the broadcasters give the public what it wants. While this may be an easy answer to the problem of mediocre programing, it cannot be defended if one takes the time to examine the statement.

In the first place, there is no such thing as "the public." Our society is composed of a number of different publics, with compositions which vary from time to time, and each of those publics has its own dreams, desires, opinions, and ideas, which may change from time to time. Thus the audiences for programing generally and for any given programs specifically, are constantly changing so far as makeup is concerned. So it is impossible to program for "the public."

Secondly, people are not able to tell just what they want. Can you want something that doesn't exist or that hasn't existed, or that you know nothing about? Could the radio listeners of the 1920's demand broadcasts that featured Benny Goodman or Orson Welles? If you were offered a choice between a steak dinner or a widget, could you make a reasoned choice? How about if you were offered $10 or a widget? Of course you could not choose wisely because you have no idea what a widget is. This same problem exists regarding broadcast programing. Would the audience rather watch "All In the Family" or a widget? There is simply no sound research method which will answer that question. One of the major problems may be that there are very few widgets in the broadcasting milieu; not enough new and different programing ideas are tried, so there is no criterion by which to judge.

Broadcasters often claim that they cannot "afford" to program good music, good literature, and serious discussion programs because the audience would not be sufficiently large to make the programs economically viable. While it is true that such programs might not draw top ratings, that does not mean that at least certain "publics" are satisfied with the present offerings. Furthermore, while it is highly unlikely that there are large audiences "out there" awaiting such programs, neither can it be claimed by the broadcasters that many people would reject the programs; they never have had the opportunity to choose. Secondly, people can and do improve their tastes, but only if their taste buds are challenged by new and different things.

4.5.2   The Rating Game

A discussion of ratings is, in fact, an extension of the discussion in the previous section. While the ratings do indicate a fairly accurate picture of how many people were watching or listening to a given program at a given time, ratings should not be considered a valid indication of actual program preference. Let us consider first some of the more pedestrian shortcomings.

The most generally accepted television ratings, Neilsen, are based on electronic recording of how many sets in the sample (we will not even consider that the sample is too small to be adequate) are tuned to each network. The fact that the set is on and tuned to a specific station does not guarantee that anyone is watching, or how many in the family are. Telephone surveys, most often used for radio, ask, "Is your radio on, and, if so, to what station is it tuned?" The same objections as were indicated for television may be raised here. In neither of these surveys is it possible to determine whether any other activity is taking place, such as conversation while the set is on. However, if these "minor" objections to the ratings systems are overlooked, there is another, more important factor that must be considered.

Let's assume that on a given evening the three networks are carrying programs A, B, and C at seven in the evening, programs D, E, and F at eight o'clock, and programs G,H, and I at nine. Let's assume a somewhat exaggerated situation for the sake of illustration. Assume that for a given "public" the actual order of program preference is: A, B, C, D, E, F, G, H. They can watch only one program at a time, based on their preferences, so they will elect to watch A at seven o'clock, D at eight, and G at nine. Thus the ratings would indicate that those were the three most popular programs which would be far from the truth.

This is an example of what often happens, because networks may well pit their best against one another in a desperate battle for ratings. An equally misleading rating results if two networks happen to schedule poor shows opposite a fair show on the third network. The audience will turn to the third program even if it is not one they particularly like, simply because it is better than the other two. In other words, rather than an absolute indication of preference, the ratings more often indicate what most of some of the viewing publics (those so addicted that they turn their sets on when they get up and often don't turn them off until they are falling asleep at night) choose to select *from what is available at that given moment.* This last point re-emphasizes what was said about widgets in section 4.5.1: If such unusual programing is not among the choices, how can one judge properly?

4.5.3   Some Observations

Unfortunately, audiences tend to behave like sheep; they will follow almost any trend, they are gullible when it comes to evaluating content of either programs or commercials, and they are not very demanding. Visitors to this country from Western European nations are often appalled at the docile manner in which the American audience accepts what the networks and stations offer. Perhaps it is because, until the relatively recent advent of public television, we simply weren't aware that there could be something else — not necessarily something *better*, but a greater degree of choice.

The broadcast media have too great an influence on us as individuals and upon our society for us not to take some action to alert the broadcasters to our needs. Interestingly enough, most broadcasters will admit that they are often influenced by just a modicum of audience response. When *Star Trek* was originally removed from the air the number of persons complaining was small by network standards (reportedly less than 100,000), but the show was returned to the air because of the *type* of person complaining. The network

indicated it was impressed with not only the intensity of the pro-
testors but with their "quality." So, it is not always necessary to
meekly accept what the broadcast media offer. A great deal needs
to be done to increase greatly the number of *critical* viewers and
listeners. This is a responsibility that needs to be shared by parents,
educators, and the broadcasters themselves.

Interestingly enough broadcasters' reaction to what they perceive
as audience desires sometimes results in skewing programing in an
unexpected direction. In recent years certain discontented, vocal
segments of the audience have complained about "under-represen-
tation." Typically, broadcasters and advertisers have again over-
reacted and have provided ready-made stereotypes for these various minorities. Consequently, much concern has been expressed
over the stereotyped images assigned to certain ethnic and racial
minorities by the media. The standard casting of Italians as Mafia
extortionists, Jews as penurious penny pinchers, Blacks as illiterate
household servants, and Chicanos as switchblade-carrying criminals
not only reinforced the public's outmoded beliefs and antiquated
prejudices, but it produced a new stereotype — the all-American,
middle-class, middle-aged WASP bigot.

However, where most whites seem to be able to laugh at Archie
Bunker's misguided attempts at perpetuating conventional racial
and ethnic prejudices (in fact, some viewers undoubtedly regard
him as a "hero" of sorts), it is doubtful that actual minority members react as laughingly to their programing counterparts. In fact,
programs dealing with minority situations, that Neilsen has ranked
highly, have often been targets of attack from the minority groups
themselves. For example, generally blacks did not approve of *Julia*,
and pressure from Catholic and Jewish factions forced *Bridget
Loves Bernie* off the air.

Such a reaction is quite understandable, and it is regretable that
broadcasters often adopt such stereotypes in the interest of pulling
large audiences; they don't worry too much about losing some of
the minorities if the whites are amused. It is up to combined audiences to bring about changes in this arena.

4.6 PRESSURE GROUPS

Action for Children's Television (ACT) is currently engaged in an effort to convince the FCC and/or Congress to ban all advertising on children's television programs. The Committee on Mass Communication and the Spanish Surnamed has been active in Colorado for the past several years in an effort to obtain more broadcast employment for Spanish-Americans and to increase the amount of programing directed at this ethnic group. Atheist groups are demanding equal time to reply to religious programing. Black organizations continually challenge license renewals in an effort to force the FCC to make more broadcast facilities available to Blacks.

These are just a few examples of the types of groups putting pressure on broadcasters, the FCC, and Congress. Their efforts vary in viability, tactics, intensity, and specific purpose. In general, these groups are more than justified in their demands. Ethnic minorities have long been deprived of the right to sufficient programing addressing their needs, and have not been given equal employment opportunities. Groups, such as ACT, who are concerned about content, are certainly justified in objecting to those aspects of broadcast programing which are detrimental to any aspect of society. And, in recent years, these "pressure" groups have been making considerable progress in their struggle.

Considering the entire nation, the struggle by the Blacks has perhaps received the greatest attention, and deservedly so. As Whitney M. Young, Jr., Executive Director of the National Urban League writes, Puerto Ricans, Mexican-Americans, and American Indians also suffer discrimination, but the Black Americans are the only group representing involuntary imigration to this country.[8] In a somewhat broader approach, Payl Ylvisaker wrote:

> If broadcasting could choose only one community
> need to serve in this . . . Second Emancipation, it
> should be to fan the small fires of self-respect which
> have been lit in the breasts of the community's ne-
> glected and disadvantaged citizens. This year (1964)
> it was the Negro; next year it may well be the Spanish-
> or Mexican-American; the year after, perhaps, it may
> be the American Indian or the mountain white.[9]

The pressure by ethnic groups has certainly resulted in some changes in programing, as was indicated to some extent in section 4.5.3. We find many more Blacks, for instance, serving as newspersons for both networks and local stations. Just this month (July, 1978) ABC announced the use of the first Black in an anchor position on network news. The question one should ask at this point is, "Since these persons obviously did not simply spring up 'overnight,' completely competent, what is the broadcasting industry's rationale for not having used such persons long before this?"

Pressure has also resulted in more programing for and about ethnic minorities. Blacks are now generally fairly well presented as normal human beings (when you discount such early attempts as *Julia* and similar programs), but there is still room for improvement. Foreign language broadcasts are much more readily available, and documentaries are beginning to deal with the real issues underlying social and economic problems of ethnic minorities, whereas formerly, such programs merely dealt with the surface manifestations of such problems and simply showed pictures of slums, rats, poorly dressed children, etc.

Just recently "pressure" of a new kind has developed. An adolescent in Florida pleaded innocent to a murder charge on the grounds that television violence had driven him to such a state mentally that he was not responsible for his violent actions. The courts found against the young man, but the case is a milestone, because it brings into very sharp focus the mounting concern, in many quarters, about the effect that violence on television may be having on society. Many studies, especially those carried out under the direction of Dr. George Gerbner, Dean of the Annenberg School of Communications at the University of Pennsylvania, lend considerable support to the contention that television violence is affective. Further corroboration may be found in *Television and Social Behavior,* a five-volume study done under the general auspices of the Surgeon General's Scientific Advisory Committee on Television and Social Behavior.

Another more recent, and perhaps more consequential, case has been filed in San Francisco as a result of an NBC telecast in 1974, starring 15-year-old Linda Blair as an inmate of a state home for girls. In one scene she was artificially raped by four other girls with the handle of a "plumber's helper." Three days later, a San Francisco girl nine-years old was similarly attacked on the beach by other girls using a bottle. The victim and her mother are suing NBC and its San Francisco outlet for $11 million, charging that NBC was negligent in showing the program, particularly during the so-called "Family Hour." A friend-of-the-court brief filed by the California Medical Association asserted that "television is a school of violence and a college for crime," adding that, because the attacking girls "acted out what they saw, we now have a real life victim with real life scars — brought to her by NBC."[10]

Regardless of the outcome of this trial, which had not yet commenced at the time of this writing, such situations as recounted above are bound to have some effect on broadcasting, and, thus, must be considered as controlling influences.

4.7 SOCIOLOGICAL AND PSYCHOLOGICAL INFLUENCES

As indicated earlier, our societal mores are changing quite rapidly, and this fact is reflected in the type of broadcast programing available to us. For example, our concerns about sex are being gradually replaced by our concerns about both violence and showing fairly the manner in which many of our minorities have to live. The programmatic results have been discussed in earlier sections of this chapter.

The trend toward the scientific "push" of the last several years is reflected in programing. "Star Trek," "Space: 1999," "$6 Million Dollar Man," "The Bionic Woman," and numerous other programs give evidence of this general move toward a more scientific society. On the other hand, this movement seems to be inherently resisted by many who feel that the rights, freedoms, and wishes of the individual are being violated by the scientific movement. Their hopes, aspirations, and fears are reflected by such programs as "Grizzly Adams," "How the West Was Won," and "Little House on the Prairie," all examples of man's struggle for individuality.

A somewhat related, but still different, sociological phenomenon is individualism vs collectivism. The idea of group activity, of "joining," of "working for the good of the whole," is becoming more dominant in our society, and, as with the scientific "push" we are finding resistance increasing. In effect, what many people fear so far as collectivism is concerned, is that they will loose the opportunities to do what they want to do with their lives, to feel that they can be what they want to be without overt pressure from some group. Programing dealing with the 1950s, in one way, tends to offer escape for these persons, as do some of the programs mentioned above.

There has been increasing concern expressed over the years that nationally broadcast election returns of predictions based on computer projections could have an effect on an election outcome, particularly where polls still remained open. For example, in 1976 an 8:00 p.m., EST prediction of a winner would have occurred while 23 states were still voting, with an electoral strength of 216 out of a total 538. This is due not only to difference in time zones but differences in the times at which polls close in the various states.

Psychologically, the question should be raised as to whether the West coast voter, having heard a prediction that his favorite was certain to lose the election, would simply not bother voting on the assumption that his vote would apparently do no good. There is the companion question about whether or not the projections made by the networks are really as reliable as is claimed when they project a winner, perhaps only minutes after the polls in the East close (or may still be open, for that matter), based on only a handful of returns. And yet those projections might influence considerable numbers of voters, either not to vote at all, or suddenly decide to "get on the winning side."

From time-to-time Congress has expressed concern over this problem, and several bills have been introduced, so far unsuccessfully (which may well be an indication of another pressure affecting broadcasting: lobbying by affected interests). Several years ago, in connection with the consideration of one piece of legislation dealing with this problem, Senator Vance Hartke directed a letter to Chairman Pastore of the Subcommittee on Communications, pertinent excerpts from which follow:

> As you know, (election) coverage by radio and television has reached a high degree of speed and interpretation. Network newsmen have computers, researchers, pollsters, and other specialists at their disposal . . . Yet, the projections based on fragmentary samples may prove incorrect. Viewers and listeners are misled — *often unintentionally* (emphasis added) . . . The late President John F. Kennedy won the 1960 election by a 112,692 plurality vote. If *one* voter in each of the 173,000 voting precincts in the United States had switched his vote . . . Nixon would have won the popular vote . . . Realistically, had there been a switch of one vote in Kennedy's favor in each of the 10,400 precincts in Illinois, plus a switch of nine votes in each of the 5,000 precincts in Texas, Mr. Nixon would have tallied the required 270 electoral votes and would have been our President. [11]

The fact that is most frightening in regard to this matter is that by his use of the phrase "often unintentional," Hartke is making quite clear his feeling that some, if not many, of the projections are *designed* to mislead the viewer and listener.

Thus, we see the very strong possibility that the pressure of "beating the competition" in broadcasting the expected election results (and thus building the confidence and numbers of both audience and advertisers) leads to projections, which may not only be erroneous but intentionally so, which, in turn, leads to a psychological impact on the audience which might have a real effect on the final election results.

4.8 TECHNOLOGICAL CONTROLS

Although directly related to governmental controls, discussed earlier, it is necessary to point out that such developments as cable television, pay television, and theater television have an effect on what we see, because these systems make more program choices available to us (discounting for the sake of this discussion the Federal limitations, as discussed in Section 4.2.5). Laser beams, fiber optics, and other technological advances will have a similar impact.

On the other hand, reticence on the part of the electronics industry, often in concert with the broadcasters, can restrict our programing opportunities. The reluctance of television receiver manufacturers to include UHF tuners in their sets not only restricted our choice, but, perhaps, was the principle cause for the failure of UHF to develop properly. By the time the law required all sets manufactured after April 1, 1964 to contain UHF tuners, the entire matter of frequency allocation and related matters had been determined. It was during those considerations that UHF needed to be made viable.

4.9 CONTROL BY OTHER MEANS

The television critic could have a considerable influence on what we see and hear if there were more critics and fewer reviewers. There are only a handful of capable, reliable critics in the country. The rest who may claim that name are merely columnists, often simply rewriting releases they receive from the networks and stations. The public would be well served if there were more critics like Jack Gould of the *New York Times* and Lawrence Laurent of the *Washington Post,* who strive to be truly critical in the hope of improving broadcast fare.

It is undoubtedly true that the broadcast media, and particularly television, have suffered greatly from a lack of motivating criticism. As Robert Landry has said of television:

> The only art medium with a universal audience, the one conduit of ideas that must be kept unclogged, is practically without any organized, extensive, general criticism. What little there is is apt to be offhand, careless, and feeble. [12]

Thus, although broadcast critics do not currently have much influence on what we see and hear, it would seem to be important that valid, learned criticism be increased. This would alert the audience to what to expect of given broadcasts, and advise them in the same manner that theater, book, and motion picture critics advise their audiences. This, in turn would put pressure both on the broadcasters and the advertisers to produce more meaningful programing.

Another controlling factor that needs to be considered briefly is the National Association of Broadcasters. This is the national trade association in the broadcasting field, and it is natural to expect that such an organization would not be very "harsh" in its attempts to "police" the industry. However, due to a fear of government-enforced programing standards, of stricter regulation of commercial message content, and of general chaos within the industry, the NAB has attempted to undertake self-policing. As far as control of programing is concerned they have established a Radio Code and a Television Code. These are designed to serve as guidelines for the industry by indicating parameters in such matters as number of commercial messages per hour, type and content of commercial messages, and guidelines for children's programing and general programing.

The industry is to be applauded for attempting to prevent programing and commercial excesses. However, from a practical point of view this type of self-regulation is extremely difficult. Only a portion of the broadcast stations belong to the NAB and only a portion of *them* subscribe to the Codes. Those who do pay an extra annual fee in order to support the activities of the Code Authorities who administer the Codes. The Code Authorities can admonish stations violating the Codes, and, as an ultimate step, could "expel" a station from being among Code subscribers. This action, however, is self-defeating because such expulsion would result in the loss of income needed to administer the Codes. So far as the public is concerned, the only visible (or aural) evidence is the display of (or announcement about) a seal indicating that the station does subscribe to the Code. A recent survey indicated that fewer than five percent of the broadcast audience even knew what it meant for a station to subscribe to the Codes. Thus, the effectiveness of the Codes as a real factor in program and commercial content seems to be questionable. (For a more complete discussion and specific examples of Code guidelines, refer to any basic text on broadcasting).

4.10 REVIEW AND FORECAST

This brief overview has indicated many, but certainly not all, of the various factors that have a controlling influence on broadcasting and what we see and hear. As indicated earlier, space did not permit more extensive discussion, but it is hoped that this information will serve as a springboard for further reading, discussion, and research.

Although the information presented in this chapter may seem to be rather pessimistic, it was not intended to discourage anyone from being a "broadcast fan," nor was it designed to dissuade anyone from constantly striving to bring about improvements in the broadcasting fare currently available to the American public. Indeed, there have been considerable strides made in recent years, and it is only by continuing, careful examination of such factors as have been discussed here that we can hope for this trend to improve and accelerate.

It is to be noted that perhaps the most "beneficial" influences have been exerted by the various so-called "pressure groups" — not universally, of course, but for the most part. On the other hand, perhaps the most deleterious influence has been that caused by the commercial nature of the media and the resultant attitude on the part of network management and commercial advertisers and their agencies. As the FCC's Office of Network Study concluded in 1965:

> . . . the policies and practices of network managers. . . tended to substitute purely commercial considerations based on circulation and "cost per thousand" for considerations of overall service to all advertisers and to the various publics, as the dominant motives in the plan and design of network schedules. In other words, network television became largely a "slide rule" advertising medium principally motivated by a commercial concept. . .

Perhaps more critical viewing and listening, causing the ratings (in spite of their weaknesses) to reflect the public's discontent with overcommercialism, would result in a lessening of control by advertisers. Direct protest to sponsors is another avenue of "protest" for those who would like to see broadcasting serve a larger number of audiences and serve them much more meaningfully.

As indicated earlier, there is at least some interest in Congress in deregulating radio programing, so that the government would have no voice in what was aired. However, the No. 2 Democrat on the Subcommittee on Communications (the one considering the legislation), Rep. John Murphy (D-N.Y.) indicated in July, 1978, his opposition to the bill, because he said that freeing broadcasters from regulation would have a predictable result:

> For most broadcasters, the ratings alone will prevail.
> In the end, there will be fewer public affairs programs,
> if that is possible. Our children and young people will
> continue to get short shrift. The cheap programing
> that fills the Saturday morning ghetto period — which
> earns high profits at low cost — will still be their stan-
> dard fare. And in adult entertainment, banality will
> continue to rule. [13]

He said further that the new provisions miss the target because they "leave untouched network control over American television."

The last point is the strongest point made, perhaps strong enough to convince other Congressmen that the proposed legislation will not be any improvement over the present law. However, the balance of his argument is not necessarily constructive. One of the main tenets of American life has been "a free exchange of ideas in a free market place." Why should broadcasting be exempted from that privilege? It seems reasonable that freedom from government regulation just might result in greater diversification. Of course, government deregulation will not prove to be a panacea, what with all of the other controls and influences currently having an impact on what we see and hear, but it would be a step in the right direction, and that freedom might lead to freedom from other pressures.

There is no point in hoping for better programing through freer choice if the first and easiest step, government deregulation, is impossible to achieve, particularly when some feel that the government regulation is there to protect us! That is too much like "Big Brother" watching over all of us. Government deregulation should be instituted in the area of cable television as well, thus permitting more local control and more access to diversified programing by the public.

In spite of the concerns expressed in this chapter about the great number of controlling factors present in broadcasting, the future appears bright — at least brighter than the past. The broadcasters and sponsors are providing us with more hours of meaningful programing than was available a few years ago (admitting that there is still a great deal of pap on the air), and the government is at least considering deregulation. And, perhaps most important, it appears that the public is beginning to demand more of a voice in what programing is presented.

REFERENCES

[1]   Aaholm, Leslie and Carol Williams, "Majority-Minority," an unpublished graduate research study, University of Colorado-Boulder, 1974, p. 1.

[2]   Friendly, Fred, *Due to Circumstances Beyond Our Control* (New York; Random House 1967) p. 265.

[3]   Gurevitch, Michael and Phillip Elliott, "Communication Technologies and the Future of the Broadcasting Profession," in *Communication Technology and Social Policy,* edited by George Gerbner, Larry P. Gross, and William H. Melody (New York: John Wiley & Sons 1973) p. 505.

[4]   Walsh, Patrick, "Commercials and Our Changing Times," in *Television,* edited by Barry G. Cole (New York: The Free Press 1970) p. 238.

[5]   Aaholm & Williams, *op. cit.,* p. 2.

[6]   Gould, Jack, "Control by Advertisers," in *Problems and Controversies in Television and Radio,* edited by Harry J. Skornia & Jack William Kitson (Palo Alto, Ca: Pacific Books, Publishers 1968) p. 418.

[7]   Yoder, Dale, *Personnel Management and Industrial Relations* (Englecliffs, NJ: Prentice-Hall, Inc. 1970) p. 396.

[8]   Young, Whitney M., Jr., "This Is Where We're At," in *Broadcasting and Social Action,* edited by the editors of *Educational Broadcasting Review* (Washington, D.C.: National Association of Educational Broadcasters 1969) p. 5

[9]   Ylvisaker, Paul, "Conscience and the Community," *Television Quarterly,* Winter, 1964, p. 11.

[10] Seib, Charles B., "Broadcasters under Attack Again," a column in the *Washington Post,* reprinted in the *Denver Post,* July 24, 1978.

[11] Hartke, Senator Vance, *Senate Report,* August 29, 1967.

[12] Landry, Robert, "Wanted: Critics," *Public Opinion Quarterly,* December, 1940.

[13] Murphy, Congressman John, in "Deregulation Would Increase TV 'Banality,' Lawmaker Says," a UPI news story, datelined Washington, D.C., appearing in the Boulder *Daily Camera,* July, 26, 1978.

Chapter 5

TELECOMMUNICATIONS AND THE CONSTITUTION:
SPEECH AND PRIVACY

Irving J. Kerner
Attorney, Lecturer in Telecommunications Law
University of Colorado
Boulder, Colorado

5.0 INTRODUCTION

The preservation of two values, free speech and privacy, is
central to the concept of a free society. The explosion in
telecommunications technology has affected the human a-
bility to experience these values. The data processing-data
communications connection, in particular, raises serious
doubt as to whether, in reality, true privacy can be experienced in this society. Thus, this chapter will deal with the
informational aspect of privacy.

This chapter will also briefly survey the history of the First
Amendment. Anyone involved in the telecommunications
industry should be sensitive to First Amendment problems
and have some familiarity with its relationship to telecommunications.

5.1 THE RIGHT TO INFORMATIONAL PRIVACY

The development of sophisticated communication systems and
technological methods of storing vast amounts of data in computer banks, raises an increased concern about the amount of
privacy an individual can expect regarding his personal life. Computers have become essentially communication networks capable of transmitting stored data from various sources to single
terminals;[1] thus government and private enterprise can now
coordinate information about an individual and create detailed
personal dossiers. Fears of misuse and inaccuracy of individual
dossiers have resulted in a theory of informational privacy; namely, the right of an individual to exercise some measure of control
over the collection and dissemination of information about himself.[2]

In times past the storage of information posed little threat to
an individual's privacy. Filing systems could not accommodate
very much accumulated information; information was difficult
to retrieve in large amounts; storage problems limited the detail of the information. Most important, the information was
largely decentralized, and agencies and businesses were unable
to coordinate the diverse data they had gathered.[3]

The computer system, however, has largely eliminated impediments to information storage and retrieval. The computer, especially since the development of the Electronic Funds Transfer System (EFTS),[4] has the ability to store volumes of information on very little space, to allow retrieval of the information in seconds, and to transmit the information through telephone networks and other transmission devices.[5] The computer has made it possible for diverse agencies and businesses to share information about individuals for the purposes of ascertaining credit worthiness, insurance risks, criminal records, or threats to society.

An individual's right to privacy is dramatically affected by such technological advances. Citizens expect to be able to exercise some control over the collection and dissemination of personal information. In addition, citizens are concerned that their complete file not be revealed to persons who need to see only part. Finally, the collection of detailed information by various governmental agencies may have a chilling effect on individual political activity.[6]

Balanced against concerns about informational privacy, however, is the benefit to society of technological advances in collection and rapid transmission of data.[7] Without these advances consumers would be denied the advantages of instant credit; crime prevention and detection would be hindered; business would suffer from inefficient information-gathering and communications systems. The development in the law of the right to informational privacy will affect the design of the new technological systems and the costs of adequate financial and security services. The type and quality of information that the law allows to be maintained and shared among credit granting institutions will affect the accuracy of the information, the cost of extending credit, and ultimately the price to consumers.[8]

The law, therefore, must offer the individual adequate protection of privacy without unduly restricting the development of new uses of electronic technology. Traditional common

law theories of the right to privacy and constitutional interpretations have not kept pace with rapidly increasing technological advances.[9] Thus, neither the common law nor the constitutional right to privacy adequately safeguard the individual's
right to informational privacy. Ultimately the appropriate balance must be struck by Congress.

5.1.1 Common Law Right of Privacy

The common law protection of privacy is found in two legal
theories: defamation and invasion of privacy.[10] Damages
for defamation can be recovered in a libel action if the plaintiff can show that information released concerning him was
false and either injured his reputation or diminished the esteem, goodwill, or confidence in which he had been held.
Thus credit reporting companies utilizing computer data
banks may be potentially liable for the release of injurious
information. Recovery of defamation in these cases, however, is uncertain at best. The company may assert the defense of truth or of qualified privilege. Qualified privilege
is available if the company shows the consumer reports are
useful as business services to those who have legitimate interests in obtaining information. The plaintiff can defeat the
qualified privilege defense only by showing that the company
acted with malice or with conscious or reckless disregard for
his rights, or by proving the company released the information
to the general public or to persons with no legitimate business
interest in the information. In most cases, an individual cannot confidently rely on the law of defamation to protect his
right to informational privacy.

The other common law theory, the tort of invasion of privacy,
originated in 1890 with the publication of "The Right to Privacy" by Samuel Warren and Louis Brandeis.[11] Reflecting
a concern about governmental and nongovernmental threats
to individual privacy, the article advocated a remedy at law
to protect individuals from unwarranted intrusions into personal life. The theory gained widespread acceptance after publication, and four separate doctrines of the right to privacy
have evolved. Dean Prosser has identified as actionable torts:
1) the appropriation of another's name or likeness, 2) the

intrusion upon the seculsion of another, including unauthorized prying into a bank account, 3) publicity that places a person in a false light before the public, and 4) publicity given
to an individual's personal life.[12] The plaintiff in these cases
must prove that the invasion of privacy would be offensive and
objectionable to an ordinary, reasonable person considering
all the circumstances surrounding the incident. He must further prove that the defendant invaded what was expected to
be private and entitled to be private. Protection extends only
to private facts, not to information of public record available
for public inspection. Voluntary disclosure of private facts by
the plaintiff will bar him from maintaining an action for intrusion upon the seclusion of another, and the plaintiff has no
cause of action based on unauthorized publicity unless the information was made available to the general public by written
or printed word or radio or television.

The common law protection of the right of privacy has not
expanded to protect informational privacy. Most injury from
misuse of records does not involve either public disclosure or
intimate fact. Most communication of information for employment or credit determinations is qualifiedly privileged since
the information is transmitted from persons with legitimate
business interests to person with legitimate business interests.
Consent is often a defense because the information is usually
given voluntarily. Even if the plaintiff manages to prove his
case, the common law provides only a remedy for injury, not
any preventive measures.[13]

Computers, moreover, raise peculiar problems that the common
law has not yet addressed.[14] An individual may be unaware
of the stored information that has caused him injury and thus
unaware of his legal claim. Proving malice or reckless disregard
of rights is becoming increasingly difficult as computer technology reaches the point where the computers themselves make
evaluative judgements concerning such matters as credit risks.[15]
In addition, retailers with on-line connection in the EFTS would
probably be considered privileged individuals with legitimate

business interests.[16] Finally, there is an inherent deterrence to litigate these injuries under common law theories because of the time, exposure, emotion, expense of lawsuits, and the uncertainty of success.

5.1.2   Constitutional Right of Privacy

Although the Supreme Court has recognized a constitutionally protected right of privacy, the Constitution, like the common law, falls short of protecting the individual's right of informational privacy. The constitutional right of privacy was first recognized and articulated in *Griswold v. Connecticut.*[17] Although no specific provision of the Constitution guarantees a right of privacy, the majority opinion in *Griswold* found the right among the penumbras of the Bill of Rights "formed by emanations from those guarantees that help give them life and substance."[18] Thus the first amendment's guarantee of the right of people peaceably to assemble, the third amendment's prohibition against forced quartering of soldiers in private homes, the fourth amendment's protection against unreasonable searches and seizures, and the fifth amendment's guarantee against compelled self-incrimination all imply that an individual has a constitutionally protected right of privacy. Without that right the amendments would be meaningless. In a concurring opinion, Justice Goldberg asserted that the right of privacy is constitutionally protected because of the ninth amendment guarantee that the enumeration of rights shall not be construed to deny or disparage others retained by the people. Justice Harlan, also concurring, found the right to privacy as part of the liberty guaranteed in the due process clause of the fourteenth amendment.

Because the *Griswold* opinion failed to identify precisely the source or scope of the right of privacy, it is unclear whether the dimensions of the constitutional right of privacy extend to informational privacy. Subsequent Supreme Court opinions show that the right to privacy includes but may not be limited to "the personal intimacies of the home, the family, marriage, motherhood, procreation, and child rearing."[19] No Supreme Court decision to date has directly addressed the issue of a constitutionally protected right of informational privacy.

Even if the scope of the right clearly included informational privacy, the right to privacy is not absolute. In *Roe v. Wade* [20] the court found that a woman's decision to have an abortion is protected under her right to privacy, but that her right to make that decision is limited as the fetus matures and the state's interest in potential life becomes compelling. Therefore, if the right to informational privacy is among the fundamental rights, the Court may still allow collection of information by the government where the government asserts a compelling reason, such as crime control or national security. If the right to informational privacy is relegated to the status of a less-than-fundamental right, the government need only show a reasonable relation between its collection of information and the governmental purpose to be achieved. "[T] here is still a long decisional path to be traversed before a constitutional right of *informational* privacy is established." [21]

Commentators have expressed doubt that the Court will extend the constitutional right of privacy to recordkeeping by the government.[22] The Court's decision in *United States v. Miller* [23] may extinguish all hopes. In that case the Court upheld the government's subpoena of copies of Miller's checks and other bank records obtained in compliance with the Bank Secrecy Act of 1970.[24] The Court held that Miller had no constitutionally protectable interest in the records. Characterizing the checks and records as negotiable instruments rather than as confidential communications, the Court stated:

> "The depositor takes the risk, in revealing his affairs to another, that the information will be conveyed by that person to the Government. . . . [T] he Fourth Amendment does not prohibit the obtaining of information revealed to a third party and conveyed by him to Government authorities, even if the information is revealed on the assumption that it will be used only for a limited purpose and the confidence of the third party will not be betrayed." [25]

The right to privacy thus did not extend to information concerning business transactions. Although bank records may not be entitled to the same expectation of privacy as confidential, sensitive information such as that transmitted by EFTS, the implications of *Miller* are that the government may be able to compell disclosure of private information about individuals from the private sector without violating any constitutionally protected right to privacy.

5.1.3 Statutory Protection of the Right of Privacy

Since the common law offers limited recourse against invasion of informational privacy and since the constitutional right to privacy may not extend to informational privacy or may be limited when the government shows a compelling interest, the only viable resolution of the problem of protecting the individual lies with the legislature. Few states have enacted effective legislation against invasion of informational privacy [26] or have included consumer protection provisions in their EFTS enabling legislation. [27] Furthermore, state legislation, should it be forthcoming, may lack the necessary uniformity to meet the needs of a mobile and sophisticated population and the national and international financial system. [28] It is essential, therefore, that Congress enact federal legislation to guarantee protection of the individual's right of informational privacy.

Congress' first response to public outcries for consumer protection was in the form of the Fair Credit Reporting Act of 1970, [29] which was designed to control information-gathering and dissemination by private enterprises such as banks, credit card companies, and other credit reporting agencies. The purpose of the Act is to ensure confidentiality, accuracy, relevancy, and proper utilization of credit information. To that end, the Act limits the purposes to which consumer reports may be put and allows other uses only by court order or with the written permission of the subject. [30] Anyone wishing to use an investigative report must disclose the request to the subject of the report and disclose the nature and scope of the investigation. The consumer has the right to see his credit file and to know why he was denied credit. He may dispute the accuracy of the report and require the agency to reinvestigate the information. The Act also provides a statute of limitations to eliminate the reporting of obsolete data. The Act has been applied to the use of computers to assure limited access to and maximum accuracy of transmitted information.[31] Commentators criticize the Act for failing to limit the kind of information that can be stored or reported and for failing to give the consumer the right to physical access to his file or to possession of a copy.[32]

The second important piece of federal legislation was prompted by the need to control *governmental* intrusions into an individual's privacy. The Privacy Act of 1974[33] recognizes the interest of individuals in government records containing information about themselves. The Act attempts to strike a balance between the need of the individual for a maximum degree of privacy over personal information he furnishes the government and the government's need for information about the individual. The Act allows the individual access to records concerning him; gives him the right to correct and challenge personal information; requires the government to file in the Federal Register the existence, nature and scope of all federal government files containing personal data; prohibits nonroutine dissemination of records without notification to individuals involved; restricts the use of social security numbers; and establishes the Privacy Protection Study Commission (PPSC) to study and make recommendations on data banks and information processing programs in the government and the private sector.

The Privacy Act attempts to protect informational privacy, but commentators criticize the Act as being "conceptually sound, but pragmatically unenforceable."[34] Citizens have no control over compilation of data by an agency, and the restriction on dissemination is weakened by broad exceptions. The Act relies on individual initiative for enforcement, but gives an individual no way to discover agency violations and no incentive to take action against violations. Most important, there is no regulation of the government's access to information maintained by the private sector. This omission becomes significant in commercial settings wherein third parties own and maintain the system through which what was formerly a private transaction must be accomplished. Thus in *United States v. Miller*[35] the government was able to compel the bank to disclose the records of an individual who conducted business by means of a checking account. The implications for governmental access to information stored in the EFTS are unclear under *Miller*. Unless the Court limits *Miller* to bank records, the Privacy Act must be amended to ensure protection against governmental acquisition of private information held by private companies. [36]

The Privacy Protection Study Commission created by the Privacy Act conducted extensive studies of interstate transfer of information by computer and other electronic or telecommunication means, of data banks and information programs which significantly affect an individual's enjoyment of privacy, of the use of universal identification numbers, and of the matching and analysis of statistical data with other sources of personal data to ascertain the impact on an individual's privacy. In its report recommending the scope of future legislation the Commission emphasized that it has three objectives: 1) the creation of the proper balance between what an individual is expected to divulge to a recordkeeping system and what he seeks in return — to minimize intrusiveness; 2) openness in recordkeeping operations in ways that will minimize the extent to which the record about an individual is itself a source of unfairness in any decision of which it is the basis — to maximize fairness; and 3) creation and definition of obligations with respect to uses and disclosures that will be made of recorded information about an individual — to create a legitimate enforceable expectation of confidentiality. In its report the commission recommended that the Privacy Act not be applied *per se* to the private sector because recordkeeping by the government is different than recordkeeping by the private sector and because behavior and management of private agencies are diverse and differently motivated. The underlying principles and philosophy of the Privacy Act should, however, be applied to the private sector in future legislation. [37]

As indicated by commentators and by the Privacy Protection Study Commission, the Fair Credit Reporting Act and the Privacy Act represent significant attempts to protect informational privacy, but both statutes leave substantial room for invasion of that privacy and abuse of personal information. Far-reaching federal legislation, however, poses its own hazards. Professor Ransom warns that the creation of federal statutes and administrative rule-making agencies may invite Big Brother into homes and offices to control abuses of information gathering processes. The exercise of federal control may cost more in terms of individual freedom and confidentiality of papers than the abuses sought to be controlled. "Let us try to be sure that the side-effects of controls or 'cures' are not more destructive to liberty than the ills caused by a data bank and dossier society." [38]

In addition to the problem of placing proper restrictions on federal intervention, Congress faces the problem of its own lack of expertise in the technological area, the difficulty in categorizing levels of sensitive or privileged information, and its fear of enacting obsolete legislation because the field of technological data gathering and transmitting is so rapidly changing.[39] The need for effective legislation that strikes the appropriate balance, however, is of vital and immediate concern. Delay in implementing new legislation may well eliminate the right of privacy that American citizens have assumed was their heritage.

5.2 RIGHT OF FREEDOM OF SPEECH: DEVELOPMENT OF THE MEANING OF THE FIRST AMENDMENT

In 1791 the First Amendment was incorporated into the Constitution. The drafters stated, in relevant part: "Congress shall make no law. . .abridging the freedom of speech, or of the press;" The constitutional right of freedom of speech had come into being. It was not until 1925, in *Gitlow v. People of State of New York*, [40] that the First Amendment was held to apply to the states as well as Congress.

The broad language of the First Amendment did not receive serious attention from the courts until more than a century after it was drafted. Shortly after World War I, the United States Supreme Court began to establish the parameters of the right of freedom of speech. [41] Since then the courts have addressed considerable attention to that constitutional right, defining both its protections and its limitations. In interpreting the

First Amendment, courts have recognized that broad protection of the right of free speech is essential in a democracy to preserve free discourse in the "market place of ideas." At the same time, courts have had enough foresight to impose limitations on that right to ensure its continued vitality and feasibility.

The First Amendment advances a number of policies on both an individual and a social level. Four important policies have been posited.[42] First, the right of free speech permits individual fulfillment by allowing self-expression. Second, the amendment encourages the advancement of truth and knowledge by permitting free interchange of ideas. Third, free speech results in broader participation in the political process. Finally, freedom of speech helps ensure the viability of our system by maintaining the resolution of conflicting policies on a peaceful level. Above all, the First Amendment reflects an optimistic view of human nature, importing confidence in the ability of rational men to learn from each other and peaceably resolve disputes by communicating with each other.

The Supreme Court's analysis of the scope of the First Amendment has resulted in the birth of related constitutional protections. The Court has recognized a right of freedom of association.[43] The right to associate freely is essential to the right to communicate freely embodied in the First Amendment. The Court has also recognized a right to refrain from speaking. In *West Virginia State Board of Education v. Barnette,* [44] the Supreme Court upheld the right of a student to refrain from saying the pledge of allegiance. By protecting the right to remain silent, the Court helped assure the vitality of the First Amendment by upholding the right to not speak that which is not believed.

In considering the First Amendment, the Supreme Court has attempted to determine under what circumstances and to what degree the right of free speech may be restricted. The inquiry has evolved on two different fronts. At first, the Court's concern was primarily with the extent of governmental interest necessary to justify a restraint on the right of free speech. More recently, the Court has begun to focus on what constitutes "speech."

The first cases addressing the circumstances under which freedom of speech could be abridged involved alleged threats to national security. Perhaps for that reason, the Court was less willing to extend protection to the right of free speech. In 1919, in *Schenk v. United States,* [45] the Supreme Court created the first important judicial standard for restricting speech. The defendants were radical speakers who had been convicted under the Espionage and Sedition Acts of 1917. The Court declared: "The question in every case is whether the words are used in such circumstances and are of such a nature as to create a *clear and present danger* that they will bring about the substantive evils that Congress has a right to prevent."[46] In the early cases following World War I, the Court acknowledged a right of free speech but always in the context of affirming convictions restricting free speech. It was not until the Depression years in the 1930's that the Court actually afforded substantive protection to a defendant's right to speak freely. Thus, in *De Jonge v. Oregon,* [47] the Supreme Court reversed the conviction of a communist under a state statute forbidding "the teaching of criminal syndicalism."

By 1950, the Court had begun to drift away from the test of "clear and present danger." In *American Communication Assn., CIO v. Douds,* [48] the court ignored the prior standard and instead applied a balancing test. Indeed, the earlier cases undoubtedly reflected a tacit aspect of balancing, but by formally acknowledging that the constitutional inquiry would focus on balancing rather than a legal phrase, the Court attempted to insure that the policies at stake would be given fuller consideration.

Although the legal standard for restricting freedom of speech may have changed, a number of limitations on that right remain. Some of the most important restraints have arisen as a means of preserving order in society. Thus, the courts have been willing to limit freedom of speech during wartime.[49] Speech which acts as an incitement to riot or contains "fighting words" may also be prohibited. In 1951, in *Feiner v. New York,* [50] the Supreme Court held that speech which acted as an incitement to riot was not entitled to First Amendment protection. In that case, the Supreme Court affirmed the conviction of Irving Feiner for "breaching the peace" by criticizing various political figures and urging his audience to fight for their rights. The Court reasoned that speech which passed the bounds of persuasion and undertook incitement to riot was undeserving of constitutional protection.

A similar limitation of freedom of speech involves "fighting words." The Supreme Court articulated this doctrine in *Chaplinsky v. New Hampshire,* [51] a case involving a speaker who was arrested for calling the chief of police a "damned fascist." The Court upheld his conviction, ruling that the First Amendment did not extend protection to such inflammatory language. "Fighting words" do little to advance the quest for truth and knowledge and threaten to remove the decision-making process from a rational to a violent level. Thus, it is doubtful that the "fighting words" limitation will seriously threaten socially important communication, and it may advance policies underlying the First Amendment.

The Supreme Court has also considered the manner in which speech is delivered as well as content in determining the scope of First Amendment protection. Speech which is delivered in an offensive manner, to the extent of interfering with personal enjoyment of life, may be prohibited. Thus, in *Kovacs v. Cooper,* [52] the Supreme Court upheld an ordinance prohibiting raucous soundtrucks from broadcasting in residential neighborhoods. In a concurring opinion, Justice Frankfurter noted that, "The various forms of modern so-called 'mass communications' raise issues that were not implied in the means of communication known or contemplated by Franklin and Jefferson and Madison."

Other major developments in defining the right of free speech have occurred in the formulation of the definition of "speech." Important developments have arisen in respect to protest marches and other symbolic gestures, obscenity, and commercial speech.

Defining "speech" has been a particularly thorny problem for the courts in cases involving symbolic acts. In 1969, in *Shuttlesworth v. City of Birmingham,* [53] the Court overturned the convictions of civil rights demonstrators who had marched

without obtaining a permit as required by local law. In so
ruling, the Court went further than it ever had before in extending protection to protest marches.[54] The case reflects
a recognition that speech could be manifested in forms other
than the spoken and written word. The Court also served notice
that laws purporting to grant wide discretion to public officials
to regulate that right could not be constitutionally enforced.

The Supreme Court has also identified other symbolic acts as
protected speech. In 1931, the Supreme Court extended the
reach of the First Amendment to include a symbolic display
of a red flag.[55] Almost thirty years later, the Supreme Court
again affirmed that speech was not limited to words. In *Tinker
v. Des Moines Independent School District*, [56] the Court upheld the right of high school students to protest the Vietnam
War by wearing black arm bands. Although it is clear that symbolic acts may constitute speech, in many cases the courts will
attempt to avoid characterizing the action as "speech."[57]
Often the result in these cases will turn upon the degree of dis-
ruption caused by the action.[58]

Significant developments in the definition of "speech" have
also occurred in respect to pornography and obscenity. When
the Supreme Court first considered the issue in 1957,[59] the
Court ruled that obscenity was not speech and therefore not
entitled to protection under the First Amendment. The Court
extended the First Amendment to protect all works except
those "utterly without redeeming social importance."[60]
Although the Court formulated its most liberal definition of
obscenity in the 1960's, less protection was afforded commercially exploitive material.[61] At the same time, Constitutional
protection was extended to material consumed within the confines of one's home.[62] Thus, the Court has considered the
manner of presentation and consumption as well as the content
of the material in defining obscenity. More recently, the Court
has again addressed the issue of obscenity in *Miller v. California*
[63] and *Paris Adult Theaters v. Slaton*, [64] resulting in more
restrictive definitions of obscenity, but permitting standards
of obscenity to be determined in reference to local mores. Although restrictions on the right to distribute obscene material

as "speech" are arguably necessary, the Court's refusal to characterize obscene material as "speech" may not be intellectually defensible. Suprisingly, the Court has been slow in extending the First Amendment to include commercial speech such as advertising. Earlier cases refused to apply the First Amendment to commercial speech at all.[65] A change in the Court's thinking was signaled in *New York Times v. Sullivan,* [66] a 1963 case in which an "editorial advertisement" was deemed speech under the aegis of the First Amendment. It was not until well into the 1970's, however, that the Supreme Court acknowledged commercial advertising as meriting attention under the First Amendment. In both *Virginia Citizens' Consumers Council, Inc. v. Virginia State Board of Pharmacy* [67] and *Bates v. State Bar of Arizona,* [68] the Court struck down laws prohibiting advertising as violative of the First Amendment. Although commercial advertising may not serve as important a role as political dialogue, the Court has belatedly acknowledged that the exchange of information inherent in advertising is also deserving of constitutional protection.

5.2.1 Prior Restraint

A person's right to express himself is not unlimited. In certain circumstances, a speaker may be subjected to criminal liability, for example, for "disturbing the peace." In other circumstances, he may incur civil liability for slander or libel. Determining appropriate limitations on the right of free speech has entailed difficult assessments of the social utility of the communication and the costs of allowing it. For the most part, however, courts have extended broad protection to the right of freedom of speech. Inherent in that broad protection is the fundamental belief underlying the First Amendment that if people can express themselves freely, society becomes a better place, socially as well as politically, in which to live. Thus, the courts have interpreted the First Amendment to prevent the federal and state governments from instituting a system of prior restraints on speech or the press.

Prior to the drafting of the First Amendment there was a long
history of censorship both in England and the Colonies. England
licensed all publications until 1695, and after that date it continued to heavily tax publications, limit the introduction of
printing presses into the Colonies, and enforce criminal sanctions against seditious libel. Suppression was aimed primarily at
publications critical of governmental operations. The law of seditious libel punished publications if they were defamatory, regardless of their truth. Thus, exposure of scandals such as Watergate would have been impossible. The drafters of the First Amendment intended to stop licensing, censorship, and punishment of
government critics.

In *Near v. Minnesota,* [69] a state statute provided for the abate-
ment as a public nuisance of "malicious, scandalous, and defamatory publications." The statute further provided that all persons
found guilty of a nuisance could be enjoined from further publication. An action was brought under the statute against a pub-
lisher, and he was enjoined from further publication on the
ground that his newspaper had accused law enforcement officials of failing to take action against organized crime. On appeal the United States Supreme Court found that the effect of
the statute was to suppress further publication, constituting a
prior restraint on the press prohibited by the First Amendment.

Reasons given for holding the prior restraint in *Near* unconstitutional included the fact that prior restraint is broader in
coverage, more uniform in effect, and more easily and effectively enforced than the proper remedy of subsequent punishment. Upholding the prior restraint would subject everything
to scrutiny, through a process geared to suppress, and even
if material was not banned time delays would render it obsolete.
Thus *Near* focused on the effect of a prior restraint on expression. However, *Near* did not establish an absolute bar to prior
restraints, leaving the door open by indicating only that prior
restraints come before the court with a heavy burden and presumption of unconstitutionality.

5.2.2 National Security and the Right to Publish

The issue of prior restraint also arises in the context of national security and the release of government information. Governments have an inherent right to protect state secrets where release would be harmful to the public interest or the survival of the state. The most obvious example is the release of troop movement information during war. However, most cases are more subtle, creating a conflict between the people's right to know and questions of national security. Even though the Supreme Court ruled that President Nixon had to turn over the Watergate tapes, it was made clear that there is an inherent right to protect secrets, especially those relating to national security and diplomatic affairs. See United States v. Nixon [70].

The problem of national security and freedom of the press becomes more acute when the media has possession of classified material and wants to publish. In *Near* the Court left open the possibility of a prior restraint on the release of military secrets during war. In the "Pentagon Papers Case," *New York Times Co. v. United States*, [71] Daniel Ellsberg released to the New York Times and other papers top secret classified documents relating to U.S. policymaking in Vietnam. After the New York Times published part of the material, the U.S. Justice Department was successful in obtaining a restraining order to prevent further publication. The Supreme Court freed the Times to continue publication on the ground that prior restraints are unconstitutional and that the government has a heavy burden not met in this case. However, the implication was that given sufficient proof of direct and immediate damage restraint would be permissible.

5.3 DIFFERENT MEDIA CREATE DIFFERENT PROBLEMS

5.3.1 Films

The idea that different media raise different problems, thus requiring a variable approach to speech, is seen in the standards for prior restraint applied to films. The Supreme Court has found that the nature of films makes it proper to subject them to pre-censorship or prior restraint, with appropriate procedural safeguards.

In *Times Film Corp. v. City of Chicago*, [72] Chicago had an
ordinance which stated that before a film could be given a permit to be shown it had to be,

> "produced at the office of the commissioner of po-
> lice for examination. . . or censorship, [if the picture]
> is immoral or obscene, or portrays depravity, crimi-
> nality, or lack of virtue of a class of citizens of any
> race, color, creed, or religion and exposes them to
> contempt, derision, or obloquy, or tends to produce
> a breach of the peace or riots, or purports to repre-
> sent any hanging, lynching, or burning of a human
> being, it shall be the duty of the commissioner of
> police to refuse such permit; otherwise, it shall be
> his duty to grant such permit."

Thus the police became a pre-censor of all films in Chicago. An
independent film distributor refused to submit "Don Juan" to
the censor, and the case went to the Supreme Court where the
issue was whether pre-censorship of films was constitutional.
By a 5-4 majority the Court held pre-censorship to be constitutional, Justice Clark saying in part that, "Chicago here emphasizes its duty to protect people against the dangers of obscenity
in the public exhibition of motion pictures. To this argument. . .
the only answer is that previous restraint cannot be justified.
With this we cannot agree. . ." Movies, the majority said, were
not,". . .necessarily subject to the precise rules governing any
other particular method of expression."

The minority, however, vehemently opposed the decision. Chief
Justice Warren said it

> ". . .presents a real danger of eventual censorship
> for every form of communication, be it newspapers,
> journals, books, magazines, TV, radio, or public
> speeches. . .[There is] no constitutional principle
> which permits us to hold that the communication
> of ideas through one medium may be censored
> while other media are immune. . ."

Chief Justice Warren reviewed many cases illustrative of the evils
involved in prior restraint of movies. The Chicago Police Commissioner had banned newsreel films of Chicago police shooting at

labor pickets. That same Commissioner had ordered the deletion of a scene showing the birth of a buffalo from Walt Disney's "Vanishing Prairie." Memphis censors had banned "The Southerner," which was about poverty among tenant farmers, because the film, "reflected on the South." Atlanta had forbidden the showing of "Lost Boundaries" because it depicted black people passing as white and would "adversely affect the peace, morals, and good order." In his review of these cases, Chief Justice Warren indicated that pre-censorship was much more harmful than subsequent criminal prosecution. Delays in adjudication would cause damage; some distributors would not or could not afford to challenge license denials; fear of the censor would deter the creation of new ideas — all resulting in a suppression of free speech.

Times Film Corp. made it clear that a general constitutional attack on a statute requiring prior censorship of motion pictures will fail. However, in *Freedman v. Maryland*, [73] the Supreme Court established procedural safeguards for the prior restraint of films. A film distributor challenged the Maryland censorship procedures as being unduly restrictive and dilatory. The Court agreed that the statute as written was an unconstitutional prior restraint. The Court stated that: 1) the burden of proving a film obscene rests with the censor, 2) the censor must act on a film within a short, specified period of time, and 3) there must be a prompt, final judicial determination. However, even with procedural safeguards, the prior restraint of films is more restrictive than the alternative of subsequent criminal prosecution.

5.3.2   The First Amendment and Radio and Television

Unlike the printed media and films, radio and television share a unique characteristic — both are limited access media. The rationale that there are only a limited number of frequencies available and access to the airwaves is necessarily limited, has justified government regulation of radio and television. [74] The stated purpose of such regulation has been to ensure that the public's right of access to a complete, "marketplace of ideas", the paramount goal of the First Amendment, is not endangered.

The regulation of the broadcast media is enforced through a system of licensing controlled by the Federal Communications Commission (FCC), an independent administrative agency of the federal government which exercises powers delegated to it by Congress. The FCC has the power to allocate the frequency spectrum and to grant and renew licenses for use of particular frequencies by local broadcasters provided they will operate or have operated in the public interest. The broadcaster has no property right in his license. A license only confers a temporary right to use a designated frequency, and the license expires within three years unless renewed.

Courts may review an FCC determination to grant, deny, or renew a license to assure that reasonable consideration was given to all pertinent facts and to assure that the FCC did not abuse or exceed its authority.

Although technological developments may affect the rationale for regulation, at this time the broadcasting industry is licensed by a government agency, and is in fact accountable to the agency for many of its programming decisions. It is established, that in its determination of whether an applicant for a license or renewal will serve the public interest, the FCC will review what the programming has been or will be.

Perhaps the most controversial and important aspect of government regulation of content unique to radio and television is the fairness doctrine,[75] which applies to all of a station's public affairs programming including news broadcasts. The most basic requirement of this doctrine places an affirmative responsibility on the broadcast licensees to provide a reasonable amount of time for the discussion and consideration of public issues. The second aspect of the doctrine requires that broadcasters must cover public issues fairly so that an opportunity for the presentation of contrasting viewpoints is allowed.

The licensee has the duty to evaluate the needs of its particular community to determine what is a reasonable amount of time to devote to the presentation of public issues. It is also the responsibility of the broadcast licensee to decide which statements are controversial and require offsetting presentations. Mere discussion of a public issue does not necessarily make it controversial. Once the licensee's obligation arises under the doctrine, the presentation of opposing views must be done at the broadcaster's expense if sponsorship is unavailable, and the programming must be obtained at the licensee's initiative if it is unavailable from any other source.

The doctrine does not mean that equal time must be given to different viewpoints. The doctrine also does not require that the various sides of an issue be communicated in a single broadcast or in the station's overall programming. Thus the doctrine does not assure that each individual listener will hear all sides of a controversial issue.

One of the most important aspects of the fairness doctrine is the "personal attack" rule which applies to both radio [76] and television. [77] The rule provides that when an attack is made upon the honesty, character or integrity of a person or identified group during the presentation of views on a controversial issue of public importance, an opportunity to respond must be provided to the person or group attacked. Attacks on foreign groups or foreign public figures, legally qualified candidates, and bona fide newscasts, interviews, and on-the-spot coverage of a news event are exempt.

5.4 CONSTITUTIONALITY OF THE FAIRNESS DOCTRINE

The following arguments have been presented to support the position that the fairness doctrine is an unconstitutionsl abridgement of the broadcasters' First Amendment rights of freedom of speech: The First Amendment protects a broadcaster's desire to use the frequencies to broadcast and exclude whatever he chooses. No one should be prevented from saying what he thinks or from refusing to give equal weight to the views of opponents. The fairness doctrine is not applicable to newspapers because of the "free press" provisions of the First Amendment, and since radio and television have "free press" protections also, to permit the fairness doctrine to be applied to radio and television is to countenance a constitutionally impermissible double standard.

These arguments were considered and rejected in *Red Lion Broadcasting Co. v. F.C.C.,* [78] in which the United States Supreme Court upheld the constitutionality of the principle of the fairness doctrine and two specific aspects of the doctrine: the personal attack rule and the political editorializing regulation,[79] which requires that when one candidate is endorsed in a political editorial, other candidates must be offered reply time.

The Court adopted the rationale that unlike the press, access to radio and television is limited, thus application of different First Amendment standards is justifiable. Most importantly, the Court stated that the rationale of the fairness doctrine and regulation of limited access media is based on the right of the listening and viewing public to be informed and to have access to ideas. "It is the right of the public to receive suitable access to social, political, aesthetic, moral and other ideas and experiences which is crucial here."[80]

The goal of the First Amendment is to guarantee the free flow of information and to remove restraints that may inhibit that flow. If the First Amendment right that the government is trying to protect through programming regulation is viewed as the public's right to this free flow of information and not the private broadcaster's right to say what he pleases or to omit an opponent's views, then it is arguable that the government regulation through the fairness doctrine might accomplish First Amendment goals.

One conceivable result of the mandatory application of the fairness doctrine is that the doctrine could be evaded by a licensee simply refusing to devote any coverage to controversial issues. In this way no time would have to be given to opposing viewpoints and the purpose of the doctrine to encourage the flow of information would be stifled. It is argued that this has not happened, but if it could be proven that it did, the Court recognized in *Red Lion* that the FCC has the power to prevent this occurence by conditioning the grant or renewal of a license on the broadcaster's willingness to give adequate and fair coverage to public issues.

5.4.1   Limitations on the Doctrine

In *Columbia Broadcasting System v. Democratic National Committee*, [81] the Supreme Court announced that the fairness doctrine and the "public interest" standard of the Communications Act do not require broadcast licensees to accept editorial advertisements. Thus there is no right of access for people who might want to purchase air time. The court reasoned that the public interest in providing access to the marketplace of ideas would not be served by a system weighted in favor of the wealthy. If this system were adopted, the views of the affluent would prevail over the views of the poor because the rich would have the power to purchase time more frequently. Even if under the fairness doctrine broadcasters were required to provide time, perhaps for free, to those with an opinion contrary to the ones expressed in the paid advertisement, the wealthy could still determine what issues were to be discussed on the air. Thus the journalistic freedom given a broadcaster to select his choice of material would be seriously infringed. The broadcaster, not private affluent groups or individuals, is given the power to determine what is broadcast and the broadcaster is answerable to the public through the licensing procedure if he fails to meet the public's needs.

In *CBS* the Court also found that the actions of a broadcaster in refusing to accept editorial advertisements, which is in effect a type of private censorship, is not state action. First Amendment constitutional limitations and obligations do not apply unless it is the government which has restrained freedom of expression. Since the broadcaster's actions are not those of the government, the broadcaster's policy of refusing to sell time for editorial advertisements could not be a violation of the First Amendment rights of those who seek to advertise.

Controversy engendered by the fairness doctrine well illustrates problems that arise out of government involvement in content regulation. Whether such involvement serves to enhance policies which underlie the First Amendment, or in fact detract from those policies, is a question not answered. In light of that, a second question arises: if the first question cannot be answered affirmitively, are there any policies to be furthered, the need for which outweigh dangers inherent in government regulation of content?

5.4.2   Cable Television

As of this writing, the extent of FCC regulatory authority over cable television is unclear. It is settled that the Commission has limited jurisdiction to regulate in this area. In *United States v. Southwestern Cable Co.* [82] the Supreme Court upheld the FCC's regulation of cable to the extent that such regulation is "reasonable ancillary to the effective performance of the Commission's various responsibilities for the regulation of television broadcasting."[83]

The Supreme Court's opinion in *Southwestern Cable* has been criticized for failing to consider either the short term or long term first amendment implications of its holding.[84] "Rather than approaching the case as one involving fundamental constitutional issues requiring careful judicial scrutiny, the Court approached the case as a pure question of administrative law requiring maximal judicial restraint."[85]

Other courts agree with such criticism. Thus in *Home Box Office, Inc. v. FCC,* [86] the Court of Appeals for the District of Columbia invalidated Commission rules[87] limiting pay cables' ability to cablecast certain popular entertainments. Such rules had been criticized in the past as having stifled cables' growth by preventing the industry from achieving the popular support necessary to expand its services,[88] thereby depriving the public of the multiplicity and diversity of voices which by its other regulations the FCC attempts to promote. Further, "by limiting the commercial development of cable, these restrictions perpetuate the conditions which arguably justify content regulation in the electronic media," [89] — that there is a scarcity of available air time necessitating government regulation of that time in the public interest. Thus, the question is raised as to whether, ". . . the constitutionality of broadcast regulation [can] be sustained if it is the government itself which perpetuates the need for such regulation."[90]

The "reasonable ancillary" rationale was again attacked in
the recent case of *Midwest Video Corp. v. F.C.C.,* [91] wherein the Court of Appeals for the Eighth circuit held that the
Commission could not require cable television operators to
set aside channels for public access. The Court ruled that the
regulations were not, "reasonable ancillary," to the Commissions' responsibilities for regulatory broadcast television, nor
does the Communications Act of 1934 grant specific authority
over cable systems. In regard to constitutional implications
the Court commented that it could not find authority for
giving the Commission, "an affirmitive duty or power to advance First Amendment goals by its own tour de force, through
getting everyone on cable television or otherwise." The Court
also stated that, "governmental interference with the editorial
process raises a serious First Amendment issue," since the access regulations, "strip from cable operators, on four of their
channels, all rights of material selection, editorial judgment
and discretion enjoyed by other private communications
media, and even by the 'semipublic' broadcast media."

In seeking Supreme Court review of the decision, the American Civil Liberties Union has argued that the case presents
an opportunity for the Court to decide, "where cable television fits into the First Amendment system of freedom of
expression." The ACLU also stated that there are important
distinctions between cable systems and newspapers — the
systems' service, for instance, depends on broadcast signals —
and technical similarities between cable and telephone systems, which are regulated. [92]

Thus, the question of the form and breadth of government
regulation of cable television is unanswered. The answer is
basic and critical to future national communications policy.

5.4.3 Satellites

Under present technology, the main use of satellites for television purposes is to allow networks to send a signal to local
broadcasters more cheaply than using traditional cable or
microwave transmissions. Future technological advances,
however, may allow a broadcaster to send a signal directly

to the home of the consumer.[93] Such a system has obvious implications for the local licensee/public trustee concept
of broadcast regulation. It also would have profound impact
upon any rationale for content regulation.

Satellites will allow the development of a truly international
television system. Inasmuch as many nations do not hold first
amendment rights in as high regard as does the United States,
many nations may be sensitive to the types of material sent
to their countries by foreign broadcasters. The United Nations
has debated whether an international agreement limiting the
content of direct satellite broadcasts should be adopted.[94]
Serious questions exist as to whether the United States, consistent with the first amendment, could participate in and
enforce such an agreement.

Two possible approaches for achieving the requisite international agreement have been suggested in this debate: consent
to program content by the receiving country and international
program standards. The "consent" approach might arguably interfere with the first amendment rights of American viewers to
receive foreign programming and of American broadcasters to
present uncensored programs, should the United States be required to tailor a broadcast to conform to the receiving country's standards.

Establishing international principles to govern the content of
international broadcasts presents the difficult question of whether the United States should agree to international standards which
legitimize restrictions on the content of information that flows
into the United States or other countries.

ACKNOWLEDGEMENTS

The author would like to acknowledge the aid of Kevin Shea,
Esq., Andrea Bloom, Douglas Bry, D. L. Hutchinson, Pam McKee,
David Steefel and particularly Jan Holladay in preparation of
this chapter.

REFERENCES

[1]   Comment, *The Computer Data Bank — Privacy Contro-
      versy Revisited: An Analysis and an Administrative Pro-
      posal,* 22 *Cath. U. L. Rev.* 628, 642 (1973) (hereinafter
      Computer Data Bank); Scaletta, *Privacy Rights and Elec-
      tronic Funds Transfer Systems — An Overview,* 25 *Cath.
      U. L. Rev.* 801, 802 (1976).

[2]   Davis, *A Technologist's View of Privacy and Security in
      Automated Information Systems,* 4 *Rut. J. Comp. & L.*
      264, 265-66 (1975).

[3]   Scaletta, *supra* note 1, at 802.

[4]   The Electronic Funds Transfer System has been defined
      as the transmission of information regarding fund trans-
      fers over communication networks, starting with input
      from a terminal at the point of sale and culminating in
      a computerized bookkeeping transaction at some central
      funds transfer computer station, which in most cases
      would be a banking institution. Bequai, *A Survey of Fraud
      and Privacy Obstacles to the Development of an Electronic
      Funds Transfer System,* 25 *Cath. U.L. Rev.* 768 (1976).

[5]   Scaletta, *supra* note 1, at 802. *See also* Bequai, *supra*
      note 4, at 768.

[6]   *Computer Data Bank, supra* note 1, at 637. *See also*
      Davis, *supra* note 2, at 266. Public concerns increased
      dramatically in the 1960's when the federal government
      proposed the National Data Center to centralize all gov-
      ernment files into one massive data bank. Scaletta, *supra*
      note 1, at 802.

[7]   Brandel & Gresham, *Electronic Funds Transfer: The Role
      of the Federal Government,* 25 *Cath. U. L. Rev.* 705, 728
      (1976); *Computer Data Bank, supra* note 1, at 629.

[8]   Brandel & Gresham, *supra* note 7, at 728.

[9]   *Computer Data Bank, supra* note 1, at 631; Bequai, *supra*
      note 4, at 791; Scaletta, *supra* note 1, at 809.

[10] *See generally, Note, Constitutional Right of Privacy and Investigative Consumer Reports: Little Brother is Watching You, 2 Hast. Const. L. Q.* 773, 786 (1975); Note, *The Privacy Act of 1974: An Overview and Critique,* 1976 *Wash. U. L. Q.* 667, 675 (1976) (hereinafter *The Privacy Act of 1974*); Scaletta, *supra* note 1, at 801; *Computer Data Bank supra* note 1, at 631; Prosser, *Handbook of the Law of Torts* (4th ed. 1971) § 117, p. 802.

[11] Warren & Brandeis, *The Right to Privacy,* 4 *Harv. L. Rev.* 193 (1890).

[12] Prosser, *supra* note 10, at 807-09.

[13] *Computer Data Bank, supra* note 1, at 631-32.

[14] *Id.*

[15] Bequai, *supra* note 4, at 791.

[16] *Id.*

[17] 381 U.S. 479 (1965).

[18] *Id.* at 484.

[19] 413 U.S. 113 (1975).

[20] 410 U.S. 113 (1975).

[21] Miller, *The Right of Privacy: Data Banks and Dossiers, Chief Justice Earl Warren Conference Final Report, Privacy in a Free Society* 72, 80 (1974).

[22] *Id.* at 81.

[23] 425 U.S. 435 (1976).

[24] 12 U.S.C. § 1829(d) (1970).

[25] 425 U.S. at 443.

[26] State statutes often deal only with the accessibility of official records and leave a great deal to the discretion of the recordkeepers. *Computer Data Bank, supra* note 1, at 629.

[27] Weber, *A Public Policy Overview of Electronic Funds Transfer Systems,* 25 *Cath. U. L. Rev.* 687, 691 (1976).

[28] *Id.*

[29] 15 U.S.C. § 1681 *et. seq.* (1970).

[30] Section 1681(b) limits dissemination to persons who intend to use the information in connection with credit transactions, for employment purposes, for insurance purposes, for determination of eligibility for a license or other benefit conferred by a governmental agency required by law to consider the applicant's financial interest, or for other legitimate business needs. Section 1681(c) provides that except for governmental use under § 1681(b) information to the government is limited to the name, address, former addresses, places of employment, or former places of employment of the individual.

[31] Bequai, *supra* note 4, at 790 quoting *Federal Trade Commission, Division of Special Projects, Bureau of Consumer Protection, Compliance With the Fair Credit Reporting Act* 8 (1970).

[32] Bequai, *supra* note 4, at 790.

[33] 5 U.S.C. § 552a (1974).

[34] *The Privacy Act of 1974, supra* note 10, at 679. *See also* Brandel & Gresham, *supra* note 7, at 727-28.

[35] 425 U.S. 435 (1976).

[36] Ware, *Handling Personal Data,* 23 *Datamation* 83 (1977). The full report of the Commission is "Personal Privacy in an Information Society," U.S. Government Printing Office (Superintendant of Documents, Washington D.C. 20402) July 1977, Stock No. 052-003-00395-3.

[37] Ware, *supra* note 36, at 83.

[38] *Chief Justice Earl Warren Conference on Advocacy in the United States, Privacy in a Free Society, Final Report* 35 (1974).

[39] *Computer Data Bank, supra* note 1, at 648.

[40] 263 U.S. 652 (1925).

[41] *See* Schenk v. United States, 249 U.S. 47(1919); Abrams v. United States, 250 U.S. 616 (1919).

[42] *See T. Emerson, The System of Freedom of Expression,* 6-7.

[43] NAACP v. Alabama, 357 U.S. 449 (1958).

[44] 319 U.S. 624 (1943).

[45] 249 U.S. 47 (1919).

[46] *Id.* at 52.

[47] 299 U.S. 353 (1937).

[48] 399 U.S. 332 (1950).

[49] *See* Schenk v. United States, 249 U.S. 47 (1919).

[50] 340 U.S. 315 (1951).

[51] 315 U.S. 568 (1942).

[52] 336 U.S. 77 (1949).

[53] 394 U.S. 147 (1969).

[54] *See* Walker v. City of Birmingham, 388 US. 307 (1967); Poudos v. New Hampshire, 345 U.S. 395 (1952).

[55] Stromberg v. California, 283 U.S. 359 (1931).

[56] 393 U.S. 503 (1969).

[57] *See* Street v. New York, 394 U.S. 576 (1969); United States v. O'Brien, 391 U.S.367 (1968).

[58] *Compare* Tinker v. Des Moines Independent School District, 393 U.S. 503 (1969) *with* United States v. O'Brien, 391 U.S. 367 (1968).

[59] Roth v. United States, 354 U.S. 476 (1957).

[60] *Id.* at 484.

[61] *See* Nishkin v. New York, 333 U.S. 502 (1966).

[62] Stanley v. Georgia, 394 U.S. 557 (1969).

[63] 413 U.S. 476 (1973).

[64] 413 U.S. 49 (1973).

[65] *See, e.g.,* Valentine v. Chrestensen, 316 U.S. 52 (1942); Breard v. Alexandria, 341 U.S. 622 (1951).

[66] 376 U.S. 255 (1964).

[67] 425 U.S. 748 (1976).

[68] 97 S. Ct. 2691 (1977).

[69] 283 U.S. 697 (1931).

[70] 418 U.S. 683 (1974).

[71] 403 U.S. 713 (1971).

[72] 365 U.S. 43 (1961).

[73] 280 U.S. 51 (1965).

[74] National Broadcasting Co. v. United States, 319 U.S. 190 (1943).

[75] This administratively created policy was given statutory force in 1959 by an amendment to the Federal Communications Act 47 U.S.C. § 315.

[76] 47 C.F.R. § 73.123(a), 73.300.

[77] 47 C.F.R. § 73.679.

[78] 395 U.S. 367 (1969).

[79] 47 C.F.R. § 73.123(c).

[80] 395 U.S. at 390.

[81] 412 U.S. 94 (1973).

[82] 392 U.S. 157 (1968).

[83] *Id.* at 178.

[84] Hagelin, *The First Amendment Stake in New Technology: the Broadcast-Cable Controversy*, 44 U. Cinn. L. Rev. 427, 502.

[85] *Id.* at 502.

[86] 567 F.2d 9(1977).

[87] 47 C.F.R. § 76.225.

[88] Note, *Cable Television and Content Regulation: The FCC, The First Amendment and the Electronic Newspaper*, 51 N.Y.U. L. Rev. 833 (1975).

[89] Hagelin, Supra Note 84, at 524.

[90] *Id.* at 523.

[91] 571 F.2d 1025 (1978).

[92] Broadcasting, June 5, 1978, p. 60.

[93] Prince, *The First Amendment and Television Broadcasting by Satellite*, 23 *U.C.L.A. L. Rev.* 879 (1976).

[94] *See, e.g.,* G.A. Res. 2916, 27 U.N. GAOR Supp. 30, at 14, U.N. Doc. A 2916 (1972).

Chapter 6

THE INFORMATION SOCIETY

Howard Higman
Department of Sociology
University of Colorado
Boulder, Colorado

6.1 INFORMATION

Information may best be defined by what it is not. Information is in contrast to noise. Noise is meaningless perturbation in the sense environment. Information, on the other hand, is any perturbation in the sense environment which can be seen to have predictable recurring consequences.

The odor emanating from an empty can of sardines tossed into the trash is information. If there is a cat in the neighborhood, we can predict that the cat will overturn the trash can. The odor from the sardine can is a perturbation in the environment with a fairly predictable consequence. A streak of lightning in the air frequently will result in an individual's avoidance of standing free in an open field or close to a tall object. The sound of the rattler of a rattlesnake to some individuals is not noise but rather information, and will result in that individual's freezing in motion.

6.1.1 Signals

The above examples refer to information conveyed by what we may call "signals." Signals deal with the here and now. One group of chickens was known to have a thirteen-word vocabulary. With signals they gave information. Among the words in their vocabulary they had a word for "There is water in the trough" and "There is corn in the dish."[1] Life with meaningful signals resulting in predictable responses may have been on the earth for a billion years.

6.1.2 Symbolic Language

Symbolic language, on the other hand, is only as old as the human race. The thing that characterizes the human being, as distinguished from his nonhuman predecessor, is his ability to refer to things that are not here and not now, and possibly not at all. The human animal can say, "I will meet you on Monday at 12:00 noon at the Kennedy Airport," when it is not Monday at 12:00 noon and we are not at the Kennedy Airport. Dogs cannot make appointments.

The information in a preliterate society is limited by the memory and the vocabulary of a single human being. Life goes on, daughter like mother, daughter like mother, and son like father, son like father, generation after generation for a million years.

6.1.3   Writing

Some six thousand years ago, somewhere between the Tigris and Euphrates rivers there was a radical revolution in the information system — the invention of writing. Writing has many consequences. The two to which we will refer are first that it eliminates the restriction of the size of a human society. Prior to writing no society could be larger than could be assembled to receive the spoken word, for *in the beginning was the word*. To survive a society must know whether it is at peace or whether it is at war. With writing, for the first time, the word could be written on scrolls and yeomen could ride into the upper and the lower kingdoms, and presumably all illiterate persons could hear what they believed to be the same words simultaneously, thus making it possible for Egypt to be formed. Civilization, then, proceeded father like son, father like son, daughter like mother, daughter like mother, for ten thousand years, until about 400 years ago.

While the invention of writing made possible the enlargement of the size of a human society, far beyond that which could be assembled to hear the verbal presentation of truth, the second effect was to make possible the tolerance for deviance and experiment since the oral presentation of odd or nonconforming recipes did not threaten the very survival of society itself, as was the case when the only storage house of truths was their monotonous repetition. As long as the recipes for survival were recorded only in conduct and in the oral tradition, any form of behavior that varied from the mores constituted a pollution of the word and could not be tolerated. With writing, however, the correct recipe for survival was safe whether or not any specific individual's conduct conformed to that norm. This partly accounts for the dynamic nature of civilization as compared with the static and stable nature of the pre-literate society.

6.1.4   Printing

In Korea in the fifteenth century printing was invented.[2] It comes down to us in the West through the movable type of the Gutenburg Bible. Prior to printing, the handwritten word was expensive and therefore the property of a small elite population.

After printing the word was available for wide, popular distribution among the masses. With printing, in contrast to manuscripts written by hand, the word was sufficiently inexpensive for there to be thousands of copies where there had been single or tens of copies of the words before. Thus "universal literacy" became possible. The monopoly of the possession of the word by the aristocrats was broken; this gave rise to the House of Commons, public opinion, and democracy. Life went on for the last 400 years father like son, father like son, father like son, until now.

6.2 DISSEMINATION OF INFORMATION

6.2.1 The Post Office

The literature, both fact and fiction, of modern times, that is to say the last 400 years, was dominated by the post office. While the remnants of feudal aristocracy still received their messages in special scrolls produced by ambassadors with couriers and attendants, the mass of humanity in the Western world was dominated by messages contained in "letters." Not only did the arrival of the letters and the messages contained in the letters play a crucial role in fiction and non-fiction in modern times, but it is also significant that the dominant political figure in the President's cabinet classically was the Postmaster General, and a dominant figure in every small town or large city was the Postmaster. The Postmaster who presided over the flow of the delivery of information from citizen to citizen or government to citizen or citizen to government vied with the clergy, whose information was presumed to be associated with a supernatural network.

6.2.2 Electronic Communication

Electronic communication, which we know as the telegraph, the telephone, the radio, the television and the computer, has introduced a discontinuity in the course of human history of equal magnitude to the three previous revolutions: symbolic language a million years ago, writing ten thousand years ago, and printing 400 years ago.

The telegram, which was of course electronic communication transformed into a sort of letter, had the aura of special urgency because of the speed with which the message was transmitted, particularly over long distances. Thus, it was associated with serious matters such as death, induction into the army, casualties from warfare, and birthdays. A serious exchange of thoughts in long-hand by letter constituted a written record between two human beings sometimes of such value as to be worthy of preservation or of being sold and bought at auction and reproduced in book form. There obviously is no counterpart in a collection of the exchange in telegrams; telegrams might be collected as a display of congratulations.

The distribution of information on a mass scale was primarly by means of the printed word in newspapers, magazines and journals. Journals got their name from the thought that there was new news each day, and thus the word "journal" derives from "day" as does journey, referring to how far one could travel on a horse in one day.

Small town editors complained that they did not indeed circulate the news; everyone in town knew it. Their problem was more that of archivist or recorder. When the editors "got the news wrong," the townspeople descended upon them to correct them. It was not the information in the newspaper that was disseminated but rather the opinions and editorials espoused by the muckraker or the crusader that dominated the role of the newspaper. Information in the newspaper that was relevant was, of course, the advertisements, which were not dominated by persuasion but were announcements of what it was that was available when, where and for how much. This fact is attested to by the relative energy spent in accuracy in newspaper advertisements in contrast with the energy spent for the accuracy in nonadvertising content. [3]

6.3 SOCIAL IMPACT

6.3.1 Propaganda

We normally credit John Milton in his essay *Areopagitica*, with the first clear expression of the ideas underlying the first amendment to the United States Constitution. This amendment, guaranteeing freedom of speech, is based on the notion that error should be free to be promulgated for the sophisticated reason that only with the promulgation of error, subsequently defeated in competition with truth, was truth itself to be defined. The idea of the difference between propaganda and facts became a major preoccupation with Western man during and after World War I. It was at this time that we began to get scholarly studies of the conscious management of the dissemination of "information," or propaganda, as a major instrument of the struggle for the maintenance of power over the people. This occurred at the same time that telecommunications in the form of radio made its major debut.

While it could be argued that the promulgation of printed opinions in no way interfered with the opportunity of others to state their printed opinions, the use of limited and scarce airwaves presented a different problem. For this reason it was initially agreed that the society itself would control, by setting up rules, who would get to say what to whom on the airwaves.

Studies in the United States during the 30's and 40's showed repeatedly that persons believed that the news in the newspaper was highly likely to be incorrect, whereas the advertisements in the newspapers were highly likely to be correct.[4] The reverse was true with radio broadcasts where it was believed that the news was highly likely to be correct, whereas the advertisements were highly likely to be incorrect. The reasons for this are fairly obvious in that the advertisements in the newspapers tended to refer to what was for sale, when, where and for how much whereas radio advertisements contained persuasion and argument. The news on the radio, on the other hand, was restricted largely under the control of government to objectively verifiable shorthand assertions of events as they occurred.

With the frightening rise of totalitarian societies in Italy, Spain, Japan and Germany, came the overtly conscious management of the dissemination of "information," along with an enormous surge of interest in what we called "propaganda." Some persons sought to define propaganda as a conscious distortion of information through the use of less than all of the facts. Others defined propaganda as the use of information with an intention to influence

the receiver of the information. Still others used the term propaganda to refer to the attempt to use information to create opinion.

Up to now telecommunications in the area of mass information
has tended to be controlled almost entirely by the mass response.
With the exception of the small effort of public broadcasting, television and radio are overwhelmingly restricted by the popularity
of the offering as measured by scientific sampling.

6.3.2   Public Opinion

Prior to 1936 public opinion was a word invoked by persons with
the hope of persuading other persons to join them. The truth of
the claim was anybody's guess. The *Literary Digest* magazine
claimed to know that public opinion insured an overwhelming
defeat of Franklin Delano Roosevelt for his second term and the
election of Alfred Landon to be President of the United States.
We know in retrospect this disastrous prediction was based upon
a simple error of what we would now call a stratified sample.[5]
The invention of the scientific public opinion poll, which we associate with Gallup, Roper and Harris, was probably as radical
an invention in terms of its social and political consequences as
the atomic bomb. The survival of political positions as well as
their promulgators has become dependent upon how they are
doing at the polls.

Electronic communication, in the form of a "face-to-face" fireside
chat between President Franklin Delano Roosevelt and the masses
of the American people, is given credit for his ability to father a
major social revolution in the American lifestyle. However, the public relations firm of Whittaker and Baxter in managing first the election of Governor Earl Warren, and later of General Eisenhower and
Richard Nixon in 1952, may be recorded by historians as conducting the first telecommunications presidential election. This marriage
of scientific sampling, market research, telecommunications and
computer projections has replaced the strong-willed, fortunate or
brilliant business tycoons in the management of giant national and
multi-national western corporations in the modern era. Names like
Harriman, John Pew, J. P. Morgan, McCormick, John Deere and
Henry Ford no longer explain the decisions of the corporate structures of the western economies.

6.4 THE TELEPHONE

Sixty years ago the cost of an individual personal telephone was approximately one third of a working man's monthly income. This, of course, accounts for the fact that he did not have a telephone, and also for the fiasco of the *Literary Digest's* selection of samples of voters from telephone books fifteen years later. Today the telephone company is the largest single corporate employer in the United States. Let us notice the contrast between life with the telephone and life with the letter. One does not visit by letter. If letters are other than an exchange of logistical information, then they are exchanges for deep thoughts and feelings and can be published, read, and reveal to us the inner person. The telephone "visit" is superficial, transitory, and is characterized by such phrases as "touching base." The telephone may be used as a thermometer to run a check or to determine the state of tension and well-being or the continuity of a previous relationship. Karen Horney [6] has observed that Americans have "a neurotic need for affection." Margaret Mead [7] observes that Americans are possessed of what she calls the "success ethic" which is measured by the responses of others. This is called the "How am I doing?" syndrome. The telephone is used by Americans as a thermometer, and conversely, as a device for sending out an alarm as well.

Secondly, the telephone compared with the letter is coercive. The voice seems to demand an immediate clear-cut response. There is frequently a loss of reflection time on the part of the respondent to the telephone call. Whereas one may lay a letter down, read it over and over, and reexamine it after his response to it, these options are not available to him in an exchange on the telephone. The bell of the telephone seems to act as a command. All of us have been infuriated by having service we were receiving in a department store interrupted by the ringing of a telephone as the person waiting on us stops to respond to the command, presumably coming from the other end of the telephone.

The telephone is active and, by comparison, the letter is passive. That the telephone is an instrument of invasion is attested to by the unlikely anomaly of the words "the number is witheld at the request of the subscriber," as well as the universal language describing the desire for rest and rehabilitation as "getting away from the telephone."

A unique advantage of the telephone call, contrasted to the letter, is the enormous potential for permitting the creative use of ambiguity. It would be exceedingly difficult to find letters in which the intent of the author is unclear. The message of the letter is clear, and the ambiguity can be preserved in face-to-face conversation. The telephone provides us with a superior mask for ambiguous exchange. This makes the telephone call a primary device for negotiation and adjudication for human differences. The phrase "may I say who is calling?" is primarily intended to facilitate the formation of strategies and the choice of tactics for what is expected to follow. The telephone call may be generative; that is, there is built into the telephone call rapid and extensive feedback which is slow or missing in the exchange of written letters, and as a result, the call itself may end up in conclusions that cannot be said to have been in the mind of either the initiator or the recipient of the telephone call.

The telephone call is "a dissolving message." This characteristic of the telephone call is attested to by the alarm with which persons greeted the advent of the tape-recorded telephone call and the insistence that the parties of the telephone call be informed that something was recording the call other than the human brain. We might well study the significant role of the dissolving message. The dissolving message is also attested to by the common practice of a subsequent letter saying, "this will confirm my telephone call," or conversely the statement, "would you kindly put that in writing and slip it in the mail to me." The phrase, "I'll get back to you," being an affirmation by denial, also attests to the general awareness of the ephemeral nature of the telephone communication. The phrase, "I won't keep you," means either "I got what I wanted" or "I guess I'm not going to get what I want."

By the end of the nineteenth century, a change in society occurred from one largely constituted around nouns by which individuals were defined primarily by their Aristotelian essences; that is, individuals were perceived as subjects of sentences or as nouns. They had a clear-cut sense of identity, a high degree of continuity in

space and time, and were characterized by such personalities as
Abraham Lincoln, Henry Ford and J.P. Morgan. In the Artistotel-
ian scheme, identities are defined as what the individuals "are."
However, with the appearance of the assembly line, the automobile, and the telephone, the contemporary mobile society defines
individuals as predicates, rather than subjects, of sentences. They
are what they are "doing." In *The Lonely Crowd* David Reisman
discusses this shift as being analogous to a shift from "gyroscopes"
to "radar." We might add that the shift from the printed (passive)
communication represented by the letter to the electronically transmitted spoken communication (active) represented by the telephone
is a shift of the same order.

In a recent interview Mr. William Baker, president of the Bell Laboratories, made these remarks which point up the magnitude of
the role played by modern telephone networks:

> "It appears that the telephone is the principal organ-
> izing element in the ordering of an informational so-
> ciety. It appears that the switch telephone system is
> as big an element as anything in reducing the entropy
> and bringing order in the broad philosophical sense. . .
> Since about 1945 the amount of information in the
> records doubles every 7 years. It has been determined
> that a weekday copy of *The New York Times* has as
> much to read as the educated individual in 16th cen-
> tury Europe absorbed in his life time. Now that im-
> poses on society a huge burden because people can't
> absorb information any faster than about 40 bits (bi-
> nary digits, the smallest measure of information) per
> second. Our evidence is that people today can't ab-
> sorb information any faster than stone age people,
> so you can see that getting information fed to you
> faster over the phone becomes critical." [8]

James Flannegan, Director of the Acoustics Research Department
at Bell Labs, has described a voice recognition system which will
let a person in a checkless society call up his bank and have it send
money from his account to a store to pay for a purchase.[9] The
computer succeeded in recognizing individual voices with 95% accuracy.

Saul Buchsbaum, Bell Labs' Vice-President for Network Planning, states:

> "There won't be such a thing as a telephone in 50 years. We'll have an instrument, and at the press of a button you'll either turn it into a telephone or it will connect you with someone else or something else like a computer. You will be able to get a video display as well as a voice. You will be able to connect to a vendor or a bank. You will have a communications terminal. When you get a busy signal your phone can monitor the line and complete the call as soon as the line is free. You will also have a display on the phone that will show what number is calling you. If you recognize it as that of someone you don't care to speak with you can ignore the ringing."[10]

In 1900 a pair of copper wires carried one message. By 1918 a twisted pair could carry 12 channels. By 1950 a microwave link could carry 1,800 voice channels. By 1970 a coaxial cable carried 32,000 voice channels. Now a helical waveguide will carry 100,000 voice channels or the equivalent. The cost of delivering messages is falling at an exponential rate. We increasingly live in a "shrinking world;" in every sense, man around the world is again similar to primitive man in a single tribe since "accessibility" of information is almost universal.

6.5 COMPUTATION

A look at the effect of the new information technology on the speed and cost of arithmetic computation illustrates the benefits derived from printed circuit and semiconductor technology. Carl Hegel calculates that one man takes one minute to do one multiplication at a cost of $12,500,000 for 125,000,000 multiplications. A desk calculator completes the multiplication in 10 seconds at one sixth the cost of a man. A Harvard Mark I completes the calculation in one second at the cost of $850,000. Today a CDC 6600 does the job at 0.3 microseconds or at a cost of $4 in contrast to $12,500,000 for one man's effort.[11]

In 1945 the labor for a million operations on a key board would take a month at a cost of $1,000, at the rate of a dollar an hour. In 1972 a computer could do a million operations for less than 6¢. By 1975 this cost had dropped to less than 0.6¢, according to the futurist John McHale.[12]

In the Organization for Economic Cooperation and Development Information Studies #3, 1973, "Computers in Telecommunications," there is a list of possible services from a broad band switched network: advertising, pictorial consumer information, alarm (burglar, power failure, fire, and so on), banking, facsimiles of documents, newspapers, emergency communications, communications between subscribers and computers, meter readings for utilities, the distribution of radio programs, shopping from homes, television originating and distribution, television-stored movies available on demand, educational television, telephone, computer-aided instruction, picture phone, video phone, and voting.[13] Buckminster Fuller has suggested the replacement of the representative system of government (Congress) with a national plebiscite on each specific issue after a public discussion of the issues conducted via telecommunications systems.

The Japan Computer Usage Development Institute suggested that Japan might set a national goal transforming the country from industrialization to informationalization by the year 2000. The major projects in the plan would be as follows: 1) formation of a nation-wide information network (1,000,000 cables); 2) formation of an administration information center for integrated policy decisions; 3) formation of a data bank by industry for up to 50 industries; 4) computer oriented education to cover elementary, secondary and high schools all over Japan; 5) remote medical care systems to service doctorless areas (currently there are 570 doctorless areas in Japan); 6) formation of a pollution information center with a warning system; 7) large-scale supermarkets which sell fresh foods for up to 1,000 stores; 8) an automatic driving transportation system for the center of Tokyo; 9) international cooperation through medical and educational networks using communications satellites.[14]

6.6 THE TWO BRAINS

Recent work[15] in socio-biology is uncovering a distinct dualism
in the human brain. The left brain is assigned the task of unilinear,
sequential, logical thought proceeding from bit to bit to the whole,
while the right hemisphere of the brain is assigned the process of a
gestalt, a holistic leap. The traditional transmission of information
through books and sentences associated with newspapers, journals
and printing is characteristically that of the left brain. On the other
hand, information transmitted by television, which is increasingly
the source of news would be right-brained, starting with a holistic
comprehension which then may be broken down into bits and
pieces by analysis by the left brain.

The recent outcry that "Johnny can't read" may in fact reflect
the replacement of the unilinear process of left brain reading by
a right brain gestalt awareness. In the pre-telecommunications era
of the newspaper, a political campaign was waged, as in the example of the Lincoln-Douglas debates, by left-brained dialectical, unilinear, sequential argument characterized by logic. On the other
hand, in the telecommunications world, the political campaign is
waged by a 47-second television aphorism associating the candidate
with either aversion or attraction. The most famous (or infamous)
of these political advertisements for television was that of the Democratic Party in the 1964 presidential campaign which showed a little girl plucking off daisy petals followed in sequence by an explosion of a mushroom cloud and a picture of the Republican candidate, Barry Goldwater.[16]

This shift from a unilinear, sequential arrangement of argument
associated with books and the left-brain by holistic, right-brained
leap to conclusions first and reasons afterward may be associated
with the widespread disappearance of departments of debate from
the college scenes of Western Europe and America in which debating
teams stated logical propositions and carried on according to unilinear arguments for or against a proposition. It has been noted that
the current so-called presidential debates in American campaigns do
not have the character of debate but rather of a projection of image
of two persons dividing the time equally. Departments of debate at
universities have been largely replaced by departments of communication in which persuasion and advertising take on a holistic or

right-brained character. Advertising in American newspapers during the 20's and 30's reflects the presentation of claims about the physical characteristics of the product being sold, whereas advertising since the advent of telecommunications seeks to create in the purchaser a feeling with reference to the product with little or no reference to the qualities or characteristics of the product itself. An advertisement for an automobile in 1928 gave its dimensions, its length, its width, its height, the amount of material used to manufacture the automobile, and the horsepower of its engine. Advertisements for automobiles in the 70's refer to what feelings the possession of such an automobile might induce in its owner.

Books are characterized by titles, chapters, paragraphs, and sentences. They have a beginning, they proceed to the end, and then they stop. Television communication has no boundary lines, no margins, no paragraphs, no sentences, no beginning and no ending.[17] A parallel can be found in the evolution of musical forms. This may be symbolized by the difference in music. In the classical music of Corelli, Vivaldi and Mozart one must learn when growing up when one should not applaud. Classical music, like chapters, paragraphs, and sentences in the written word, has a rigid and conventional form. Rock concerts, on the other hand, are lacking in the highly stylized form of classical music, and instead are free-flowing "happenings" characterized by the romantic form of the amoeba with no particular beginning or end.

6.7 "THE NOTICERS"

In a quick examination of the three component parts of a university, we determined that in 1938 the university was possessed of x number of students, y number of professors and teachers, and z number of persons who were neither teachers nor students but facilitated the work of both.[18] Forty years later, in 1978, the same institution had multiplied its number of students by 7, but it had multiplied the number of faculty by 5, thus decreasing the teacher-student ratio. The striking fact, however, was that the number of persons associated with the university who were neither teachers nor students was multiplied by 17. We might call these persons "noticers," for they were neither teaching nor studying.

They were noticing who was teaching, who was studying, what was being taught, what was being learned, and represented an interface with a larger community that funded and tolerated the academic activity. This gigantic rise in the number of "noticers" is directly correlated with the growth of relevance of information as opposed to the growth of activity in modern society as it is emerging.

The same observation could be made with reference to the growth of "noticers" if we can apply that title to the persons on the staffs of the elected officials in the House of Representatives and in the Senate of the United States, or to the "noticers" currently employed in the White House as opposed to the line officers in the administration of the government. The same development would be characteristic of the difference in the Ford Motor Company in 1978 and that at the time of Henry Ford's personal dominion.

It is significant that in an interview in Paris between a manager of a Soviet corporation and a manager of a multi-national Netherland-based corporation, each revealed that with the advent of computer inventory their two jobs were converging and becoming surprisingly similar. With the ascendancy of telecommunications in management and decision making, ideologies such as those of Adam Smith, on the one hand, and those of Karl Marx or Engels may become increasingly difficult to recognize as having a bearing on the conduct of an industry. Cybernetics in a rug factory in a socialist society and in that of a so-called free enterprise system may be, in fact, identical.

Much is said about "information overload." We predict that with a strong instinct for survival the individual human animal will maintain his/her sanity by a simple refusal to engage with the available information in the environment and consequently will carve out small, but habitable, communities. These communities in the future will lack spatial definitions. A person's community may not be a block in a small town, but rather a list of frequently telephoned individuals regardless of their physical location.

6.8 TELECOMMUNICATIONS AND THE FUTURE

Perhaps the most optimistic and hopeful projection one can make about the impact of telecommunications in the information society of the future is the appearance, only now, of the technical ability to monitor through satellites the activities of groups and subgroups organizing for the expression of hostility by a use of arms. Although disarmament has been a goal spasmodically asserted from time to time by various groups in the world, only now is the mechanical means available to make universal world disarmament feasible because the primary logical reason for avoiding disarmament is the gnawing fear that the potential opponent has not done so. Man has for the first time the technical ability, if he chooses to use it, to agree to disarm and to be able, by satellite surveillance, to know that his potential enemies are keeping the bargain.

It has been argued that the new information society will destroy privacy and individuality. This may be incorrect for it may be argued that privacy was never a reality in the history of man. The fact that data banks know everything there is to know about an individual puts them in probably the same category as the brain of one's spinster aunt at the turn of the century; in the pre-telecommunications world of 50 years ago, most persons lived in small towns and rural communities, and in a town of 4,000 the two local bankers knew all that there was to know about everybody in town that could be stored in a telecommunications data bank today or tomorrow.

Not only may the widespread universal availability of knowledge make possible a rational allocation of the increasingly scarce resources of the earth's crust and eliminate disease and poverty, but it may also eliminate the need for irrational ideological pursuits based upon a zero-sum-game.[19] The technology of the Industrial Revolution produced a stingy material world described by David Ricardo's "Iron Law of Wages" which holds that most persons would have to live at a bare subsistence level as indeed is the case in the Third World today.[20] That is a zero-sum-game. The technology of the information society, on the other hand, is producing an exponential growth of food, goods, and services making possible an ever-enlarging size of the pie providing, in the words of R. Buckminster Fuller, "more and more" for "less and less." For all to win it is no longer necessary for some to lose.[21] That is a non-zero-sum-game. The information society holds out the possibility that the non-zero-sum-game can accommodate all persons around the world.

REFERENCES

[1] Sebeck, Thomas Albert, *Perspectives in Zoosemiotics*, The Hague, Mouton, 1972.

[2] Carter, T. L., *The Invention of Printing In China and Its Spread Westward,* Ronald Press, New York, 1955.
While printing in the form of medals or short inscriptions had existed, the first book-length printed document is one in 868 A.D. in China.

[3] La Pierre, Richard, *Collective Behavior,* McGraw-Hill, New York and London, 1938, Chapter 13.

[4] *Ibid.*

[5] Likert, Rensis, "Pubic Opinion Polls, " *Scientific American,* CLXXIX, 1948, No. 6, pp. 7-11.

[6] Horney, Karen, *The Neurotic Personality of Our Time,* Norton, New York, 1937, p. 281.

[7] Mead, Margaret, *And Keep Your Powder Dry,* Morrow, New York, 1943.

[8] *New York Times,* May 28, 1978.

[9] *Ibid.*

[10] *Ibid.*

[11] Hegal, Carl, *Computers, Office Machines and New Information Technology,* MacMillan Company, New York, 1969. p. 96.

[12] McHale, John, *The Changing Information Environment,* Westview Press, Boulder, Colorado, 1976, p. 7.

[13] *Computers and Telecommunications,* Organization for
      Economic Cooperation and Development Information
      Studies #3, Paris, 1973, p. 121.

[14] *The Plan for Information Society,* The Japan Computer
      Usage Development Institute, Tokyo, Japan, 1973.

[15] Ten Houten, Warren D. and Charles D. Caplan, *Science and
      Its Mirror Image: A Theory of Inquiry,* New York, Harper
      and Row, 1973. See also "Neurolinguistic Sociology," *Socio-
      linguistics Newsletter,* Warren Ten Houten, Vol. VI, No. 2,
      pp. 4-9, July 1975.

[16] Wright, Quincey, *A Study of War,* University of Chicago
      Press, 1942, p. 31.

[17] Cater, Douglas, *Television and the Thinking Person: A
      Policy Paper,* The Aspen Institute of Humanistic Studies,
      Series on Communications, Palo Alto, 1975.

[18] Unpublished paper by Howard Higman, University of Color-
      ado, Boulder, 1978.

[19] Bagel, Medard, *Energy, Earth and Everyone,* with introduc-
      tion by Richard Buckminster Fuller, Straight Arrow Books,
      San Francisco, 1975.

[20] Ricardo, David, *Principles of Political Economy and Taxation,*
      Everyman, London, 1817.

[21] Ferkiss, Victor C., *Technological Man: Myth or Reality,*
      Braziller, New York, 1969.

Chapter 7

Robert J. Williams
Department of Engineering Design and Economic Evaluation
University of Colorado
Boulder, Colorado

7.1 Economic Analysis a Prerequisite to Action in the Telecommunications Industry

    7.1.1 Co-determinants for Decision-making

    7.1.2 Economics as a Design Parameter

7.2 Interest and Return

    7.2.1 Economy Must be Measured over the Life of the Capital Investment

    7.2.2 Interest

    7.2.3 Simple and Compound Interest

    7.2.4 Equivalence

    7.2.5 Symbols and Terms

    7.2.6 Nominal and Effective Interest

    7.2.7 Uniform Series

    7.2.8 Examples of the Series Notations

    7.2.9 Derivation of the Time Value Factors

    7.2.10 Examples of Applications of the Time Value Factors

    7.2.11 Continuous Compounding

7.1 ECONOMIC ANALYSIS A PREREQUISITE TO ACTION IN THE TELECOMMUNICATIONS INDUSTRY

The technologies of the various communication processes are in the midst of changing society much as printing did five centuries ago, or as broadcasting did in the first half of this century. Not only are there many types of processes, as indicated in earlier chapters, but each has been developed to a sophisticated point almost undreamed of a relatively short time ago. Then, too, processes can be joined in unique ways, and through coupling with the logic of computers, the collection, storage, manipulation, and transmission of information literally knows no bounds. Voice signals, video signals, reference data, and control signals are typical of information which can be reduced to the common denominator of electronic digital data. Rarely are there significant technical limitations. The limits are now those of regulation and of economics.

To design today's complex telecommunications systems, to buy even "off-the-shelf" equipment, and to keep the entire system operating with trained personnel, adequate inventory, and within the various regulations, requires, of necessity, a centralized function in most organizations. The manager of this function is usually a highly skilled technical individual with considerable formal training. Technical decisions by this individual as to procurement of equipment through purchase or lease generally results in long-term financial commitments by the organization. Once such commitments are made and facilities or equipment are built or purchased, the decisions are frequently irreversible. In other words, it does no good to change one's mind once this point of no return has been reached.

7.1.1 Co-determinants for Decision-making

Each project, system, or major individual item of equipment under consideration for procurement is normally evaluated in two distinctly different ways—whether it meets the required technical specifications and whether it is economically viable. Those alternatives which are technically acceptable generally each present a different set of economic consequences, requiring an economic feasibility study. Economic evaluations may be intuitive, may be by the "squeaky wheel" method, by the "necessity" process (wait 'til the old one breaks down), or by a comprehensive life-cycle cost analysis. The latter method considers all of the costs and all of the anticipated savings over the economic life of the proposed investment. It responds to the question of what are the future economic consequences of a decision to be made now. At one time in history the engineer was content with doing the technical design and the technical feasibility studies. The physical environment was considered to be more well-ordered, and the complexities of the economic aspects were left to "someone else." Surely, management could always find the funds to pursue a "hot" idea. The result was that an effective optimization of resources of time, facilities, and money was seldom achieved.

The concepts of what has come to be called Engineering Economy have been developed to assist engineers in presenting the economics of technical alternatives. Other terms, including managerial economy, have been used, but the most generally accepted one is engineering economy. This chapter is intended to provide some understanding of the principles, techniques, and reasoning by which a technical person may be guided in his economic analysis relating to long-term investments in capital equipment. A number of references are cited for those who wish to pursue the topics in more detail. In teaching his courses in engineering economy the author has found it difficult to keep current with the rapid technical and cost changes in the broad telecommunications field. He has found it more desirable to present general examples to illustrate the fundamental concepts. This approach avoids the question of whether the data in the examples is technically and economically up-to-date and permits more emphasis on learning the principles. Students readily adapt the concepts to their own special fields. This chapter emphasizes the same approach.

7.1.2 Economics as a Design Parameter

While economic analysis frequently follows a technical feasibility study (will it work? will it deliver the promised output?), we now find that economy has become a criterion of design. This means that the engineer and the communications manager, from the first conceptual efforts, must actually consider economics in parallel with function and reliability. Or, to express it another way, there is now a growing tendency in design to emphasize the joint achievement of performance objectives and product or service costs. It is apparent at the present time, for example, that telephony is no longer an unique art. Groups outside the telephone business with experience in electronics and information theory are reaching in to specified narrow segments of the business because they recognize a profit potential. They have also been assisted by the 1968 Federal Communications Commission (FCC) Carterfone decision which permitted non-monopoly equipment to be connected to telephone company lines. Cost parameters within these segments are first established, and then a system is specifically designed to operate within these parameters. *Rocky Mountain Industries* magazine in October, 1977 examined the growth of the interconnect business, stating that "in the Denver area about 2,000 private interconnect systems have been installed," with "the cost of owning a private telephone system as much as 15 to 35 percent less than the monthly rental of a comparable system from Mountain Bell, depending on size and type of equipment."[1] The net result of such cost-effective designs is to drive down unit costs. Many functions of interactive communications media, however, tend both to require and to generate increasing quantities of information, so the capital investment necessary to produce lower unit costs is at the same time increasing.

7.2 INTEREST AND RETURN

Effective use of capital as a limited resource is essential to the profitability of any company. While the "bottom line" is viewed as the real measure of the effective use of capital, it cannot be an optimum unless each of the individual units of invested capital is optimized. Timing, in terms of the moment, as well as the duration of commitment of individual units, is an essential part of this optimization. Capital expenditure decisions generally involve an investment of funds now in anticipation of greater future receipts. An evaluation of the effectiveness (or return on investment) of the expenditure requires comparisons of future sums of money, or receipts, at various points in time. Often the capital expenditures for a particular project also come at various time intervals. An understanding of the time value of money (interest) is essential in order to measure the effectiveness of the investment.

7.2.1 Economy Must be Measured over the Life of the Capital Investment

The cheapest equipment to buy or build, the yardstick of the purchasing agent, is not always the cheapest equipment to own. Savings in construction materials, components, and complete units of equipment may be wiped out by continuing high operating and maintenance costs. What may appear to be expensive refinements as reflected in a higher first cost may prove over the life of the equipment to actually reduce operating and/or maintenance costs. Operating people who have to live with the equipment have a constant fight against costly down-time and always push for the best in the original procurement. They find it difficult to express their justification in economic terms, but know they want the higher priced alternative. Tools of engineering economy use figures to determine which of several possible alternatives costs the *least in the long run*. These tools are *not* used to prove that a preconceived solution or alternative is the only correct one. The trend toward consideration of costs over the long run is recognized by the term life cycle costing.

7.2.2 Interest

The term interest is used as the cost of capital, or the rental amount
charged for the use of money. It represents a practice which is as
old as the recorded history of man. Time is one of the key variables,
and is a measure of the financial gain anticipated by the owner
through his lending, and of the benefit which the borrower or user
anticipates by having capital available. The economic gain through
the use of money by either party is what gives money its time
value. "The growth of money in time must be taken into account
in all combinations and comparisons of payments," [2] so adequately stated in the early development of engineering economy, still
holds today. Regardless of the source of funds, whether equity or
debt, it is necessary in economy studies to consider that invested
capital must gain compensation for its owners. Measuring the
amount or the degree of compensation of invested capital forms
the basis for economic decision analysis. Funds which are idle are
not gaining compensation, and thus are actually costing money,
commonly known as the cost of foregone opportunities. (The
effects of inflation/deflation will not be considered in this chapter.) Thus, an investment of one dollar which showed an earning
of $0.15 at the end of the year would be said to yield a return
investment of 15%. The earning may be due to an increase in
sales, or in savings in operations or maintenance.

7.2.3 Simple and Compound Interest

Interest may be either simple or compounded. In simple interest,
the principal and the interest become due only at the end of the
loan period. Since this application is seldom found in typical
business situations, it will be discussed no further here. All
applications in the remainder of this chapter will assume compound interest. When money is borrowed for a length of time
equal to several interest periods, interest is calculated at the end
of each of those periods. There are a number of loan repayment
plans, ranging from paying the interest when due at the end of
each period to paying no interest until the end of the total loan
period. Interest due but not paid thus becomes an addition to the
loan, and interest on this addition is charged at the end of subsequent periods. Interest in this case is said to be compounded.

Assume, for example, that $1,000 is borrowed to be paid in 3
years with interest at 9%. Table 7.2.1 shows the year-by-year
activity when interest is paid at the end of each period of one
year.

Year	Amount owed at beginning of year	Interest due at end of year	Amount owed at end of year	Amount paid by borrower at end of year
1	$1,000	$90	$1,090	$90
2	1,000	90	1,090	90
3	1,000	90	1,090	90

Table 7.2.1: Loan of $1,000 on which interest is paid annually.

The other extreme, payment of no interest until the end of the
total duration of the loan, is illustrated in Table 7.2.2

Year	Amount owed at beginning of year	Interest due at end of year	Amount owed at end of year	Amount paid by borrower at end of year
1	$1,000	$ 90	$1,090	$ 00.00
2	1,090	98.10	1,188.10	00.00
3	1,188.10	106.93	1,295.03	1,295.03

*Table 7.2.2: Loan of $1,000 on which accumulated interest is paid
at expiration of loan.*

The interest is compounded in Table 7.2.2, based on an interest
charge payable on interest previously due but not paid. The force
of compound interest is considerable. At the 9% illustrated, the
principal amount doubles in just eight years.* A third arrangement, shown in Table 7.2.3, is more frequently used, and is typical of mortgage payments. Interest is paid when due, and the
sum of the payment of the interest due together with a portion of
the principal is constant each year.

*The number 72 can be used as the product of the interest rate and the time in years
to determine the doubling time. For instance, money at interest at 8% will double in
value in 9 years. Similarly, a country whose population is increasing at a rate of 4% per
year will have its population doubled in 72/4 or 18 years.

Year	Amount owed at beginning of year	Interest due at end of year	Amount owed at end of year	Amount paid by borrower at end of year
1	$1,000	$90	$1,090	$ 395.06
2	694.94	62.54	757.48	395.06
3	362.44	32.62	395.06	395.06

Table 7.2.3: Loan of $1,000 in which year-end payments are equal.

7.2.4 Equivalence

Each of the above illustrations satisfies the payment of a loan of $1,000 for 3 years at 9% compounded annually, and are said to be equivalent to each other. The out-of-pocket amounts do vary from one to the other because they are made at different dates. *The concept of equivalence is the key to understanding engineering economy.* It provides a means by which elements of an investment, each perhaps made at a different time, may be compared with the earnings which may occur periodically over the total life of the investment. The evaluation of any resulting gain over the life may be thus expressed as a "return" on the investment.

7.2.5 Symbols and Terms

Several symbols and terms are used in describing interest relationships:

i represents an interest rate per period. Interest rates may be for a one-year period, a six-month, or quarterly, or other lesser periods.

n represents the number of interest periods.

P represents a present sum of money at time zero.

F represents a sum of money that occurs at some time, n periods, other than time zero.

A represents a uniform end-of-period payment or receipt made each period. It is known as an annuity.

G represents a uniform increase or decrease per period in the value of A, and is known as a gradient.

*The concept of equivalence, with its implication of the time value
of money at some interest rate, permits translation of any amount
of money at any date into an equivalent amount at any other date.*
Sums of money at different dates can only be added or subtracted
when they are translated into a common date. The simplest
translation would be to determine the future value, F, n periods
hence, of a present sum of money, P, placed in a savings account
with an interest rate of i per period. Or, knowing an amount, F,
desired n periods in the future, one could determine the present
amount, P, which must be deposited, at interest rate i per period.
These two translations are illustrated in figure 7.2.1.

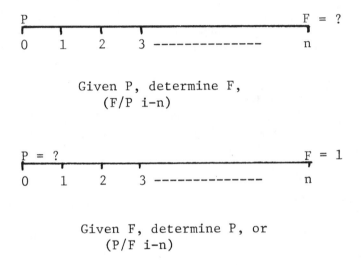

Given P, determine F,
(F/P i-n)

Given F, determine P, or
(P/F i-n)

*Figure 7.2.1 Translation of Present and Future Values of a
Single Amount.*

The notation, (F/P i-n) is called the future from a present amount,
at some rate of interest, i, per period for n periods. Similarly,
(P/F i-n) is called the present from a future amount, at some rate
of interest, i, per period for n periods. These notations, together
with those for a uniform series derived later, were endorsed several
years ago by the Engineering Economy Division of the American
Society for Engineering Education, and are now used in most
writings in engineering economy. They follow very closely the

system designed for the 1962 edition of *Engineering Economy,* published by the American Telephone and Telegraph Company. This *Green Book,* [3] as it has been commonly called, is now in its third edition and is a comprehensive work on Engineering Economy.

7.2.6 Nominal and Effective Interest

Reference has been made several times to the role of interest per period. A period may be any length of time up to and including a year. An interest rate is commonly specified in terms of a year and is known as a nominal rate. However, when the compounding intervals are less than a year, the nominal rate must be divided by the number of compounding intervals, or periods, to determine the rate per period. A nominal rate is always qualified accordingly. Thus, a rate stated as 6% compounded quarterly, would actually be 1-1/2% per quarter. Six percent would be called the nominal rate, and the qualifying phrase following it would indicate the frequency of compounding, or the number of periods.

When the compounding intervals are less than one year, interest is then earned on earlier interest, thus exceeding the nominal amount. The actual interest earned, or paid, may be expressed as the effective interest rate. If a person deposited $100 at a rate of 6% compounded semi-annually, the amount of accrued interest at the end of one year would exceed $6. At the end of the first six months, or the first period, interest of $3 would accrue as 3% of the $100. During the second interval, the original $100 plus the $3 interest, or $103, would earn 3% or $3.09. The total at the end of the year, or at the end of the two periods, would thus be $106.09. The actual interest rate, or effective rate, would be 6.09%. A dramatic distinction between nominal and effective interest rates can be drawn in the case of a typical personal loan where it turns out that the true cost of the loan is considerably higher than the nominal or stated rate indicates. Such loans have a nominal rate of interest applied to the amount borrowed, with this total amount divided by the number of months of the loan to determine the monthly amount. The effective rate can be readily calculated, and for any nominal rate is almost double.

7.2.7 Uniform Series

Other notations are used to describe the four basic translations of
a uniform series. A uniform series, A, may be translated into either
a present single amount, P, or a future single amount, F. Conversely,
a present single amount, P, or a future single amount, F, may be
converted into a uniform series, A. *The six translations, two for
single amounts and the four for a uniform series, become the basis
for determining equivalent amounts which in turn are the key to
determining the effectiveness of capital investments.* The notation,
(P/A i-n) is called the present value of an annuity, or the present
value of a uniform series, A, at some rate of interest for n periods.
The single future amount, F, equivalent to a past annuity, is called
the future worth of an annuity, (F/A i-n). A present single amount
may be translated into a uniform series using the notation (A/P i-n)
or annuity from a present amount. A future single amount is translated into a uniform series using the annuity from a future amount
notation, (A/F i-n).

7.2.8 Examples of the Series Notations

The present value of an annuity, (P/A i-n), may be used to translate
a series of potential annual savings which are expected to result
from the installation of a new unit of equipment. The present worth
of the savings may then be compared with the investment, with i
representing the minimum return expected on the investment. Such
a procedure is also known as the discounted cash flow method.

The future value of an annuity, (F/A i-n), is helpful in calculating
the amount one would have saved over a number of years by setting
aside uniform amounts every month or year.

An annuity from a present amount, (A/P i-n), may be used to determine the periodic amount one may withdraw over a given period
of time from the accumulated savings in the previous paragraph. A
common application in engineering economy studies is to determine
the annual allocation to costs of an investment in capital equipment.
The annuity or uniform amount thus resulting is a combination of
interest or return on the investment plus a portion of the original
sum or principal. The notation, (A/P i-n), is then called a capital
recovery factor. By including the return on the invested capital as

a cost, the engineer makes provision in his analysis for a minimum profitability which the accountant, during and at the conclusion of the investment, will express in dollars of profit on the capital. The telephone industry refers to the portion of principal thus allocated as capital repayment, or the annual apportionment of the original sum of money invested in plant and equipment representing a cost to the firm which must be recovered from the customer.[4]

The annuity from a future amount, (A/F i-n), may be used to determine the amount which must be set aside periodically to pay the principal amount of a bond issue due at maturity. The notation is then called a sinking fund factor. An individual who wishes to accumulate some predetermined future amount may use the notation to determine the necessary periodic amounts which must be deposited.

Mathematical formulas and tables are available to translate any amount occurring at any particular time into an equivalent amount at some other time. A representative table is shown in Table 7.2.4.

7.2.9 Derivation of the Time Value Factors

A cash flow diagram aids in understanding the necessary translations by providing a graphical description of each problem situation. A horizontal line represents a time scale divided into intervals or periods. Arrows signify cash flows, with downward arrows representing receipts or a positive (+) cash flow, and upward arrows representing expenditures or a negative (–) cash flow. Perhaps it is easier to distinguish the direction of the arrows if one considers that funds that flow downward into the cash "pot" enrich it, and funds that move out (upward) decrease it. A vector-type plot, perhaps more easily understood by some, would reverse the arrows (arrows pointing upward are positive and proportional to cash inflows, while those pointing downward are negatives representing cash outflows).

n	Single Payment		Equal-Payment Series				Uniform Gradient-Series Factor
	Compound-Amount Factor	Present-Worth Factor	Compound-Amount Factor	Sinking-Fund Factor	Present-Worth Factor	Capital-Recovery Factor	
	To Find F Given P $F/P\ i,n$	To Find P Given F $P/F\ i,n$	To Find F Given A $F/A\ i,n$	To Find A Given F $A/F\ i,n$	To Find P Given A $P/A\ i,n$	To Find A Given P $A/P\ i,n$	To Find A Given G $A/G\ i,n$
1	1.080	0.9259	1.000	1.0000	0.9259	1.0800	0.0000
2	1.166	0.8573	2.080	0.4808	1.7833	0.5608	0.4807
3	1.260	0.7938	3.246	0.3080	2.5771	0.3880	0.9487
4	1.360	0.7350	4.506	0.2219	3.3121	0.3019	1.4038
5	1.469	0.6806	5.867	0.1705	3.9927	0.2505	1.8463
6	1.587	0.6302	7.336	0.1363	4.6229	0.2163	2.2762
7	1.714	0.5835	8.923	0.1121	5.2064	0.1921	2.6935
8	1.851	0.5403	10.637	0.0940	5.7466	0.1740	2.0984
9	1.999	0.5003	12.488	0.0801	6.2469	0.1601	3.4909
10	2.159	0.4632	14.487	0.0690	6.7101	0.1490	3.8712
11	2.332	0.4289	16.645	0.0601	7.1390	0.1401	4.2394
12	2.518	0.3971	18.977	0.0527	7.5361	0.1327	4.5956
13	2.720	0.3677	21.495	0.0465	7.9038	0.1265	4.9401
14	2.937	0.3405	24.215	0.0413	8.2442	0.1213	5.2729
15	3.172	0.3153	27.152	0.0368	8.5595	0.1168	5.5943
16	3.426	0.2919	30.324	0.0330	8.8514	0.1130	5.9045
17	3.700	0.2703	33.750	0.0296	9.1216	0.1096	6.2036
18	3.996	0.2503	37.450	0.0267	9.3719	0.1067	6.4919
19	4.316	0.2317	41.446	0.0241	9.6036	0.1041	6.7696
20	4.661	0.2146	45.762	0.0219	9.8182	0.1019	7.0368
21	5.034	0.1987	50.423	0.0198	10.0168	0.0998	7.2939
22	5.437	0.1840	55.457	0.0180	10.2008	0.0980	7.5411
23	5.871	0.1703	60.893	0.0164	10.3711	0.0964	7.7785
24	6.341	0.1577	66.765	0.0150	10.5288	0.0950	8.0065
25	6.848	0.1460	73.106	0.0137	10.6748	0.0937	8.2253
26	7.396	0.1352	79.954	0.0125	10.8100	0.0925	8.4351
27	7.988	0.1252	87.351	0.0115	10.9352	0.0915	8.6362
28	8.627	0.1159	95.339	0.0105	11.0511	0.0905	8.8288
29	9.317	0.1073	103.966	0.0096	11.1584	0.0896	9.0132
30	10.063	0.0994	113.283	0.0088	11.2578	0.0888	9.1896
31	10.868	0.0920	123.346	0.0081	11.3498	0.0881	9.3583
32	11.737	0.0852	134.214	0.0075	11.4350	0.0875	9.5196
33	12.676	0.0789	145.951	0.0069	11.5139	0.0869	9.6736
34	13.690	0.0731	158.627	0.0063	11.5869	0.0863	9.8207
35	14.785	0.0676	172.317	0.0058	11.6546	0.0858	9.9610
40	21.725	0.0460	259.057	0.0039	11.9246	0.0839	10.5699
45	31.920	0.0313	386.506	0.0026	12.1084	0.0826	11.0447
50	46.902	0.0213	573.770	0.0018	12.2335	0.0818	11.4107
55	68.914	0.0145	848.923	0.0012	12.3186	0.0812	11.6902
60	101.257	0.0099	1253.213	0.0008	12.3766	0.0808	11.9015
65	148.780	0.0067	1847.248	0.0006	12.4160	0.0806	12.0602
70	218.606	0.0046	2720.080	0.0004	12.4428	0.0804	12.1783
75	321.205	0.0031	4002.557	0.0003	12.4611	0.0803	12.2658
80	471.955	0.0021	5886.935	0.0002	12.4735	0.0802	12.3301
85	693.456	0.0015	8655.706	0.0001	12.4820	0.0801	12.3773
90	1018.915	0.0010	12723.939	0.0001	12.4877	0.0801	12.4116
95	1497.121	0.0007	18701.507	0.0001	12.4917	0.0801	12.4365
100	2199.761	0.0005	27484.516	0.0001	12.4943	0.0800	12.4545

Table 7.2.4 8% Interest Factors for Annual Compounding Interest

If an amount, P, is deposited now at an interest rate i per year, how much principal and interest will be accumulated at the end of n years? The cash flow diagram, Figure 7.2.2, illustrates the time line.

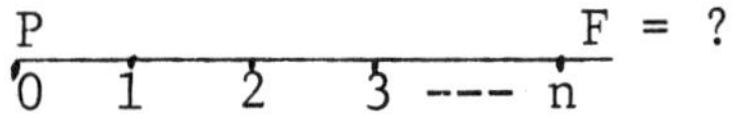

$$P \qquad\qquad\qquad F = ?$$

$$\overline{0 \quad 1 \quad 2 \quad 3 \; --- \; n}$$

Figure 7.2.2 Single Payment Present Amount and Future Amount.

The future value, F, or compound amount, of an investment of P dollars may be developed as shown in the Table 7.2.5. The derived *single payment compound amount* factor is designated (F/P i-n).

Year	Amount at Beginning of Year	Interest Saved During Year	Compound Amount at End of Year	
1	P	Pi	$P + Pi$	$= P(1+i)$
2	$P(1+i)$	$P(1+i)i$	$P(1+i) + P(1+i)i$	$= P(1+i)^2$
3	$P(1+i)^2$	$P(1+i)^2 i$	$P(1+i)^2 + P(1+i)^2 i$	$= P(1+i)^3$
n	$P(1+i)^{n-1}$	$P(1+i)^{n-1} i$	$P(1+i)^{n-1} + P(1+i)^{n-1}_i$	$= P(1+i)^n = F$

Table 7.2.5: Derivation of Single Payment Compound Amount Factor

The single payment present worth factor, designated (P/F i-n), may be solved from the derivation in Table 7.2.5.

$$P = \frac{F}{(1+i)^n} \quad \text{or} \quad P = F \left[\frac{1}{(1+i)^n} \right]$$

As stated earlier, there are four basic translations involving a uniform series. Figure 7.2.3 illustrates the future value, or the accumulation at some future date, of a uniform series of year-end payments or annuities.

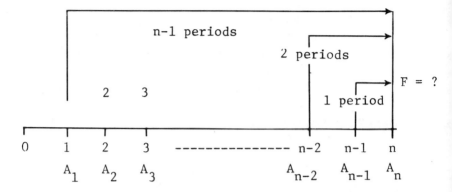

Figure 7.2.3 Uniform Series Compound Amount

Each single payment, A, draws compound interest for a different number of periods.

Payment A_n, at the end of the period n draws no interest.

Payment A_{n-1}, at the end of the period (n-1) draws interest for 1 period.

Payment A_{n-2}, at the end of the period (n-2) draws interest for 2 periods.

Payment A_1, at the end of the period 1 draws interest for n-1 periods.

Each A payment thus corresponds to a single payment present worth amount, and each individually will amount to a corresponding single payment compound amount, F.

The sum of each of the individual F amounts becomes:

$$F = A(1) + A(1+i) + A(1+i)^2 + \text{------} + A(1+i)^{n-1} \qquad (7.1)$$

Multiplying both sides by (1+i):

$$F(1+i) = A(1+i) + A(1+i)^2 + A(1+i)^3 + \text{-----} A(1+i)^n \qquad (7.2)$$

Subtracting (7.1) from (7.2):

$$F(1+i) - F = A(1+i)^n - A$$

$$\text{or } F = A \left[\frac{(1+i)^n - 1}{i} \right]$$

The factor $\left[\dfrac{(1+i)^n - 1}{i}\right]$ is known as the *uniform series compound amount factor,* and is designated (F/A i-n).

If each A value in the time line in Figure 7.2.3 is considered as a future worth F in a single payment present worth factor, then the present worth values may be summed to derive the *uniform series present worth factor.*

$$P = A\left[\dfrac{1}{(1+i)^1}\right] + A\left[\dfrac{1}{(1+i)^2}\right] + \text{------} + A\left[\dfrac{1}{(1+i)^{n-1}}\right] + A\left[\dfrac{1}{(1+i)^n}\right]$$

This can subsequently be expressed as:

$$P = A\left[\dfrac{(1+i)^n - 1}{i(1+i)^n}\right] \qquad \text{which is designated (P/A i-n)}$$

Through rearranging, A can be expressed in terms of P:

$$A = P\left[\dfrac{i(1+i)^n}{(1+i)^n - 1}\right] \qquad \begin{array}{l}\text{which is designated (A/P i-n) and is also} \\ \text{known as the } capital\ recovery\ factor.\end{array}$$

Similarly, A can be expressed in terms of F to derive the annuity from a future amount factor, (A/F i-n), also known as the *sinking fund factor.*

$$A = F\left[\dfrac{i}{(1+i)^n - 1}\right] \qquad .$$

There are frequent times when annual charges are not uniform, as for instance maintenance charges as equipment becomes older, or when receipts increase as an investment becomes more profitable. When such amounts increase by a relatively constant amount each period, such arithmetic progression is called a gradient. A gradient conversion factor may be used to find an equivalent uniform annuity and to determine the present or future worth of a gradient series.

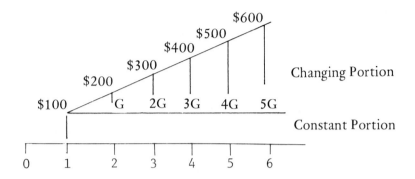

Figure 7.2.4 Cash Flow Diagram for Disbursements Increasing at a Uniform Rate.

The future worth of the complete series illustrated in Figure 7.2.4 is the sum of the constant portion and the changing portion. Similarly, the present worth of the series may be determined, and also the annuity over the total period which is equivalent to the changing portion of the arithmetic progression. More complete books on engineering economy may be consulted for the derivations of the respective gradient factors as well as tables for each of the three at varying rates of interest.

A careful review of the various interest factors derived together with a display of the available data on a time scale will indicate the following points which must be kept in mind:

1. P is at the beginning of a year at a time considered as being the present, or time 0. F is at the end of the nth year.
2. An A occurs at the end of each period. The P of a series is always one period prior to the first A. Exceptions are termed an annuity due, where the first A is at time 0. An appropriate P or F may be calculated by a shift in the time line. The F of a regular series occurs at the same time as the last A.

7.2.10 Examples of Applications of the Time Value Factors

Assume in each of the examples, unless otherwise noted, than an i of 8% is used.

1. A deposit of $1,000 will accumulate to what amount at the end of 10 years?
   $F = P \ (F/P \ 8\text{--}10)$
   $= \$1,000 \ (2.1589)$
   $= \$2,158.90$
2. What deposit must be made now if a sum of $1,000 is desired 6 years from now?
   $P = F \ (P/F \ 8\text{--}6)$
   $= \$1,000 \ (0.6302)$
   $= \$630.20$
3. If $1,000 is set aside at the end of each year, what will be the amount accumulated in the fund at the end of 10 years?
   $F = A \ (F/A \ 8\text{--}10)$
   $= \$1,000 \ (14.487)$
   $= \$14,487$
4. A community has successfully sold a $1,000,000 bond issue to develop a cable TV system. The bonds mature in 20 years. What amount must be set aside at the end of each year to (a) pay the interest due the bondholders, and (b) to eventually pay off the bonds at maturity? Assume the sinking fund earns interest at 5%.
   (a) $I = Pi$
   $= \$1,000,000 \times (0.08)$
   $= \$80,000$ interest due each year
   (b) $A = F \ (A/F \ 5\text{--}20)$
   $= \$1,000,000 \ (0.03024)$
   $= \$30,240$ deposited each year to retire bond issue at maturity
   Total annual cost of bond issue:
   $= \$80,000 + \$30,240$
   $= \$110,240$
5. What would be the total annual cost of the bond issue in problem 4 if the community desired to retire the bonds in 10 years rather than 20, and was able to earn 6% on the annual funds set aside?

Interest due each year would remain the same.
Amount to be set aside in a sinking fund would be:
$A = F (A/F\ 6{-}10)$
   $= \$1{,}000{,}000\ (.07587)$
   $= \$75{,}870$
Total annual cost, 10 year basis:
     $= \$80{,}000 + \$75{,}870$
     $= \$155{,}870$

6. A new piece of equipment, with an installed cost of \$65,000, is proposed to augment an existing system. What would be the equivalent annual allocation or charge for this equipment if its economic life is estimated to be 10 years?
$A = P (A/P\ 8{-}10)$
   $= \$65{,}000\ (0.14903)$
   $= \$9{,}687$

7. A cable TV company is considering purchasing a new van costing \$6,000 for use in installation and repair work. It would like to finance the purchase by borrowing money from a local bank at 9%. What would be the monthly payments if the loan is to be paid off in 3 years?
Since the payments are monthly, the interest rate per period would be 9/12, or 3/4% per month for 36 months.
$A = P (A/P\ 3/4{-}48)$
   $= \$6{,}000\ (0.0318)$
   $= \$190.80$

8. Assume that in problem 7 the company purchased the van and later found it had a larger than anticipated cash flow in its operations during the year following the purchase. Immediately after making the 12th payment the company desired to pay off the remaining amount due. What single payment would be equivalent to the remaining 24 payments? Assume no pre-payment penalty.
$A = \$190.80$ from problem 7

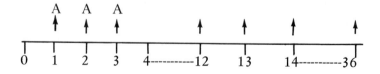

The end of month 12 becomes point 0 for the 13th through the 36th payments. The equivalent single payment thus is the present worth, at a new 0 point, of the remaining 24 payments. Since each of the $190.80 payments is part interest and part principal, the present worth at the end of the 12th year is the sum of the principal portions of the remaining payments.

P = A (P/A 3/4–24)

= $190.80 (21.8883)

= $4,176.29

Note that the remaining payments are not multiplied by 24, as to do so would ignore the time value of money and equivalence concepts.

9. Parents of a young daughter are concerned that they will have sufficient funds for a college education for her. They estimate in 1978 they will need a minimum of $6,500 per year for four years beginning at age 18 in 1992. If the daughter has just had her fourth birthday, what annual amounts, beginning now and continuing through her 17th birthday, would have to be set aside in a special fund? Assume 6% interest.

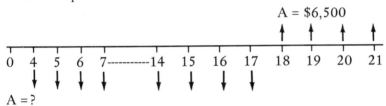

The present worth of the four $6,500 amounts must be determined first. The last paragraph in Section 7.2.9 gives a reminder that the P of a series occurs one period before the first A. Thus, the present worth as of the 17th birthday will represent the amount necessary to begin the $6,500 payments at the end of that period or the 18th birthday.

PW age 17 = A(P/A 6-4)

= $6,500 (3.465)

= $22,523

This amount at age 17 is now the future amount of the required deposits. The yearly deposits into the fund may be determined:

A = F (A/F 6–14)

  = $22,523 (0.04758)

  = $1,071.64

10. Assume in problem 9 the parents found shortly after making the payment on the daughter's 14th birthday, that they could no longer make the special deposits. However, they would keep the accumulated fund intact. If no further payments were made, what four equal amounts would be available for the daughter beginning on her 18th birthday?

To determine the amount in the fund on the 14th birthday:

F = A (F/A 6–11)

  = $1,071.64 (14.972)

  = $16,044.60

This single amount will accumulate at compound interest until the 17th birthday. That date, as indicated in problem 9, is the P for the yearly withdrawals.

F age 17 = P(F/P 6-3)

        = $16,044.60 (1.1910)

        = $19,109

The equal annual withdrawals, beginning on the 18th birthday, can now be determined:

A = P (A/P 6–4)

  = $19,109 (0.28859)

  = $5,514.67

11. Assume in problem 9 that the parents in their original planning desired to have $6,500 available for the first year of college, with an increase to $7,000 for the second year, $7,500 for the third, and $8,000 for the fourth. What would be the necessary amount deposits beginning at age 4 and with the last payment at age 17? The amounts necessary for college now represent a $6,500 annuity together with a $500 gradient. The amount necessary in the fund on the 17th birthday thus becomes:

P = A (P/A 6–4) + G (P/G 6–4)
  = \$6,500 (3.465) + \$500 (4.9455)
  = \$22,523 + 2,473
  = \$24,996
The annual deposit may be found as in problem 9:
A = F (A/F 6–14)
  = \$24,996 (0.04758)
  = \$1,189.31

7.2.11 Continuous Compounding

The payments and receipts described so far have all been assumed to occur discretely at the end of an interest period. The fact is that in any business funds generally flow continuously — whether in or out. Some analysts insist that continuous compounding must therefore be used in order to achieve greater accuracy. On the other hand, the estimates of cash flow over the life of a projected investment are approximations which will not be made more accurate with continuous compounding. Readers are directed to some of the references for a further analysis of continuous compounding.

7.3 EQUIVALENT UNIFORM ANNUAL COST COMPARISONS

The time-value translations introduced earlier emphasized that cash flows can be converted into equivalent amounts at any time using a predetermined interest rate. This equivalency concept provides the basis for the three principal methods of economic analysis which may be defined as follows:

1. *Equivalent uniform annual cost*, or annual cost method. Cash flows are converted to an equivalent uniform yearly amount over the economic life of n periods at interest rate i. May be thought of as representing a leveled, year-by-year amount.
2. *Present worth*, or discounted cash flow. Cash flows over the economic life are "discounted" or converted to a single equivalent amount, P, at time 0, at an interest rate i.
3. *Rate of return.* The negative and positive cash flows are equated. The rate of interest which makes the two equivalent represents the extra compensation gained, or the "return."

Another method commonly used is the benefit-cost analysis. Operationally it is very similar to the present worth method, and is used primarily in evaluating public projects.

A typical investment in plant and equipment involves a present expenditure, occasionally a supplementary capital expenditure, followed by a series of annual disbursements for operation and maintenance, a series of savings or receipts generated by the equipment, and ending with the disposal for some net realizable salvage value. The cash flow pattern is illustrated in Figure 7.3.1.

Investment

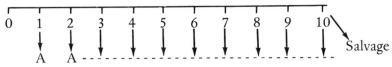

A = Excess of receipts over disbursements

Figure 7.3.1 Cash Flow Diagram of Typical Equipment Investment

Example 1. Assume that in Figure 7.3.1 the investment is $29,000 with an estimated salvage value of $2,000. The annual excess of receipts over disbursements for the 10-year life is estimated to be $4,000. Would the investment be justified if the minimum rate of return is 10%?

Using an equivalent uniform annual cost basis, the annuity cost equivalent would be:

EUAC = ($29,000 – 2,000) (A/P 10–10) + 0.10 (2,000)
     = ($27,000) (0.16275) + 200
     = $4,594

The annual excess of receipts over disbursements of $4,000 is not sufficient to cover the equivalent uniform annual cost of $4,594. On the other hand, if the rate of return were 6% rather than 10%, the EUAC would be $3,688 + 120 or $3,788, and the investment would be justified.

7.3.1 Treatment of Salvage Values

In the diagram in Figure 7.3.1, a salvage value was indicated. This is an estimate of the net realizable value of an asset at the time it is removed from service at the end of its life. The net realizable value is the dollar amount which might be received from the sale or trade of the asset, either as a used piece of equipment or for scrap, less the estimate of the cost of removal. Costs of disconnecting utilities, removal of a base, and the moving of the asset to the point where it may be disposed of, are all included in the cost of removal. Since the net realizable value represents recoverable funds, that portion is not allocated on an annual basis. Thus, in Example 1, the $2,000 is subtracted from the total investment. However, the $2,000 is part of the investment as far as the earning of a return is concerned, and for 10%, this would amount to a necessary earning of 0.10 ($2,000) or $200 per year.

It is frequently extremely difficult to estimate net salvage at the end of an economic life. Technological obsolescence plays a major role in contributing to the difficulty in determining what the used market might be in 10, 15, or 20 years. For this reason, many companies, including those in any way related to investment in telecommunications equipment, are disregarding any salvage values. If there finally is some realizable amount, it is added to income that year as a gain on the sale of an asset. Similarly, if there is a loss because of the high cost of removal, that is charged against current income. Such adjustments in accounting are allowable when item, but not group, accounting is used. The corresponding tax treatment provides for the benefit of taxation at half-rate if there is a net gain, and full deduction for a net loss. Since gains and losses of a given year must be offset against each other, it is generally to the taxpayer's advantage to group losses one year and gains the next.

7.3.2 Comparison of Alternatives on an Annual Cost Basis

Example 2. It has been determined that the present manual method of handling a certain expanding operation is no longer adequate. Two proposals for mechanization have been submitted. Each will perform the required operations within the established technical limits. Plan A has a first cost of $20,000. Annual disbursements for labor and labor extras (social security, paid vacations, various other employee fringe benefits) are estimated to be $9,000. Annual payments for maintenance, power, property taxes, and insurance are estimated to be $1,800. Salvage value is estimated at $1,000.

An alternative proposal, Plan B, has a first cost of $28,000, with an estimated salvage value of $2,000. This plan is considered to be more efficient and more reliable, resulting in lower labor and maintenance costs. Labor and labor extras are estimated to be $7,000 and other payments are estimated to amount to $1,500. It is expected that the need for this operation will continue for 10 years. The minimum required rate of return is 10%. Which is the more economical plan?

	Plan A	Plan B
Capital recovery:		
$\quad = (\$20,000 - 1,000) \, (\text{A/P } 10\text{-}10) + 0.10 \, (\$1,000)$		
$\quad = (\$19,000) \, (0.16275) + 100$	$\$\ 3,192$	
$\text{CR} = (\$28,000 - 2,000) \, (\text{A/P } 10\text{-}10 + 0.10 \, (\$2,000)$		
$\quad = (\$26,000) \, (0.16275) + \$200$		$\$\ 4,432$
Labor	9,000	7,000
Maintenance, power, etc.	1,800	1,500
$\qquad$ Equivalent Uniform Annual Cost	$\$13,992$	$\$12,932$

The equivalent uniform cost method is commonly used when the individual or agency to be "sold" is accustomed to considering financial matters in annual blocks of time. All three basic methods will select the same preferred alternative. The choice of method depends upon which might be more easily understood by the person or agency receiving the analysis and also on the form of the data. In public projects, for example, there are frequently large initial investments with relatively smaller annual amounts, usually for maintenance, and the present worth method may be more straightforward.

7.3.3 Asset Life

Mention has been made of asset life in this chapter without any definition. There are several ways to describe the life of an asset, and the following are among the more common.

Physical life implies a period of time over which an asset remains physically sound with reasonable care and maintenance. This period may extend beyond the time for which there is a useful function for it to perform.

Accounting life is based primarily on tax considerations as regulated by the Internal Revenue Service, and is the basis for depreciation charges. It may or may not correspond to its physical life or its economic life.

Economic life is an optimum one over which the costs of the asset are either minimized or its benefits are maximized. It is strongly influenced by technological obsolescence. The economic life is the most appropriate one to consider in engineering economy studies.

An asset such as a solid state device may be as physically sound after several years of use as the day it was purchased. On the books of account it may have a number of years yet remaining before it is fully depreciated. However, its economic life may be in jeopardy if a more profitable device becomes available as an alternative.

7.3.4 Evaluation of Assets with Unequal Lives

It frequently happens that alternatives under consideration in a study do not have the same economic lives. Some adjustments are thus necessary to avoid skewing the results. The primary consideration is the anticipated life of the plant or larger unit of which the alternatives represent some component. What is the continuing requirement? In no case would it be valid to consider a component as having a longer life than its larger system. If the continuing requirement is for 35 years, for example, and the life of one alternative A is seven years, and that for alternative B is five years, then five identical alternative A units and seven units of B, would be considered. The assumption is that each identical replacement would have the same stream of costs. There may be other situations in which the life of the shorter lived alternative may well be taken as the study period, or conversely, the life of the longer-lived alternative. There may be other study limits, but whatever the analysis, it remains important to carefully examine the consequences of the varying lives, possible differences in salvage value, and the options available subsequent to the study period.

7.4 COMPARISON OF ALTERNATIVES BY THE PRESENT WORTH METHOD

The same data presented in Example 2 in section 7.3.2 may be used to determine the preferred investment by the present worth method.

Example 1:

PW Plan A = $20,000 + 9,000 (P/A 10-10) + 1,800 (P/A 10-10)
$$-1,000 \text{ (P/F 10-10)}$$
$$= \$20,000 + 10,800 \text{ (6.144)} - 1,000 \text{ (0.3855)}$$
$$= \$85,969$$

PW Plan B = $28,000 + 7,000 (P/A 10-10) + 1,500 (P/A 10-10)
$$-2,000 \text{ (P/F 10-10)}$$
$$= \$28,000 + 8,500 \text{ (6.144)} - 2,000 \text{ (0.3855)}$$
$$= \$79,453$$

Plan B has the lower present worth cost and thus is the preferred alternative as previously determined in Example 2. The equivalent uniform annual costs from Example 2 in section 7.3.2 can readily be converted into present worth equivalents:

PW Plan A = $13,993 (P/A 10-10)
$$= \$13,992 \text{ (6.144)}$$
$$= \$85,967$$

PW Plan B = $12,932 (P/A 10-10)
$$= \$79,454$$

(Such slight differences are due to the rounding off variations in individual calculators including those used in constructing the tables.)

The choice of using the present worth method, as in the case of the equivalent uniform annual cost method, depends on who will be receiving the study or the preference of the analyst. Either method gives the same results.

7.4.1 Present Worth Represents a Financial Commitment Over Project Life

The use of the present worth, sometimes called present value, reduces the difficulty of comparing expenditures at varying periods of time by converting all such expenditures into a single value for comparison with other single values. Such single amounts represent the *total* commitment, at time 0, of all future expenditures. This method frequently finds acceptance in large public projects involving deferred investments, in a variety of valuation situations, and in capitalized cost considerations. The latter is not be be confused with the accountant's "capitalizing" an expenditure in order to record it as an asset rather than a current expense. In engineering economy studies the concept of capitalized cost represents a single amount, a present worth, which at some given rate of interest will be equivalent to a cash flow of equal annual payments extending to infinity. The subsequent interest earned from the present worth would provide the required annual annuity indefinitely.

The present worth amount may be that of an excess of receipts over disbursements as well as expenditures alone. It is important that all alternatives be expressed in a similar fashion and that they all cover the same study period.

7.4.2 Importance of Proper Rate of Return

The specified interest rate, or the rate of return, is a crucial element, particularly when comparing expenditures whose timing may vary considerably from alternative to alternative. Careful consideration must be given to use a rate which is consistent with others used in the organization, considering the risks involved. Projects which will show a favorable excess of receipts over disbursements under one return may well show a negative present worth under another return. It is well to remember once again that in calculating a present worth of a series of negative and positive cash flows, a positive present worth means that the stipulated rate has been exceeded. Similarly, a negative cash flow means that the excess of receipts over disbursements has failed to equal the specified rate.

Consider the following example.

Example 2: A firm is exploring the possibility of replacing a present piece of equipment with one of two alternatives. Option A has an initial cost of $35,000. It is anticipated that receipts with Option A will exceed disbursements by $6,500 the first year, and will gradually increase by $400 per year over its economic life. Option B has an inital cost of $50,000 with an excess of receipts over disbursements estimated at $7,800 the first year with a gradual increase of $500 per year. Neither will have a net salvage value. With a minimum rate of return of 10% over a 10-year study period, which alternative is the more economical?

(In calculating a present worth of a combination of expenditures and receipts, it is advisable to use a minus (–) sign to denote all expenditures, and a plus (+) sign for all receipts.)

For Option A:

$$PW_A = -\$35,000 + 6,500 \ (P/A \ 10\text{-}10) + 400 \ (P/G \ 10\text{-}10)$$
$$= -\$35,000 + 6,500 \ (6.144) + 400 \ (22.8913)$$
$$= +\$14,093$$

For Option B:

$$PW_B = -\$50,000 + 7,800 \ (P/A \ 10\text{-}10) + 500 \ (P/G \ 10\text{-}10)$$
$$= -\$50,000 + 7,800 \ (6.144) + 500 \ (22.8913)$$
$$= +\$9,369$$

Both options will provide the minimum required 10%, and thus either might be acceptable. However, Option A has a larger positive present worth. This means that not only will the $35,000 investment provide a 10% return, but $14,093 above that amount — an amount greater than the excess with Option B.

Now assume that a 15% return had been specified rather than 10%:

For Option A:

$$PW_A = -\$35,000 + 6,500 \ (P/A \ 15\text{-}10) + 400 \ (P/G \ 15\text{-}10)$$
$$= -\$35,000 + 6,500 \ (5.019) + 400 \ (16.9795)$$
$$= +\$4,416$$

For Option B:

$$PW_B = -\$50,000 + 7,800 \ (P/A \ 15\text{-}10) + 500 \ (P/G \ 15\text{-}10)$$
$$= -\$50,000 + 7,800 \ (5.019(+ 500 \ (16.9795)$$
$$= -\$2,362$$

Option B has now failed to meet the minimum required rate of return. Option A still meets the minimum plus an additional sum.

In the calculations using 10%, it is obvious that Option A is the preferred alternative. The interpretation of the amounts may not always lead to an obvious choice. In that case the question may well be raised as to the earning power, expressed as a rate of return, of the extra $15,000 initial investment required by Option B. That investment will earn $1,300 per year more in addition to a gradient of $100 per year. This may be expressed by setting the present worth equal to zero. Some rate of interest will then make the receipts equivalent to the initial investment.

PW = 0 = $15,000 + 1,300 (P/A i-10) + 100 (A/G i-10)

The rate which makes these amounts equivalent is 2.67%. This calculation shows that the additional investment required by Option B fails to meet the minimum required rate of return. It would be better to invest the extra $15,000 elsewhere in the firm where it could earn at least the minimum.

7.4.3 Comparison Between Immediate and Deferred Alternatives

Example 3: A company is planning for a new operation which will require a special heat pump. Pump A, a small unit with a first cost of $5,000, will handle the work for four years, after which the volume is expected to increase such that a second unit will be required. Pump B is a larger unit and would meet the anticipated increase with no difficulty. Its price of $7500 appears to offer some economy, even though not all its capacity would be needed right away. The annual operating and maintenance costs of A are estimated at $800 per year, while B's costs are estimated at $1,150. Pump A has an estimated salvage value after 8 years of $450, and Pump B at 8 years an estimated $750. Use the present worth method to determine whether it is best to buy the larger unit now or to select the smaller one and then purchase a similar additional unit four years hence. Use a 10% minimum rate of return. The study period is 8 years, after which it is likely the operation will be abandoned. In that case the second unit of A would have an estimated salvage value of $2,500.

PW of Pump A = -\$5,000 - 800 (P/A 10-8) - 5,000 (P/F 10-4)
$\qquad$ -800 (P/A 10-4) (P/F 10-4) + 450 (P/F 10-4)
$\qquad$ + 2,500 (P/F 10-4)
$\qquad$ = - 5,000 -800 (5.335) -5,000 (0.6830) -800
$\qquad$ (3.170) (0.6830) + (450 + 2,500) (0.3855)
$\qquad$ = \$13,278

PW of Pump B = -\$7,500 - 1,150 (P/A 10-8) + 750 (P/F 10-8)
$\qquad$ = - 7,500 - 1,150 (5.335) + 750 (0.3855)
$\qquad$ = \$13,351

The present worth equivalents of all the estimated costs for each pump are within \$73 of each other, and thus either one could be selected. The final determination will depend upon a close examination of the intangible factors. The differences in outcome of present worth studies is generally considered to not be of any consequence if the net present worth between the two best plans is less than 5%.

Example 4. Radio relay stations in areas of extreme heat environment usually have high battery failure rates. A proposal is to provide air conditioning for the stations and thereby increase battery life from the present maximum of 6 years to a more normal life of 15 years. Assume a station's complement of batteries costs \$16,000. Equipment to air condition the stations to maintain recommended temperatures has a first cost of \$15,000. Annual testing, repairs, and ad valorem taxes on the batteries are estimated at 10% of first cost. Maintenance and ad valorem taxes on the air conditioning equipment are estimated at 3% of first cost. Assuming a 30-year life for the air conditioning equipment, and an 8% minimum rate of return, compare the present worth of the two alternatives. Salvage on batteries and air conditioning equipment is assumed to be negligible. (Note: To reduce some calculations, it may be easier to find some equivalent uniform annual costs and convert these to present worth.)

$\quad$ No air conditioning:

PW of batteries $\quad$ = \$16,000 (A/P 8-6) (P/A 8-30)
$\qquad$ = \$16,000 (0.21632) (11.258)
$\qquad$ = \$38,965

PW of annual charges = 0.10 ($16,000) (P/A 8-30)
                     = 0.10 ($16,000) (11.258)
                     = $18,013
Total PW   = $56,978

Air Conditioning:

PW of batteries   = $16,000 (A/P 8,15) (P/A 8-30)
                     = $16,000 (0.11683) (11.258)
                     = $21,044

PW of annual charges
for batteries   = $18,013 (same)

PW of air condition-
ing equipment   = $15,000

PW of annual charges
for air conditioning = 0.03 ($15,000) (P/A 8-30)
                     = 450 (11.258)
                     = $5,066
Total PW   = $59,123

The difference in the present worth of the two alternatives is $2,145 or 3.7%. Since the difference is also less than 5% as in the previous example, the consideration of intangibles will play a significant role. Intangible benefits of air conditioning the stations include improved transmission qualities because of longer battery life and reduced equipment deterioration. Access to the remote areas would also be a factor.

7.4.4 Use of Present Worth in Determining Valuations

Another common use of the present worth method is in determining valuations. Problem 8 in Section 7.2.10 is one example. Another is in determining a single figure equivalent to a series of payments on the use of a patent. Still another common application is in establishing the amount one would pay for a bond to achieve a particular yield. The term *yield* is the same as a rate of return, and is commonly used in the commercial stock and bond market. The value of a bond, or the amount one would seek to pay to achieve a given return, is the present worth of all the subsequent interest payments plus the present worth of the par or face value of the bond.

Example 5: Assume that the bond market appears somewhat depressed, meaning that a number of bond issues are selling at less than their value at maturity. An investor is willing to risk investment if he can achieve an 8% yield. What price should he pay for a $1,000, 6% bond with interest payable semiannually and which matures in 15 years?

The interest payments will be 3% per period or $30. Thus:

PW = 30 (P/A 4-30) + 1,000 (P/F 4-30)
   = 30 (17.292) + 1,000 (0.3083)
   = 518.76 + 308.30
   = $827

The 6% bond rate is only used to determine the annuity, or amount of interest, per period.

7.5 COMPARISON OF ALTERNATIVES USING THE RATE OF RETURN METHOD

In the two prior methods, the equivalent uniform annual cost technique, and the present worth process, some minimum rate of return was assumed as a cost, or the necessary return. In the rate of return method, the measure of efficiency, or the rate, is the unknown factor. It is that interest rate which makes a series of expenditures and a series of receipts equivalent to each other.

7.5.1 Analysis of the Concept

The use of the rate of return as a measure of performance is frequently confusing because the term describes two different concepts. In the widespread accounting applications, it translates the net income or profit realized during the accounting period as a percentage or index of the "investment" or book value necessary to achieve that profit. In the decentralization of profit responsibility, it is a useful measure of management stewardship. This application of the phrase "rate of return" as a measure of management stewardship is not satisfactory as a planning tool, however, because it is viewed as an overall return on the overall investment in the profit center. In planning, it is important to examine the profitability of each increment of investment such as each additional amount proposed for investments in specific units of equipment, process facilities, communications systems, buildings, and other assets. In this sense, "rate of return" reconciles the financial or the accounting viewpoint with the measurement of efficiency logic of the engineer.

The return on investment concept as generally expressed in financial statements does not give weighted consideration to the time period over which the investment earns its return. The time value of money concept differentiates very clearly the significance of immediate and near-future receipts from those some years away.

Some authors and some companies have used the term "profitability index," describing such an index as a number which equals the annual percent of compound interest which money invested in facilities and the like will yield over the life of the project. Thus, the index is usually a synonym for rate of return or return on investment.

The rate of return is also known as the *internal rate of return*, because it is devoid of external influences, and is completely dependent upon the internal cash flow consequences of the proposed investment.

7.5.2 Calculating the Rate of Return

The three principal elements are the investment, the savings or excess of receipts over disbursements, and the duration of the study period. The interest rate is the unknown, and the usual way is to solve the appropriate equation by trial and error. An estimate of the rate of return is made first to achieve some approximation.

Example 1: Company X is investigating the possibility of purchasing System Y with an installed cost of $84,000 and an estimated economic life of 6 years. A careful examination of the potential net savings estimates them at $26,500 per year. It has also been determined there will be no net realizable value at the end of the six years. What is the rate of return before income taxes?

Note that such problems may be set up on either the annual cost or present worth basis:

$$PW = 0 = -\$84,000 + \$26,500 \, (P/A \; i\text{-}6)$$

$$(P/A \; i\text{-}6) = \frac{\$84,000}{26,500} = 3.169$$

Examination of the appropriate interest tables indicates that the factor of 3.169 falls somewhere between that of 20% (3.326) and of 25% (2.951). While the derivation of the interest formulas earlier in the chapter indicates a geometric progression, straight line interpolation may be used between relatively small intervals for all practical purposes. The error resulting from linear interpolation will usually be much less than the errors in the estimates of cash flows over a period of years. Thus, the rate, i, may be found:

20%		25%
⊢	┼	⊣
3.326	3.169	2.951

$$i = 0.20 + .05 \, \frac{3.326 - 3.169}{3.325 - 2.951}$$

$$= 0.20 + 0.021 \text{ or } 22.1\%$$

7.5.3 Interpreting the Results

Now the question is that of the significance of the 22.1%. If Company X has experienced a before-tax return on investment of 10%, it might be tempted to give this proposal a high priority. Similarly, Company X may be tempted to rate this proposal highly if it considers this return in terms of the cost of the various kinds of capital available to it. Both temptations may lead to difficulty. Investment proposals should compete with one another on comparative merit alone — independent of the company's "return on investment" and the source of funds. A basic concept of economics states that all resources, including monetary funds, are limited, and it is therefore important to use those funds which will be most productive. System Y must be compared with other demands for corporate funds before a decision can be made. The technical person with a "hot" idea frequently cannot understand why funds are not immediately available to implement his request. He or she must recognize the need for understanding profitability calculations as a prerequisite to action.

7.5.4 Income Tax as a Factor in Rate of Return Calculations

Previous examples have avoided consideration of income tax. Section 7.8 will deal briefly with the impact of taxes. The complexities of the Internal Revenue Service requirements, as experienced not only in the United States but also in most other countries of the free world, are best left to the accountants. However, the technical person must understand income tax as a cost of doing business, and he must include this cost in his calculations. The most common consideration occurs when a choice is to be made between a high-cost investment in an asset, and a lower cost alternative. Generally, the only reason to invest in the higher-cost alternative is when the resulting savings "earn their way." In terms of individual projects, the lower annual costs of the higher-cost alternative may well result in lower deductions for tax purposes, and thus have a higher income tax cost. This fact should be kept in mind regardless of which of the evaluation methods is used.

Example 2: Company Z is considering two proposals for improved operations. Alternative A involves a present investment of $100,000 with anticipated annual savings of $25,000. Alternative B is more expensive, with a first cost of $140,000, but will have lower operating costs such that the annual savings are anticipated to be $29,500. It is expected there will be no remaining value with either alternative after 20 years. Estimated annual income taxes with alternative A are $10,000 with the corresponding tax with Alternative B as $11,250. Compute the rate of return on each.

For Alternative A:

$$PW = 0 = -\$100,000 + \$25,000 \, (P/A \; i-20) - \$10,000 \, (P/A \; i-20)$$
$$= -\$100,000 + \$15,000 \, (P/A \; i-20)$$
$$i = 13.9\%$$

For Alternative B:

$$PW = 0 = -\$140,000 + \$29,500 \, (P/A \; i-20) - \$11,250 \, (P/A \; i-20)$$
$$= -\$140,000 + \$18,250 \, (P/A \; i-20)$$
$$i = 11.6\%$$

Assuming that the minimum after-tax rate of return for any preliminary consideration by the company is 10%, which alternative should be selected? One would select Alternative A.

Both alternatives exceed the minimum of 10%. A further step, however, again utilizing the rate of return approach, dramatizes one of the values of this method and clarifies the differences. The incremental rate of return is a useful device whether comparing two alternatives or a number of them. Does the extra savings of $3,250 with Alternative B justify the extra $40,000 investment?

PW = 0 = -$40,000 + $3,250 (P/A i-20)

i = 5%

Without hesitation now, one can clearly state that in no way is the higher cost alternative justified.

Example 3: The two previous examples were straightfoward and required no trial and error calculations because there was only one unknown.

Assume that an investment of $50,000 with an estimated salvage value of $8,000 at the end of 10 years results in a year-by-year excess of receipts over disbursements as shown below. Calculate the expected rate of return.

End of Year	Cash Flow
0	$50,000
1	+7,500
2	+8,000
3	+8,000
4	+8,000
5	+8,000
6	+8,000
7	+7,600
8	+7,600
9	+7,600
10	+7,600
11	+8,000

The resulting equation would be set up as follows:

PW = 0 = -$50,000 + $7,500 (P/F i-1) + $8,000 (P/A i-5) x
(P/F i-1) + 7,600 (P/A i-4) (P/F i-6) + $8,000 (P/F i-10)

This may also be written as:

PW = 0 = -$50,000 + $8,000 (P/A i-10) - 400 (P/A i-4) x
(P/F i-6) - 500 (P/F i-1) + $8,000 (P/F i-10)

Before attempting a trial and error solution, a brief examination indicates an essentially level rate of about $7,700 in addition to the salvage. A perusal of the tables indicate that the present worth of an amount 10 years hence varies between 1/3 and 1/2 over a wide range of interest rates.

The equation to give an approximate rate of return is then:

PW = 0 = -$50,000 + $7,700 (P/A i-10) + $3,000 (P/A i-10)

$$= \frac{\$47,000}{\$\ 7,700} = 6.10; \text{ try } 10\%$$

for 10%:

PW = 0 = -$50,000 + $7,500 (0.9091) + $8,000 (3.791) x
(0.9091) + $7,600 (3.170) (0.5645) + $8,000 (0.3855)
= +$1,073

for 11%:

PW = 0 = -$50,000 + $7,500 (0.9009) + $8,000 (3.696) (0.9009)
+ $7,600 (3.102) (0.5346) + $8,000 (0.3522)
= -$1,184
i = 10.48%

While the calculations for the above trial and error solution may seem somewhat tedious, the use of computer programs tailored to individual company parameters permits a number of very rapid computations and comparisons. There are clues as to where to begin as illustrated above.

The rate of return method has a definite advantage where used to evaluate incremental investments, because it readily identifies just how efficient are any additional amounts over the required minimum. Comparisons are thus readily facilitated.

A major problem, perhaps, is in interpreting the rate of return which has been calculated. It should certainly be no lower than the cost of capital to be minimally acceptable. When studies are made for the purpose of rationing capital to the most profitable proposals, lower limits are usually quickly established among competing alternatives.

7.6 BENEFIT-COST ANALYSIS

The benefit-cost (B/C) analysis method is most widely used in the
public sector. It requires essentially the same data as the other
evaluation techniques. Presumably, mention of an excess of benefits over costs is more palatable to the taxpayers of the world than
to speak of a rate of return. The latter implies that the government
is earning a profit on money extracted through taxes. Taxpayers
are either in revolt or on the verge of it in many countries, resulting in an increased need for some better measure of the use
of government funds. Taxpayers, the World Bank and similar
agencies, and foreign country assistance, do not represent an end-
less source of funds, and thus the use of funds which are limited
presumably must be optimized as in the private sector. The benefit-cost analysis is not a new method, but is enjoying widespread
and accelerating popularity because governments and development institutions, as well as private investors, are requiring a much
more sophisticated analysis before committing funds. Even so,
there are many related problems, primarily because the people receiving the benefits are frequently not those paying the costs, and
vice versa.

7.6.1 Quantification of Benefits and Costs

Each proposal must first attempt to identify those who are to receive the benefits (also known as the users) and the agency which
is to pay for them. All of the benefits less any disbenefits (additional costs to the users) are included in the numerator and all of
the agency costs less any savings to the agency are placed in the
denominator. A proposal should have a ratio of at least 1.0 to
justify expenditures of funds. Again, since public funds are
limited, competition for funds would favor those projects with a
much higher ratio.

As in the other methods, all benefits and all costs must be expressed in equivalent terms — either in an equivalent uniform
annual basis, or in the single figure present worth mode. Occasionally benefits can only be expressed as differences in the costs to
the user.

Expressing benefits and costs in equivalent terms implies using some rate of interest. Public projects frequently suffer from a wide variation in interest rates, expected project lives, and operational charges. Variations are particularly widespread when a public agency is attempting to prove its case vis-a-vis a privately financed competitor. Lower interest rates tend to favor projects which might otherwise be undesirable. The present worth of benefits on a long term project is very sensitive to the interest rate used. On federal government projects, a minimum rate has been established based on the current rate for long term treasury obligations as reported in the *Treasury Bulletin* [5]. The use of this rate carries the assumption that there is no capital rationing, or, in other words, money for all purposes can be borrowed at the same rate of interest. While this is a common practice, it does not consider the degree of risk and the opportunity cost (the interest rate on the best opportunity foregone) as is commonly followed in the private sector.

7.6.2 Advantages of the Benefit/Cost Method

There are no inherent advantages of the benefit versus cost approach as compared with rate of return, present worth, or annual cost method. However, the use of the B/C approach, particularly with regard to local governments, offers some gains which might not be obtained otherwise:

- It requires the agency to think in terms of alternatives rather than advocating a single possibility.

- It helps to separate unprofitable programs or activities from those with a greater potential for productivity.

- It focuses attention on long range costs and consequences rather than on only the immediate control outlay.

- Measurable benefits can be weighed against measurable costs. Then unmeasurable benefits can be examined separately to see whether they are impressive enough to justify the necessary measurable costs.

7.6.3 Calculations

Before a ratio can be computed, all of the measurable benefits and all of the measurable costs must be identified and expressed in monetary units, either on a present worth or equivalent annual basis. There are two methods of establishing the numerator and the denominator. One is expressed above in section 7.6.1. An alternative is to subtract annual agency costs from the benefits. There is no decision problem with either method as long as the analyst is consistent.

Example 1: An emergency phone system is proposed for a certain section of remote highway where breakdowns, for some unknown reason, seem to occur more frequently than elsewhere. The cost of installing the system is $100,000, and annual operating and maintenance charges are estimated to be $5,000 during its 10-year useful life. After that time a major re-routing and improvement program should be completed and the emergency system will not be needed. If the system is installed, it is estimated that annual road user costs due to breakdowns will be reduced to approximately $7,000 from the $30,000 average at present. Savings will result primarily from reduction in down-time for commerical vehicles along with the reduction of service and towing charges. If an 8 percent cost of capital is used, what is the benefit to cost ratio for the proposal?

$$\text{Benefits} = \text{Road user savings} = \$23,000$$
$$\text{PW 10 yrs} = \$23,000 \; (P/A \; 8\text{--}10)$$
$$= \$23,000 \; (6.710)$$
$$= \$154,330$$
$$\text{Costs} = \$100,000 + \$5,000 \; (P/A \; 8\text{--}10)$$
$$= \$100,000 + 33,500$$
$$= \$133,500$$
$$\text{B/C} = \frac{\$154,330}{\$133,500} = 1.15$$

Benefit-cost analysis provides a quantitative basis for making decisions in relatively subjective areas which otherwise might be treated by intuition alone. In the above example, other alternatives might be provided, and could be compared. The method lends itself to incremental analysis similar to that used in the rate of return method.

7.7 OTHER YARDSTICKS

Different methods of evaluation are frequently used for each type
of investment proposal in order to prove a previously formed
opinion. Personal preferences may be viewed with some suspicion
when that preference switches with each new situation.

7.7.1 Payout Period

One of the common "yardsticks" is the payout or payback period,
commonly defined as the length of time required to recover the
initial investment from the net cost flow produced. A frequently
used maximum period is three years. A proposed investment is not
considered justifiable if the investment divided by the annual net
cash flow exceeds three. This is the period of time after which the
project presumably could be abandoned with the investment fully
"paid back."

Among the shortcomings of such a method are the failure to consider the time value of money and the failure to recognize the effect
of any achievements past the early years. No salvage values are included. Although there are some modifications of the simple
formula, none really addresses the two major shortcomings.

The payout period probably enjoys widespread use for two reasons.
It is simple to use, requiring a minimum of calculations. For those
who are anxious about risk and the liquidity of their positions, it
seems to favor quick-profit projects.

7.7.2 MAPI Formula

An elaborate analysis procedure has been devised by George Terb-
orgh of the Machinery and Allied Products Institute (MAPI). The
package provides a variety of assumptions, graphs, and worksheets
for use in replacement investment proposals. The trade association
represents firms doing business in the replacement market. This
bit of knowledge may make it easier to understand the emphasis
on factors for "accumulated inferiority" of present equipment.
Many of the assumptions are unrealistic, and the MAPI formula
provides no basis for direct, quantitative comparison with other
alternatives. The MAPI method has consequently fallen into general disfavor.

7.8 DEPRECIATION AND INCOME TAX

Obsolescence and physical wear and tear are two major concerns of the owners of physical assets. Obsolescence may be regarded as functional depreciation, and is due to changes in technology, demand, or requirements. It is of particular concern in high technology equipment where new developments may lead to retirement of many units practically overnight; as such, it is of far more concern than physical depreciation.

7.8.1 Definitions of Depreciation

Considerable confusion surrounds the use of the term *depreciation*, especially when various "official" definitions are examined. The Federal Power Commission System of Accounts for electric power companies and the FCC System of Accounts for telephone companies are very nearly alike in interpretation:

> " 'Depreciation,' . . . . means the loss in service value not restored by current maintenance, incurred in connection with the consumption or prospective retirement of . . . . plant in the course of service from causes which are known to be in current operation and against which the utility is not protected by insurance."

This is the "wear and tear" definition adopted by the commissions following hearings on depreciation of railroads and telephone companies during the 1930's. This definition was essentially affirmed by the Internal Revenue Code of 1954 which said: "General Rule . . . . There shall be allowed as a depreciation deduction a reasonable allowance for the exhaustion, wear and tear (including a reasonable allowance for obsolescence) . . . . (1) of property used in the trade or business, or (2) of property held for the production of income."

The effects of inflation have given rise to other interpretations. Following World War II, the U.S. Steel Company found that it could not replace worn out facilities except at greatly inflated prices as compared to the original cost, and so viewed depreciation as a process by which funds are raised to pay for the replacement of plant and equipment. In its 1964 Annual Report the American Telephone and Telegraph Company urged this same interpretation:

"We continue to urge also that both taxing and regulatory
authorities should allow depreciation charges that recognize
the lower purchasing power of today's dollar. Although in
recent years the decline in the dollar's value has slowed
considerably, much of the total investment in industry
was made when each dollar was worth much more. Hence
depreciation allowances limited to the number of dollars
originally invested cannot recover the true cost of the investment being consumed. In our judgment the tax law
should permit recovery, through depreciation, of the full
purchasing power of each dollar invested."[6]

In a case involving telephone rates, the Indiana Public Service
Commission did authorize the accrual of depreciation upon the
cost of property repriced in current dollars.[7]

Perhaps the word depreciation, so variously defined, should be
eliminated from tax and accounting terminology. In the meantime,
an interpretation found in Accounting Terminology Bulletin 1,
issued by the American Institute of Accountants, is gaining wide
approval, and is the current interpretation for U.S. federal government income tax purposes.

"Depreciation accounting is a system of accounting which
aims to distribute cost or other basic value of tangible capital
assets less salvage (if any), over the estimated useful life of
the unit (which may be a group of assets) in a systematic
and rational manner. It is a process of allocation, not of
valuation."

The key term is that of *allocation* in which a periodic charge is
made to allocate or amortize as a cost of current business, for income tax purposes, a previously incurred capital expenditure.

7.8.2 Depreciation Methods

Several methods are available by which these allocations may be
made. They are each based on time and thus are independent of
use; they are those specifically permitted by the U.S. Treasury
Department. Each of the methods — straight line, sum-of-the-years
digits, and double rate declining balance — has unique features
which appeal to different management philosophies. The latter

two, known as accelerated methods, provide for the allocation of the bulk of the investment during the nearly years of the asset's life. This provides a hedge against obsolescence and other sudden changes which could render the assets less valuable. Regardless of the method, taxes are not avoided, only the timing is changed.

7.8.3 Straight Line Depreciation

The straight line method is the simplest to understand and to apply. A constant depreciation charge is made each year. The total amount to be depreciated, installed cost less salvage, if any, is divided by the estimate of useful life in years. The book value is the difference between the installed cost and the product of the number of years of use and the annual depreciation charge. It was essentially the only method used between the major changes in tax policy of the U.S. Treasury Department in 1934 and the accelerated methods allowed beginning in 1954. Many assets do experience more rapid decrease in value during the early portions of their lives, and the accelerated methods allowed for this beginning in 1954. Although many assets do experience more rapid decrease in value during the early portions of their lives, straight line depreciation does not recognize this condition. Even though its use is declining, many firms, expecially public utilities, do continue to follow it.

Example 1:   Consider an interconnect device with a first cost of $35,000, an estimated life of 10 years, and an estimated salvage value of $3,000.

$$\text{Straight line rate} = \frac{\$35,000 - 3,000}{10} = \$3,200 \text{ depreciation per year.}$$

7.8.4 Sum-of-the-Years Digits

Both the sum-of-the-years digits (SYD) method and the double rate declining balance (DRDB) method are known as accelerated methods. Both permit allocation for tax purposes of most of the value of the asset during the first half of its life. The SYD method takes its name from the calculation procedure. The annual allocation is the ratio of the digit representing the remaining years of life to the sum of all the digits for the original entire life. The sum-of-the-years digits from 1 to n is:

$$SYD = \sum_{j=1}^{n} j = \frac{n\,(n+1)}{2}$$

Example 2:   Using the same data as in Example 1 in Section 7.8.3, determine the first year depreciation charge using SYD.

$$SYD = \frac{10(11)}{2} = 55$$

$$D_1 = \frac{10}{55}\,(\$35{,}000 - \$3{,}000) = \$5{,}818$$

The book value after five years with the straight line method would be:

$35{,}000 - 5\,(\$3{,}200)$

$= \$19{,}000$

The unallocated amount at the same time by the SYD method would be:

$$BV_5 = \$35{,}000 - \frac{40}{55}\,(\$35{,}000 - \$3{,}000)$$

$$= \$11{,}727$$

7.8.5 Double Rate Declining Balance Method

The third major method is the declining balance where a constant depreciation rate is applied to the remaining book value of the asset. The rate allowed by the Internal Revenue Service depends on the type of personal property. Most industrial and business assets qualify for the double rate which is twice the straight line rate. The salvage value is not subtracted from the installed cost before applying the depreciation rate.

Example 3: If the double rate declining balance method were to be used for the interconnect device in Example 1, Section 7.8.3, the first year depreciation charge would be 2 x 1/10 or 20%.

$D_1 = .20 \times \$35,000$

$= \$7,000$

The new book value at the beginning of the second year would be

$\$35,000 - \$7,000 = \$28,000$

The depreciation allocation at the end of the second year would be

$D_2 = .20 \times \$28,000$

$= \$5,600$

The book value at the end of the 5th year is

$BV_5 = \$35,000 (1 - .20)^5$

$= \$11,469$

A comparison of selected depreciation amounts and book values for Example 1 is of interest to note the differences in write-off:

	Straight Line	SYD	DRDB
Installed cost	$35,000	$35,000	$35,000
Depreciation, 1st year	3,200	5,818	7,000
Depreciation, 2nd year	3,200	5,236	5,600
Book value, end of 5th year	19,000	11,727	11,469

It may be noted that about two-thirds of the installed cost has been depreciated by the end of the first half of the asset's life when the accelerated methods are used. The relative size of the salvage value will cause some variations with the SYD method. The more rapid allocations result in a higher rate of return on invested capital. High depreciation charges and consequent low taxes in the early years provide the advantage because of the time value of money. When the accelerated methods provide for smaller allocations than the straight line method in the later years, the income tax charge will be higher. By the end of the asset's life the amount of the tax paid will be the same. The ability to defer federal income taxes from the use of accelerated depreciation does provide for interest-free funds to invest in additional assets or as desired.

7.8.6 After-tax Evaluations

In many engineering economy studies it will be found that an alternative which appears to be most economical in a before-tax cash flow analysis will also be the choice after income tax considerations. However, a clear picture of any differentiation of proposals can only emerge when all significant costs have been gathered, including income tax. Section 7.5.4 pointed out that more efficient alternatives will frequently have lower deductions for tax purposes, and thus must include a higher tax charge as an appropriate cost.

The calculation of corporate federal income taxes is complicated, and requires specialists. Computations are further complicated by investment tax credits, permitted under the Internal Revenue Code for specified investments in new plant facilities.

Consider the following example of an after-tax analysis:

Example 3: An $85,000 investment in new equipment is proposed. Preliminary studies indicate a potential saving in labor and other charges of $21,500 a year over the next 10 years. The investment will be depreciated for income tax purposes by the sum-of-the-years digits method assuming a 10-year life and zero salvage. The effective income tax rate is 50%. A 10% investment tax credit will be taken at year zero. What is the prospective rate of return after taxes?

A year-by-year tabular cash flow display will be necessary.

$$SYD = \frac{10\,(11)}{2} = 55.$$

End of Year	Cash Flow Before Inc. Tax	Depreciation		Taxable Income	Income Tax	Cash Flow After Income Tax
0	-$85,000	$ ---		$ ---	+$8,500	-$76,500
1	+ 21,500	$\frac{10}{55}$	-15,455	6,045	- 3,023	+ 18,477
2	+ 21,500	$\frac{9}{55}$	-13,909	7,591	- 3,796	+ 17,704
3	+ 21,500	$\frac{8}{55}$	-12,364	9,136	- 4,568	+ 16,932
4	+ 21,500	$\frac{7}{55}$	-10,818	10,682	- 5,341	+ 16,159
.	.	.	.	.	.	.
.	.	.	.	.	.	.
10	+ 21,500	$\frac{1}{55}$	- 1,545	19,955	- 9,978	+ 11,522

The after-tax rate of return may be determined by equating the net outlay ($76,500 with the tax credit) to the net annual savings after income taxes. Note also that the difference between the year-by-year after tax cash flows appears to be a constant ($773) and thus a gradient factor may be used.

PW = 0 = -$76,500 + $18,477 (P/A i-10) - $773 (P/G i-10)

The unknown i may be estimated by temporarily ignoring the gradient amount:

$$(P/A \text{ i-10}) = \frac{76,500}{18,477} = 4.14$$

From the tables, (P/A 20-10) = 4.192. Thus 20% may be used as the first assumption:

PW = 0 = -$76,500 + $18,477 (P/A 20-10) - $773 (P/G 20-10)
= -$76,500 + $18,477 (4.192 = $773 (12.8871)
= -$ 9,007

The savings failed to generate a 20% return by falling short $9,007. Assume i = 15%:

PW = 0 = -$76,000 + $18,477 (5.019) - $773 (16.9795)
= +$ 3,111   Thus 15% falls too low.

Interpolating between 15 and 20%:

$$\text{ROR} = 15\% + (.05) \frac{3,111}{3,111 + 9,007}$$

= 16.28%

Example 4: Consider the same problem as in example 3, but with straight line depreciation.

End of Year	Cash Flow Before Inc. Tax	Depreciation	Taxable Income	Income Tax	Cash Flow After Income Tax
0	-$85,000	$ ---	$ ---	+$ ---	-$76,500
1-10	+ 21,500	-8,500	-13,000	-6,500	15,000

$$PW = 0 = -\$76,500 + \$15,000 \ (P/A \ i\text{-}10)$$
$$(P/A \ i\text{-}10) = \frac{76,500}{15,000} = 5.1$$

from the tables, i - 15.5%

In the early years, the larger after-tax cash flows resulting from the accelerated method as compared with the straight line method, have a greater impact than the lesser ones toward the end of the life. The result is greater efficiency in the use of the investment, or a better return.

The impact of using accelerated depreciation, and thus deferring federal income taxes, was highlighted in 1968 during the California Public Utilities Commission rate hearings of Pacific Telephone and Telegraph Company. Prior to that time the Bell System used straight line depreciation for both book and tax purposes. The California PUC held the view that the larger tax deductions for depreciation in the early years of a depreciable asset would pro-duce tax savings, and that these savings should be treated as a reduction in expense and allowed to flow through to earnings. Pacific Telephone and Telegraph Company argued that while this method would temporarily improve earnings the taxes that were deferred would ultimately have to be paid. The PUC responded that as long as the company's growth continued, there would be no cause for concern. Other regulatory agencies, in-cluding the Federal Power Commission, had ordered the flow through principle as early as 1964.

7.9 MULTIPLE ALTERNATIVES

Preceding sections have dealt primarily with two alternatives. Frequently there are more than two possibilities, or multiple alternatives. When the selection of one alternative precludes the selection of others, the alternatives are termed mutually exclusive.

There are a number of approaches, including the basic equivalent uniform annual cost and present worth methods. They are straight-forward, once it has been determined which rate of return to use. Both the rate of return and benefit/cost approaches lend themselves more readily to the incremental concept where each additional portion of investment above the minimum required can be tested for its efficiency. Since more capital can be invested in other acceptable alternatives, the incremental investment must be justified. For the benefit/cost method of analysis the alternative to be selected is the one requiring the largest investment and whose incremental investment over another acceptable alternative meets or exceeds a B/C ratio of 1. Similarly, for the rate of return approach with mutually exclusive alternatives, the one to select is again that requiring the largest investment and whose incremental investment over another acceptable alternative meets or exceeds the minimum required rate of return.

7.9.1 Incremental Rate of Return

The rationale for investing in any alternative greater than the minimum necessary to do the job is that the extra investment is productive in the sense that it provides at least the minimum re-quired rate of return. If not, only the basic investment is justified and the extra amount is utilized elsewhere, where it can be at least minimally productive.

The specific steps to be used when evaluating mutually exclusive multiple alternatives may be summarized as follows:

1. Rank the alternatives according to increasing size of initial investment.
2. If potential savings or cost reductions are indicated for each, calculate the rate of return for each investment, using either the EUAC or present worth methods over the economic life. Eliminate any which do not meet the minimum required rate of return. If only operating and similar annual costs are specified, determine the incremental rate of return of the second alternative with the least investment. If the incremental rate of return *does not* meet or exceed the minimum, eliminate that alternative and go on to the next highest and check its in-cremental return, comparing it with the lowest. If the incremental

rate of return of the second alternative *does* meet the minimum requirement, use the second alternative as a new base and compare the third alternative with it.

3. Repeat the comparison of pairs until all alternatives have been evaluated. It is important that the correct pairs be compared. An alternative whose incremental rate of return fails to meet the required minimum is no longer a valid alternative and must be eliminated.

4. Select the alternative requiring the largest investment whose incremental rate of return meets or exceeds the minimum required.

Example 1: Consider the following five alternatives which have been found to meet the required technical specifications, including a 10-year useful life. If the minimum rate of return is 12%, which single alternative should be selected?

Alternative	A	B	C	D	E
Installed cost	$4,000	$3,500	$8,000	$7,500	$6,000
Annual Operating and Other Expenses	1,100	1,200	450	500	750

The first step is to rearrange in terms of increasing size of initial investment:

Alternative	B	A	E	D	C
Installed cost	$3,500	$4,000	$6,000	$7,500	$8,000
Annual Operating and Other Expenses	1,200	1,100	750	500	450

This rearrangement portrays a typical situation in which frequently the more expensive alternative does require less maintenance, or less labor, or has longer intervals between overhauls or major outages, etc. The real question is, then, is the added expense justified?

In comparing A with B, an extra $500 initial investment results in savings of $100 annually. Over the 10-year period the rate of return is:

$500 = $100 (P/A i-10)

(P/A i-10) = 5 or approximately 15% from the tables.

There is justification in spending the extra $500.
The next question is whether alternative E is viable.
In comparing E with A:

$2,000 = $350 (P/A i-10)

$$(P/A\ i\text{-}10) = \frac{\$2,000}{\$\ 350} = 5.7 \text{ or between 12\% and 15\%.}$$

Alternative E is a valid one.

Next, compare the extra investment of $1,500 required by Alternative D with its savings of $250 annually as compared with E.

$1,500 = $250 (P/A i-10)

(P/A i-10) = 6 or between 10% and 11%. The extra investment required by D is not justified and D is discarded an an alternative.

Alternative C is now compared with E, the largest *valid* alternative so far:

$2,000 = $300 (P/A 1-10)

(P/A i-10) = 6.33 or between 9% and 10%. Alternative C is not acceptable.

The largest investment whose incremental rate of return meets or exceeds the required 12% is the $6,000 represented by E. Extra funds that might be spent for alternatives D and C would be better utilized elsewhere for other projects meeting the 12% requirement.

The incremental concept provides a valuable tool in evaluating multiple alternatives through selective pairing. Care must be exercised in following the correct procedure, or the wrong alternatives may be selected. The largest rate of return does not optimize the selection. A relatively small investment with a very high rate of return may not generate savings or earnings of the magnitude of a larger investment which earns the required rate.

There are many other applications of the incremental concept. In evaluating the economics of leasing versus purchasing, the same technique may be applied. The question becomes: "Does the extra investment now in ownership, with its presumably lesser annual charges, provide the required minimum return over the annual costs of leasing?" Subtracting the leasing costs from the ownership costs provides a basis for calculating the rate of return. The rate must then be interpreted as to how it compares with other requests for funds.

Example 2: A company can either buy a small warehouse or lease it on a 20-year lease. The purchase price is $85,000. The lease cost if $5,000 payable at the beginning of each year. In either case the company would have to pay all maintenance and property taxes. It is estimated the building could be sold at the end of 20 years for $105,000. What rate of return before income taxes (including capital gains tax) would the company receive by purchasing rather than leasing?

The difference between the two alternatives represents the extra investment in ownership and the consequent savings each year. These incremental amounts form the basis for determining the rate of return.

$$PW = 0 = \$85,000 - \$5,000 - \$5,000 \ (P/A \ i\text{-}19)$$
$$+\$105,000 \ (P/F \ i\text{-}20)$$
$$i = 6.9\%$$

Whether or not a 6.9% rate is acceptable depends upon the minimum return this particular company expects to receive on its investments.

7.9.2 Decision Criteria

Whether choices are to be made among mutually exclusive capital investment alternatives, or from those which are not, there are tools and techniques of analysis which assist the technical person in making a decision. Pure judgment and intuition are helpful but in a greatly diminished sense. Capital is limited, and capital can be productive. The essential element in decision making is to choose that alternative which will be the most profitable in the long run.

7.10 ACCOUNTING AND ENGINEERING ECONOMY

Although the accountant and the engineering economy analyst may work with the same financial data, they do so for far different purposes. The major difference lies in the way each views a firm's profitability. The accountant is primarily concerned with evaluating the results of past decisions and operations to determine how profitable the firm has been. The engineering economy analyst attempts to predict what the profitability of a current or future decision might be. The time line used in the earlier portion of this chapter might serve to illustrate the differences by assuming that both the accountant and the engineering economy analyst are standing on it at time zero and with their backs to each other. As they move along in the progression of time, the accountant is continually classifying, recording, measuring, and summarizing various financial transactions that have already occurred. The accountant looks for the dollars of profitability. On the other hand, the engineering economy analyst, using some of the accountant's data, quantifies the expected future differences in costs and returns of alternative engineering proposals. He or she includes a capital recovery factor as a cost, which, if everything proceeds satisfactorily, the accountant will eventually find as a profit.

Accounting records provide a valuable source for the engineering economy analysis, but the technical person must be alert to the fact that accounting data is frequently collected, measured, summarized, etc., in a different form. For example, average costs may be summarized whereas the engineering economist may need to determine incremental costs.

7.10.1   The Accounting System

Basic documents for the financial management of a business of any size are the income statement (or profit and loss statement) and the balance sheet. The income statement may be expressed as a basic formula:

Profit = Revenue – Costs.

The income statement reports how well a company has done over a specified period of time such as a month, quarter, or year. With the aid of computers, some firms make this determination as frequently as weekly, or in a few instances, daily, to aid internal decision-making.

The balance sheet may be likened to a snapshot of the company at any one time in which the assets, liabilities, and net worth are displayed. The corresponding basic formula is:

Assets = Liabilities + Net Worth

or:

Assets − Liabilities = Net Worth

The change in a company's position over a period of time may be determined by comparing a succession of balance sheets. More will be said later about some significant financial ratios which are used in such comparisons.

Accounting uses a double-entry system for every transaction, and this system may be viewed as a series of equations, debits = credits, which must always be kept in balance. The terms *debit* and *credit* are conventional, and have no particular significance. One is advised to merely memorize them rather than reasoning through to a definition.

The purchase of a piece of equipment costing $6,000 and with a down payment of $2,000 and the rest charged, would result in entries as follows:

debit equipment $6,000
credit cash 2,000
credit accounts payable 4,000

The debits = credits equation continues to balance, as does Assets = Liabilities + Net Worth. The accounts, equipment, cash, and accounts payable are simply storage terms for a particular asset or liability.

Assets are everything of value owned by a firm or owed to it. All cash assets plus those resources which might normally be converted to cash within the normal operating cycle (no more than a year) of the firm are termed current assets. Fixed assets are those such as land, buildings, and equipment which are not intended to nor can normally be quickly converted into cash.

Liabilities are debts or obligations of the firm which must eventually be paid. Current liabilities consist of those debts which must be paid within the normal accounting period (a year or less). Long-term debt or long-term liabilities extend over a period longer than a year and are normally represented by bonds.

Net worth is the total of all assets minus all liabilities. This is really the ownership portion of a firm, or equity. Equity is made up of the preferred and common stock, the surplus from the sale of stock above its par value (capital surplus), and the earned surplus or re-tained earnings. Earned surplus is the accumulation of past earnings not paid out in dividends but reinvested in the business.

In the equation, Assets = Liabilities + Net worth, the right hand side may be viewed as the source of all funds within the firm. Some funds are thus available through short and long-term borrow-ing (a credit card charge is a short term loan), some through the sale of stock, and others are available through previous earnings which have been "plowed back" into the company. The left side of the equation indicates the distribution or the form in which the funds are to be found. Thus, some funds are in cash and marketable securities, others in inventory, equipment, land, buildings, patents, etc. It can be readily seen that in the double entry system any transaction does affect at least two accounts.

7.10.2 The Balance Sheet

The balance sheet for the Telecom Company shows the major accounts as of a particular time, December 31, 19xx, and con-forms to the fundamental equations, Assets = Liabilities + Net worth.

Telecom Company
Balance Sheet
December 31, 19xx

Assets		Liabilities	
Cash	$151,800	Accounts payable	$ 5,500
Accounts receivable	8,300	Notes payable	26,000
Inventory	49,200	Accrued taxes	3,400
Land	15,000		$34,900
Building	91,000	Net worth	
Equipment	39,600	Capital Stock	$200,000
		Earned surplus	120,000
			$320,000
	$354,900		$354,900

The amount of detail displayed depends on the purpose to be served by the statement.

7.10.3   The Income Statement

An understanding of the financial situation of a company also requires an income statement. For the Telecom Company, the following represents the major activity during the preceding year.

Telecom Company
Income Statement
Year ending December 31, 19xx

Gross income from sales		$360,080
Cost of goods sold		189,160
Net income from sales		$170,920
Operating expense:		
Salaries	$87,100	
Depreciation	9,300	
Advertising	8,200	
Insurance	2,400	
Utilities	19,500	$126,500
Net profit from operations		44,420
Interest expense		2,300
Federal and state income tax		21,100
Net Profit		$ 21,020

Profit must come from operations, which are the producing and selling of goods and services. The first item on the statement represents
the amount of money taken in from the sale of goods and services
during the year. Next, all the costs and expenses of doing business
must be subtracted. In the Telecom Company these included "cost
of goods sold" which represents all the production, material, and
labor costs associated with making the goods and services, salaries,
promotional expenses, etc. After subtracting all the operating costs,
a profit from the operations activities can be determined. The interest
on the notes payable (see balance sheet) is a tax deductible expense,
so is deducted before income taxes are calculated. After determining
the income taxes, the remaining amount is the "net profit" which
is the amount available for dividends and/or retained earnings.

Many of the items of expense are difficult to determine in an ongoing firm. There are problems in determining what costs are
appropriate for a particular period of time and how they should be
charged. Depreciation is one of these, and as noted earlier, can be
calculated by several different methods over some estimate of
time. "Overhead" costs present another difficulty. The light, heat,
and insurance expenses usually cannot be related to a particular
product or service. Accountants concern themselves with ways of
allocating such costs. There are also company-wide costs such as
engineering, research, personnel, procurement, and the like. They
each make a contribution to the production of goods and services,
but such contributions are often difficult to measure and to apply
to the appropriate good or service.

7.10.4   Fixed and Variable Costs

Everything that a firm does costs money. Some charges occur
whether anything is done or not. Different expenses have different
effects on a firm's profitability, depending on the variations in
operations and the timing of the operations.

Fixed costs are those which are independent of any production or
sales activity. Taxes and insurance, security expenses, interest on
debt, and administrative salaries are among those costs which
continue to accumulate regardless of the level of activity.

Variable costs vary directly with output and comprise, for the most part, the two major items of direct labor and material. The charges for each of these two categories are made directly to the product.

An engineer frequently fails to understand these costs. An error commonly made is to believe that a reduction in labor costs will result in a corresponding proportionate decrease in overhead costs.

Because the major fixed and variable cost accounts are summaries of a wide variety of lesser accounts, it is frequently difficult for the engineering economist to ascertain certain labor, maintenance, tax, and other expenses required in a study. Considerable care must be exercised to trace back through the accounting records until the necessary particular information is found. Only then can true incremental costs and incremental benefits be determined. Cost data that gives average values are frequently not adequate and may be seriously misleading.

7.10.5 Financial Ratios

There are a number of key financial ratios which may be determined by analyzing the two basic financial statements. Some are of significance to the outside investor. Others are of concern internally. In either case, it is the trend of such ratios which provides indicators of the strengths and weaknesses of a firm.

Ratios are generally classified into several fundamental types:

1. Liquidity ratios measure the ability of a firm to meet its short-term obligations.
2. Leverage ratios measure the contributions of the owners' equity as compared with the financing provided by borrowing.
3. Activity ratios measure how effectively the firm utilizes its resources.
4. Profitability ratios measure management's effectiveness as shown by the profit (or return) on sales and investment.

Each of the ratios can be compared with previous ratios in the same organization and also with those from the same type of industry. Considerable data is available for the latter comparisons and may be found in the financial reference sections of many libraries.

Another way of classifying the ratios to utilize the information which is supplied by the accountant may be for the engineer to view them as follows:

1. Scorecard ratios — am I doing well or badly in the areas for which I'm responsible?
2. Attention-directing ratios — what problems should I be looking into?
3. Problem-solving ratios — of the several ways of doing the job, which is best?

There is a growing trend, as stated initially in section 7.1.2, toward utilizing costs as a joint parameter in design. An analysis of one's own firm's costs, or an analysis of others' costs in specified segments of business, frequently identifies the area in which a new or an improved technical design is needed.

The ability of the engineer or related technical person to read and analyze financial statements can be a distinct advantage, both professionally and personally. Perhaps the advice should be presented even more strongly, for the ability to read and interpret financial statements is almost essential to survival, both professionally and personally.

7.11 SUMMARY OF CONCEPTS IN ENGINEERING ECONOMY

Eugene L. Grant, recognized as a pioneer in engineering economy, and now Professor of Economics of Engineering, Emeritus, at Stanford University aptly summarized the concepts of engineering economy at a conference some years ago. Although the conference dealt with highway planning, the concepts summarized are sufficiently general in nature as to have wide application. [9]

1. All decisions are among alternatives; it is desirable that alternatives be clearly defined and that all reasonable alternatives be considered.
2 Decision-making should be based on the expected consequences of the various alternatives. In comparing investment alternatives, it is desirable to make the consequences commensurable with the investments insofar as practicable. Money units are the only units that make consequences commensurable with investments.

3. Only the differences between alternatives are relevant in their comparison.
4. It is necessary to have a criterion for decision-making (or possibly several criteria). The criterion for investment decisions should recognize the time value of money and related problems of capital rationing.
5. In looking at the predicted consequences of various alternatives and in establishing criteria for decision-making, it is essential to decide whose viewpoint is to be adopted.
6. Insofar as possible, separable decisions should be made separately.
7. In organizing a plan of analysis to guide decisions, it is desirable to give weight to the relative degrees of uncertainty associated with various forecasts about consequences. In this connection, it is helpful to judge the sensitivity of the decision to changes in the different forecasts.
8. Decisions among investment alternatives should give weight to any expected differences in consequences that have not been reduced to money terms as well as to the consequences that have been expressed in terms of money.
9. Decisions among investment alternatives must be made at many different levels in an organization. The implementation of rules aimed at rational decision-making may appropriately be different at different levels.

REFERENCES

[1] Thomas, V. "Interconnect: Anti-competitive?", Rocky Mountain Industries, Vol. 10, No. 8, October 1977, p. 6.

[2] Fish, J.C.L., *Engineering Economics,* 2nd Ed., McGraw-Hill, N.Y. 1923, p. 20.

[3] American Telephone and Telegraph Company, *Engineering Economy,* 3rd ed., McGraw-Hill Book Company, New York, 1977.

[4] *Ibid* p. 149.

[5] Circular No. A-76, Executive Office of the President, Bureau of the Budget, March 3rd, 1966, p. 5.

[6]   American Telephone and Telegraph Company, Annual Report, 1964 (New York: American Telephone and Telegraph Company, 1965), p. 6.

[7]   Indiana Telephone Corporation (Ind. 1957) 16 PUR 3d, 490, 497.

[8]   Emerson, C.R., and W.R. Taylor, *An Introduction to Engineering Economy,* Cardinal Publishers, Bozeman, Montana, 1973.

[9]   Grant, E.L., *Concepts and Applications of Engineering Economy,* HRD Special Report 56, Workshop Conference on Economic Analysis in Highway Programming, Location and Design (Washington, D.C.: Highway Research Board, 1959).

[10] *Engineering Economist, The,* Quarterly journal of the Engineering Economy Divisions of the American Society for Engineering Education and the American Institute of Industrial Engineers, Richard S. Leavenworth (editor), American Institute of Industrial Engineers, 25 Technology Park, Norcross, Georgia 30092.

[11] Fabrycky, W.J. and G.J. Thuesen, *Economic Decision Analysis,* Prentice-Hall, Inc., Englewood Cliffs, N.J., 1974.

[12] Grant, E.L., W.G. Ireson, and R.S. Leavenworth, *Principles of Engineering Economy,* 6th ed., the Ronald Press Company, New York, 1976.

[13] Mayer, R.R., *Capital Expenditure Analysis for Managers and Engineers,* Waveland Press, Inc., Prospect Heights, Ill., 1978.

[14] Riggs, J.L., *Engineering Economics,* McGraw-Hill Book Company, New York, 1977.

[15] Smith, G.W., *Engineering Economy: Analysis of Capital Expenditures,* 2nd ed., Iowa State University Press, Ames, Iowa, 1973.

[16] Tarquin, A.J., and L.T. Blank, *Engineering Economy* McGraw-Hill Book Company, New York, 1976.

Chapter 8

TELEPHONE RATE STRUCTURE: THEORY AND ISSUES

Wesley J. Yordon
Department of Economics
University of Colorado
Boulder, Colorado

8.0 Introduction

8.1 The Logic of Marginal Cost Pricing

8.2 Adjustments for Irrationality and Externalities

8.3 The Pricing Dilemma in Decreasing-Cost Activities

8.4 Value-of-Service Pricing

8.5 Historical Sketch of Telephone Rate Structure

8.6 Towards Usage-Sensitive Pricing

8.0 INTRODUCTION

For a single-product firm such as featured in most economics texts there is no distinction between price level and price structure. As described in the preceding chapter, public utility regulation in the U.S. aims at establishing a rate level for the regulated firm which will approximate the competitive level; the margin between aggregate operating revenues and aggregate operating expenses should yield a fair return on investment, and fairness is judged by comparison with the rate of return in competitive industries.

But public utility regulation (or government ownership in some cases) normally arises as the result of a judgment that competition is unfeasible or undesirable, and one of the factors which commonly leads to such a judgment is that the industry in question sells numerous services to different classes of customers and thereby achieves economies which would be unattainable if the different services were provided by separate entities. This circumstance generates the problem of rate *structure* as distinct from rate *level*; if a firm sells a number of services in noncompetitive markets there are countless alternative price structures which could yield the same aggregate rate of return.

As a matter of practical procedure, the agent or agency responsible for setting prices for a multiproduct enterprise often finds it expedient to regulate the margin between operating revenues and operating expenses by changing prices by a uniform percentage; such across-the-board adjustments must be made frequently during periods of general inflation. Other changes in the price for a specific product or service may also occur fairly frequently on a piecemeal basis, but a thorough review of the entire price structure is such a complex and time-consuming task that the preference will be to make such a review only rarely, in the context of long-term planning. Of course the questions of rate level and rate structure are related, but as a first approximation one may think of the former as involving $\Sigma p_i q_i$, where p_i is the price of the i-th service and q_i is the quantity sold, and of the latter as involving p_i/p_j, the ratio between the prices of services i and j.

Economic theory provides abstract rules for an optimal rate structure given any specified relationship between aggregate revenues and aggregate expenses. For ease in exposition and because most readers are North Americans, our discussion of these principles will assume that the specified relationship is one which yields a fair return on investment (in accord with the telephone pricing policy which prevails in the U.S. and Canada). However, the analysis is easily extended to situations where the telephone system is owned and operated by the government and where the fair-rate-of-return relationship is unimportant because the fisc (government treasury) may absorb excess profit or subsidize deficits. (With some thought the analysis may also be extended to the rate structure for television advertising in noncompetitive markets, but this exercise is left to the interested reader.)

The first sentence of the preceding paragraph must be clarified to avoid semantic confusion. For reasons of their own, economists have defined their terms in such a way as to concede that a truly optimal ("first best") price structure would automatically generate a specific relationship between revenue and expense, probably a deficit in the case of telephone service. However, they recognize that the tax policies needed to offset such a deficit would conflict with optimal pricing elsewhere in the economy, thus making it impossible to attain a "first best" solution and forcing economists to devise rules for what they refer to as "second best" or "quasi-optimal" solutions. [1] Further discussion on this aspect appears below; at this point it is important to understand only that the theory of quasi-optimal pricing presented in this chapter is applicable in any situation where rates must be structured in such a way as to satisfy some exogenous constraint concerning the relationship between aggregate revenues and aggregate expenses, where the aim is to attain the best feasible rate structure.

The exposition of rate-structure theory aims at shedding light on historical and current issues concerning telephone rate structure in the U.S., with attention to questions such as: Does the existing rate structure discriminate against some users and favor others? Will rates for residential local exchange service increase sharply if telephone companies are forced to compete with independent suppliers in the market for bulk business services? Why are telephone companies moving to radically change their pricing structures by introducing charges for services which often have been "free", such as directory assistance, installation, unlimited local calls for a flat monthly fee, etc.? Why have rates been reduced for bulk users of long-distance service while local rates were increasing?

To analyze such issues requires some knowledge of the economic theory of quasi-optimal pricing and of telephone technology, costs, and data-gathering systems. This chapter attempts to provide this in a general fashion, but cannot describe existing practices in detail because they vary widely even within the U.S. The circumstances which have brought these issues to prominence in the U.S. are largely associated with the nearly unique American policy of providing telephone service through privately-owned enterprise, but the essential issues also exist, albeit with less fanfare, where telephone service is provided through government enterprise. Readers from nations where the latter practice prevails should be able to profit from the American experience.

I must warn against overexpectations, however. In the words of experts charged with the task of analyzing these matters:

> The rate structure for myriad telephone services is only haphazardly, if at all, related to the structure of costs. In telephony, as in transportation, the rate structure which has evolved is the product of vaguely conceived ratemaking principles and historical happenstance. The present structure, contained in thousands of tariffs and exceptions, is based in part on cost considerations, in part on demand considerations, but for the most part is inexplicable. [2]

8.1 THE LOGIC OF MARGINAL COST PRICING

Almost everyone intuitively understands the functions of a price system as it customarily operates, but a more analytical approach is usually necessary in order to reason about the merits of changing the *status quo*. For example, most readers probably feel that it is logical that long distance phone rates are lower at night than during the day, but few will have wondered why taxicab fares are not lower when the weather is pleasant than when it is inclement. The exposition of economic theory here will start at an elementary level, skim over some material which is customarily covered in introductory and intermediate theory courses, and extend to material which is not usually treated in such courses because it involves the awkward case of a multiproduct enterprise with a high proportion of joint and common costs.

Consider a person who is picking apples to satisfy his hunger. The first apple or two is intensely desired, but as the hunger pangs are eased the value of subsequent apples declines. Conversely, the first few apples are within easy reach, but subsequent ones are attained only with greater effort. How many apples should he pick and eat?

Not so many that the pain of picking the last one exceeds the satisfaction which it yields. Not so few that the nearest unpicked apple would yield more satisfaction than the pain of picking it. In the jargon of economics, the optimal quantity is reached when marginal utility equals marginal cost. (The word "utility" is used in economics to designate the satisfaction achieved from consuming a good, but economic usage does not imply usefulness as opposed to pleasurability. The word "marginal" used to designate the borderline unit; it is the economists' quaint term for what mathematicians call a first derivative.)

To express this proposition mathematically, let U equal the utility achieved from eating apples, let D equal the disutility of picking them, and let A equal the number of apples picked and eaten. The person wants to maximize net benefit (U-D), and the necessary condition is $d(U-D)/dA = 0$, or $dU/dA = dD/dA$. (See Appendix.)

In an economy where there is division of labor the consumer of a commodity is generally not its producer. Suppose that the society consists of a picker, an eater, and a benevolent central planner whose objective is to maximize social welfare. Assuming that the planner can compare the utility of consumption with the disutility of production, he should direct that the number of apples to be picked and eaten be such as to satisfy the condition $dU/dA = dD/dA$. If instead of a planner who directly controls quantities there is a regulator whose task is to set an optimal price, he will maximize social welfare by setting a price such that $(dU/dA)/\lambda_1 = P_a = (dD/dA)/\lambda_2$ where P_a is the price of an apple, and λ_1 and λ_2 are the marginal utility of money to the consumer and to the producer respectively. (As a mathematical formality it is evident that λ_1 and λ_2 provide comparability among the three terms of the equation; the economic significance of this will be discussed below.) According to the theory of consumers' behavior, a rational consumer will choose to purchase the quantity which satisfies the left-hand equality, so the regulator's task is to bring about the right-hand equality. The value $(dD/dA)/\lambda_2$ may be viewed as the marginal cost of providing an apple, so the regulator will maximize social welfare if he sets price equal to marginal cost and orders the producer to supply the quantity which the consumer wishes to purchase at that price. (If marginal cost varies with output, as in the initial illustration, the regulator must take this effect into account, but since the complications associated with rising or falling marginal cost will be treated at length below it may be helpful at this point to simplify by temporarily assuming that marginal cost is constant, unaffected by the volume of production.)

The discussion above sketches in simplest form the logic behind the important proposition that maximization of social welfare calls for the price of a commodity to be equal to its marginal cost of production. A number of qualifications have been glossed over for the moment, but before turning to these some readers may appreciate a slightly more elegant version of the proposition which does not rely so heavily on the concepts of utility and disutility. It can be shown that a more rigorous expression of the necessary condition for maximization of social welfare is

$$\frac{dU_1/dA_1}{dU_1/dB_1} = \frac{dU_2/dA_2}{dU_2/dB_2} = \frac{V_A}{V_B} \tag{8.1}$$

where A and B represent any two commodities, the subscripts designate any two persons, and V_A and V_B may be thought of as the respective marginal costs of A and B. (Thus dU_1/dA_1 designates the marginal utility which person #1 receives from his consumption of A, etc.; properly the derivatives should be partials but we dispense with this refinement.) This form involves only ratios of utilities and costs and thus avoids interpersonal comparisons of pleasure and pain; in the language of economic theory the ratios on the left are termed marginal rates of substitution and the one on the right is called the marginal rate of transformation.

For present purposes the latter concept is important. Many would have reservations in thinking about the cost of providing telephone service in terms of the disutility of someone's labor, but monetary cost is not necessarily a correct measure either. Marginal rate of transformation is a better concept. It is an expression of the *opportunity cost* of commodity A in terms of commodity B; i.e., V_A/V_B indicates how many units of B must be foregone by society in order to obtain a unit of A. Thus, if it requires 10 minutes of labor to produce a unit of A and 20 minutes to produce a unit of B, and if labor is the only resource used, $V_A/V_B = 1/2$; this indicates that production of a marginal unit of A requires society to forego the production of 1/2 unit of B, all other things being equal.

Again there is a presumption that each consumer will purchase goods or services in quantities such that $(dU/dA)/(dU/dB) = P_A/P_B$, so the regulator's task is to set prices such that $P_A/P_B = V_A/V_B$. If the monetary costs of A and B reflect the real cost of the resources (labor, land, and capital) used up in the production of commodities A and B, then the monetary cost ratio is the marginal rate of transformation, but as we shall see much of the problem of rate structure concerns this "if".

The point of this subsection has been to sketch the logic of the proposition that for optimal allocation of resources the price of a good or service should be set equal to its marginal cost of production, and to suggest that in this context "marginal cost" means the opportunity cost of the resources used, the cost of which may be difficult to measure since it is not necessarily identical with monetary expense as reported by an ordinary accounting system. Fuller expositions of the logic of marginal-cost pricing may be found in most intermediate microtheory texts, but for present purposes elaboration of the logic of the rule is less important than a number of limitations on its validity and applicability, namely questions about consumers' rationality, externalities, and equity.

8.2 ADJUSTMENTS FOR IRRATIONALITY AND EXTERNALITIES

The argument for marginal-cost pricing sketched above assumes that a consumer purchases goods in quantities which maximize his or her individual utility (in the face of given income and prices), and that a change in a consumer's utility changes social welfare by an equivalent amount (since the individual is part of society). In theory the first assumption is treated as concerning the consumer's rationality and the second as concerning externalities, but in practice the two issues often are difficult to separate. For example, it may be argued that if gin were sold at a price equal to its marginal cost some consumers would "irrationally" drink more than the amount which maximizes their utility, hence that it is desirable to offset this tendency by imposing a high tax and thus holding the price of gin well above its actual cost. Others may think that the drinker does maximize his own utility, but object to the fact that his intoxication imposes disutility or costs on others.

The question of consumer rationality does not appear to be a serious issue in telephone ratemaking. Reservations may be held concerning a subscriber's knowledge of technical matters, rates, and options, or about his susceptibility to advertising and salesmanship, but the appropriate remedy would be provision of better information rather than manipulation of rates.

Externality issues have been important. When the telephone system was in its adolescence Theodore Vail (President of AT&T, 1885-7 and 1907-19) argued that the addition of a subscriber brought external benefits to existing subscribers, and thus justified a rate structure featuring low rates for residential service. [3] To illustrate numerically, suppose that Mr. Smith would gain $5/mo. in utility from phone service, but declines to subscribe because the monthly cost is $10. But 10 of his acquaintances would each gain $1/mo. in utility if they could call Smith, so the total increment in social utility would be $15 compared to a cost increment of $10. In this circumstance marginal-cost pricing fails to maximize social welfare, and society would gain if service were offered to Smith at $5/mo., with each of his 10 acquaintances contributing 50¢/mo. to offset the telephone company's loss, say by leasing an extension cord from the company at a rate which exceeds its actual cost. Now that Vail's aim of universal service has been essentially achieved in the U.S. (95% of households having service) his argument has lost force here, but it remains important in less affluent nations.

It has been argued that a similar externality exists with respect to telephone calls, because the callee as well as the caller may gain utility from the call. If outgoing calls are priced at their marginal cost this will inhibit calls for which the marginal utility to the caller is less than the charge, but it is possible that the sum of the caller's and callee's marginal utilities exceeds the marginal cost of the call and that social welfare would rise if the call were made.[4]

This argument is less persuasive than the first because the caller-callee relationship is likely to be close enough so that one party can take into account the benefit to the other, and reciprocity would tend to even out the financial burden of a series of calls.[5]

Another externality issue of great importance has concerned the attachment of "foreign" equipment to the system; i.e., equipment not provided by the telephone company. The companies (especially those in the Bell System) have argued that such attachments are likely to impose harm on other subscribers by jeopardizing the quality of the telephone network. On the basis of this reasoning telephone companies prohibited attachment of foreign equipment until recently, and thus were able to exclude competition from the market for station equipment (see discussion in section 8.5 below).

8.3 THE PRICING DILEMMA IN DECREASING-COST ACTIVITIES

The arguments above concern efficiency, not equity; marginal-cost pricing (in the absence of irrationalities and externalities) is a necessary but not a sufficient condition for maximizing social welfare. We now turn to considerations of equity for buyers and sellers.

With respect to a buyer, it was stated above in section 8.2 that economic theory postulates that a rational consumer purchases the quantity of commodity A which makes $(dU/dA)/\lambda = P_A$. But if the consumer is poor, λ (the marginal utility of money to the consumer) will be large and he will be forced to restrict his purchase of A even though the utility from an additional unit would be large. If society judges such a situation to be unjust there are two possible remedies: grant him more money income, or reduce the price of A below its marginal cost. The latter is preferable only if irrationality and/or externalities are acting in conjunction with his poverty to limit his consumption to less than the socially optimal quantity; such an argument is commonly applied to education, housing, preventive medicine, and even to food. Telephone service is not viewed in this light; on the contrary, prior to 1968 the New York City Welfare Department regarded telephone service as an unwarranted expense in the budget of families on welfare. Opinion has shifted so that concern is now expressed over the inability of low-income families to purchase telephone service (especially the ill or the elderly for whom it may be vital in case of emergency), and a form of inexpensive ("lifeline") service has been proposed to meet this need. In general, however, such proposals are not intended to set price below cost but only to provide an option which links low price with low usage (and hence low cost) so as to avoid forcing everyone to buy the type of service which is appropriate for the average family. In sum, general opinion in the U.S. does not regard consumption of telephone service as an activity which should be promoted by subsidizing it from general taxation.

The discussion above may seem to belabor the obvious, but in fact certain technical characteristics of the telephone industry generate a pricing dilemma; apparently the principle of marginal-cost pricing can be applied only if accompanied by subsidization from general tax revenues, but social considerations apparently do not justify such a subsidy. This dilemma arises because telephone service appears to be a decreasing-cost activity rather than an increasing-cost activity, and since most presentations of economic theory focus on the latter it is desirable to elaborate the distinction and the implications thereof.

Consider the extraction of petroleum as a good example of an increasing-cost activity. For clarity, imagine that petroleum may be extracted from any one of 10 sites and that the rate of production obtainable from any given well is a constant of nature, say one million barrels per year. Due to differences in the depth of the oil, the cost of drilling and extraction vary: oil may be obtained from site #1 at a cost of $1/bbl., from site #2 at a cost of $2/bbl., etc. Suppose that the demand for oil is such that 8 million bbls/yr. will be sold if the price is $8/bbl., then demand and supply will be equated if sites #1 to #8 are exploited and the oil is sold at $8/bbl., as depicted in Figure 8.1.

Given these conditions, site #8 is the "marginal" well, and the cost at site #8 ($8/bbl.) is the marginal cost of producing a barrel of petroleum. With a uniform price equal to marginal cost the total revenue generated by selling 8 million barrels is $64 million, while the total cost is $36 million ($1 million at site #1 plus $2 million at site #2 plus . . .), so there is a surplus of $28 million. If the oil wells are owned by the government this surplus provides the

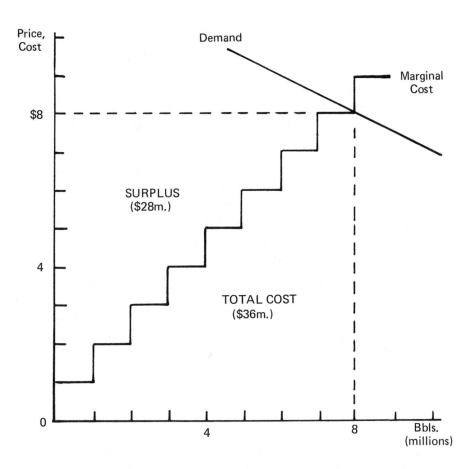

Figure 8.1. MC Pricing Where Cost Increases by Steps

government with funds which may be used for various purposes—among them perhaps to provide fuel oil for needy families. If the wells are privately owned the surplus is nevertheless available to the government through taxation. Oil from site #1 may be taxed at the rate of $6.99/bbl., oil from site #2 at the rate of $5.99/bbl., etc., without causing any diminution of production, and the total tax take will be $27.9 million, ignoring collection costs.

If production costs increase smoothly rather than by discrete steps the task of capturing the surplus through taxation may be more difficult, but the essential point remains: marginal cost exceeds average production cost, and marginal-cost pricing will generate total revenue in excess of total production cost. The case of smoothly increasing cost is depicted in Figure 8.2. If price is set equal to marginal cost at the output level which satisfies demand (P*, Q*) total revenue will be P*Q* or the area of rectangle OP*EQ*. Total cost of production will be the area of rectangle OBFQ*, and the surplus will be the area of rectangle BP*EF, or the area of triangle AP*E.

Parenthetically, to avoid confusion for those who have been exposed to an introductory course in economics, it should be noted that most economics texts assume that the surplus is captured by the land owner; i.e., that it becomes a rent or royalty which the producer must pay. Accordingly it appears as a financial cost to the producer, so that production cost plus rent equals total revenue, and average *total* cost equals price. A graph of average total cost would be represented by a U-shaped curve which reaches a minimum at point E, indicating that the producer just breaks even when price is P* and output is Q*. However, the standard treatment tends to obscure what is essential for present purposes; namely, that selling a product at a price which equals its marginal cost of production generates a surplus in activities where marginal cost rises with output.

For decreasing-cost activities the analysis is symmetrical, and the opposite conclusion follows, namely that selling a product at a price equal to its marginal cost of production will generate total revenue less than total production costs, resulting in a deficit. This

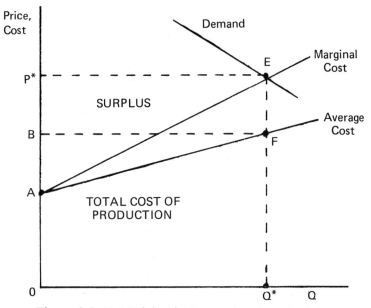

Figure 8.2. MC Pricing in Increasing Cost Activity

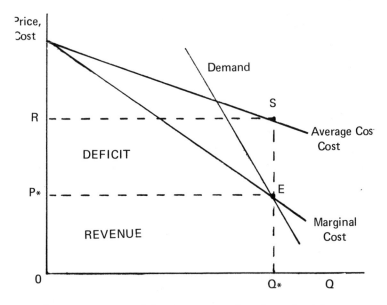

Figure 8.3. MC Pricing in Decreasing Cost Activity

situation is depicted in Figure 8.3, where total revenue is the area of rectangle OP*EQ*, total production cost is area ORSQ*, and the deficit is area P*RSE.

In the common intermediate case where average cost is constant and equal to marginal cost, marginal-cost pricing generates exactly the revenue needed to cover total cost.

Local telephone service was viewed as an increasing-cost activity by some analysts writing in the first half of the century. This view is based on the law of combinations, which states the number of paired connections between N telephone stations is given by the formula $N(N-1)/2$; e.g., 6 for 4 subscribers, 15 for 6 subscribers, etc.[6] This was a sensible argument when connections were made manually, but its validity became dubious with the spread of mechanical switching (dial systems). A recent study shows that access investment cost per subscriber falls with the number of subscribers in a given area (about $450 for 1,000 subscribers, $290 for 50,000 subscribers, and $260 for 100,000 subscribers; Figure 8.4 roughly reflects this relationship.[7]

Where average cost is falling, marginal cost must be less than average cost; so if marginal-cost pricing were applied to this dimension of service, revenue would not cover total cost, and the deficit would have to be offset with surplus revenue obtained from other activities.

For "long lines" the presence of decreasing costs has long been evident. The costs of right-of-way, poles (or trenching) are about the same for 2,000 circuits as for 2, and the cost of cable does not increase proportionally with the number of circuits. With coaxial cable, optic fiber, or microwave transmission decreasing costs are dramatic.[8]

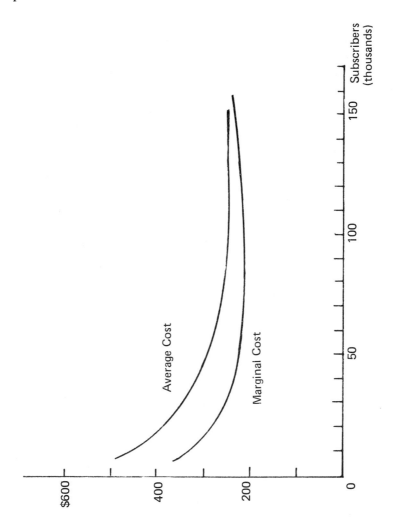

Figure 8.4. Investment Per Telephone Subscriber
(Access Equipment)

It must be stressed that analysis of telephone costs is a formidable task. Until recently the concern of citizens and regulatory agencies in the U.S. focused on the overall rate level rather than on rate structure, so there was little motivation to identify disaggregated costs. The Uniform System of Accounts (use of which is required by law) is out of date by half a century, and while reform is in progress it will be some time before data become available in the quantity and form needed for a tolerably accurate estimation of the cost functions for the numerous dimensions of telephone service. Consequently the cost functions described should be viewed only as illustrations of the probable tendency in some facets of telephone service, and it is possible that some other aspects are subject to increasing costs. It will be argued below that a refined rate structure which prices various dimensions of telephone service according to their actual costs would substantially mitigate the problem described here. However, it seems probable that the overall tendency toward decreasing costs would remain a problem even with refined cost analysis and ratemaking, and in any case it is important to understand that the telephone rate structure has historically been shaped by the assumption that decreasing costs predominate.

Some economic theorists have argued that the ideal of pricing at marginal cost should not be abandoned simply because it would result in deficits in decreasing-cost industries; they argue that the deficit should be offset by a subsidy from the fisc.[9] But where will the fisc obtain the necessary funds? Possibly by borrowing in the short run, but sooner or later only from taxation of other activities. If there are increasing-cost activities of sufficient magnitude it might be possible to extract from them surplus enough to subsidize the deficits in the decreasing-cost activities; but if not then the attempt to subsidize in order to maintain marginal-cost pricing in the decreasing-cost activity will cause prices to exceed marginal costs in others, and therefore be self-defeating. The former condition may exist in a few nations such as those whose income arises mainly from extractive industries such as petroleum, but the latter prevails in most industrial nations such as the United States where less than 5% of national income originates in agriculture and extractive industries. Given the magnitude of other demands on government funds which in the U.S. have pushed taxes to about 40% of national income it is unrealistic to assume that there is an untapped reservoir of non-distorting tax revenues which could be used to subsidize telephone service. Consequently the discussion will proceed using the assumption that telephone service as a whole is to be unsubsidized, with only a reminder that the alternative approach is not ruled out for nations where the extractive sector dominates and provides distortion-free tax revenues in excess of high-priority demands for government spending.

8.4 VALUE-OF-SERVICE PRICING

The question of how to structure prices in situations where simple marginal-cost pricing fails to generate sufficient revenues came to prominence in the 19th century with the growth of roads, canals, and railways. It was evident that the *use* of such facilities imposed small real cost, and that a pricing system which charged fees equal only to the cost of use would not generate enough revenue to cover construction cost. In 1844 Jules Dupuit, a French engineer, provided an excellent analysis of the problem and pointed out the merit of basing prices on value of service rather than on cost of service:

> Thus when a bridge is built and the state establishes a tariff, the latter is not related to cost of production: the heavy cart is charged less than the sprung carriage even though it causes more wear to the timber of the carriageway. Why are there two different prices for the same service? Because the poor man does not attach the same value to crossing the bridge as a rich man does, and raising the charge would only prevent him from crossing. Canal and railway tariffs differentiate between various classes of goods and passengers, and lay down markedly different rates for them although the costs are more or less the same. In drawing up these tariffs in advance the legislator merely defines certain features and characteristics which seem to him to indicate a greater or lesser degree of utility in the same service rendered to different people. [10]

The similarity between value-of-service pricing and marginal-cost pricing with respect to efficiency in resource allocation should be evident; in both cases the aim is to set a price which permits a person to consume a commodity up to the point where the marginal utility to the person equals the marginal opportunity cost to society. If a facility exists and use of it imposes negligible cost in terms of wear or congestion, use of it should not be parsimoniously rationed. On the other hand, if it is possible to segregate users who will not be deterred by a high fee they can be made to bear the financial burden of the construction cost without curtailing the use of the facility. Such a pricing system is sometimes described as "charging what the traffic will bear," but it may also be viewed as a system of "*not* charging what the traffic will *not* bear." [11]

It is also commonly described as price discrimination, but for reasons which will become clear this term is likely to generate more heat than light. At this point it is important to understand that value-of-service pricing by no means ignores the cost of service. For convenience of exposition, writers commonly choose examples where actual cost appears to be negligible, but it is clear that they do not envision a charity program which renders services at prices lower than the opportunity costs of providing them. In fact, with the aid of modern economic theory it becomes evident that in some cases value-of-service pricing coincides with a sophisticated form of marginal-cost pricing, but more generally it is a system where cost of service establishes the minimum charge and a variable markup is added to collect additional revenue from those who value the service highly.

Up to this point the terms "cost" and "marginal cost" have been used without specifying whether these refer to the short run or to the long run. It is my belief that the popular introductory and intermediate economics texts provide more confusion than help on this matter for present purposes. The problems are that they focus on a single-product enterprise, assume nearly perfect information, and usually illustrate with reference to wheat farming (which is an increasing-cost activity with no long-lived or lumpy investments). Samuelson has argued that the rule for optimal pricing is to equate price with short-run marginal cost, which will be equal to long-run marginal cost if and only if the capacity of the facility is optimal.[12] But in a world where there is uncertainty and where projects are lumpy rather than infinitely divisible there is ambiguity in the concept of optimal capacity.

Where need varies stochastically what capacity is required to handle the load? Does a roof have excess strength if no known snowfall has ever been enough to break it? If a facility is deliberately built oversize in relationship to current need because it is uneconomical to rebuild it as need increases does such a facility have excess capacity in any meaningful sense? Some economists would answer affirmatively and argue that price should be held down while the capacity is excessive, then raised as growing demand makes it necessary to ration use. But if long-term plans are based on current prices such a policy will induce erroneous planning,[13] and for public utilities such as telephone service it would be administratively awkward to vary rates from month to month in accord with changing relationships between demand and capacity.

Consequently it is impractical to base telephone rates on short-run marginal costs, and the relevant cost base more nearly resembles long-run marginal cost. However, the concept of long-run marginal cost in abstract theory is derived on the assumption that *all* inputs are variable (i.e., that there is no existing plant or equipment) whereas in the telephone industry some inputs have such long lives (e.g., a right of way) that the time will never come when the industry will have a fresh start with a clean planning slate. Hence the relevant cost base is not quite the same as theoretical long-run marginal cost, but with this qualification the discussion here will refer to the relevant cost base as long-run marginal cost because we envision a planning period which is long enough to permit substantial variation in plant and equipment.

For some pricing decisions, however, it is appropriate to assume that plant and equipment are fixed. For example, the relevant cost of placing a call at night is low because the facilities to handle it are already in place, and are under-utilized. Some economists view this as a situation where short-run marginal cost is low and is the correct base for price. It is more accurately viewed as a situation which involves joint cost; i.e., the same facilities provide two essentially different services, and since the demand for daytime

service is greater than the demand for nighttime service the latter
may be viewed as a byproduct of the former. This is not really a
short-run phenomenon because the difference between daytime and
nighttime demand presumably will endure for the foreseeable
future, but in any case the conclusion is valid that the relevant
cost is lower for nighttime calls than for daytime calls, and that
the rate structure should reflect the difference. Generally this will
imply that calls during the offpeak period should be priced at less
than *average* cost, so there is a strong similarity between the problem of joint cost and the problem of decreasing cost. Either case
calls for value-of-service pricing; in the former, value of service is
used to identify the relevant costs and the resulting price structure
may be viewed as marginal-cost pricing; in the latter, value of service is used to vary the margin between price and marginal cost
so as to satisfy the revenue requirement as efficiently as possible.
In practice it is virtually impossible to distinguish between these
two cases, but we proceed with a simplified example which starts
with a pure case of joint cost.

Imagine that a municipality contemplates construction of a fully
automated telephone exchange. (For simplicity we assume that
each subscriber will provide his own station equipment and line;
the municipality will provide only the service of switching calls
among the lines.) A referendum is held to ascertain whether residents favor such a project, but it is unanimously defeated because
each resident deems the cost to be excessive. Then 1,000 merchants
contract to pay $1 per month apiece for service (a sum sufficient
to fully compensate the municipality for the cost of constructing
and operating the exchange) and the exchange is built and placed
in service. At this point in time the situation is similar to the theory
described in section 8.1 — one "consumer" has declined to purchase something because its marginal cost exceeds its marginal
utility, while another "consumer" has agreed to the purchase
because marginal utility exceeds marginal cost, but here the
consumers are groups rather than individuals and the indivisible
nature of the project makes the term marginal somewhat inappropriate since the purchase involves a large rather than an infinitesmal
increment.

After the exchange is constructed it becomes evident that the merchants use it only during the day and that it might be used at night for residential service. Since construction cost has already been covered, the marginal cost of nighttime use is low, so the municipality can allow residential subscribers to connect with the exchange for a fee much lower than $1 per month if they place calls only at night. The extra revenue will permit a reduction in the fee charged to business subscribers (which may induce more merchants to subscribe, a complication treated below) but they still pay more than the residential rate. Nevertheless, they are better off than they would be if there were no residential sub-scribers, so the arrangement seems eminently fair.

The above example all too conveniently assumes that the decisions took place in a certain sequence which revealed that the telephone exchange was more intensely desired by the merchants than by the residents. Let us now modify the assumptions by postulating that the decisions are to be made simultaneously, in the light of perfect information about the demand intensities of the two groups. This and other assumptions are admittedly unrealistic, but we gain generality and rigor. The analysis may seem tedious and abstract, but there is no other way to convey an understanding of the main issues in telephone ratemaking.

Suppose that telephone subscribers conveniently fall into two groups: business users who make calls only during the day, and residential users who make calls only at night. It is not feasible to meter calls, but experience indicates that one switching unit is needed in the exchange for each 100 subscribers in order to avoid catastrophic failure of the system. The monthly capital recovery expense (interest and depreciation) of each switching unit is $100; i.e., $100 per month per switching unit provides a "fair return" on the investment in the exchange. If an exchange were built for the exclusive use of either business or residential subscribers it is evident that the appropriate monthly fee would be $1 per subscriber, but since they will jointly use the same facility the $1 is a joint cost which must somehow be split. It might seem fair to split it 50-50 but if (as is almost certain to be the case) there is a difference in the intensities of demand for the two types of service an even split would be unfair and inefficient. Figure 8.5 illustrates the problem with hypothetical demand functions B = 2000 – 1000P$_b$ and R=1450 – 1250P$_r$, where B and R are the numbers of business and residential subscribers and P$_b$ and P$_r$ are the respective monthly fees. A common fee of $0.50 would attract 1500 business subscribers but only 825 residential subscribers, generating revenue of $1,162.50 and requiring an exchange with 15 switching units (at a cost of $1,500) having excess capacity at night. Hence a uniform fee of $0.50 is neither optimal nor feasible, but this amount may be used as a reference point in calculating the optimal price structure, so it is plotted in Figure 8.5 and labeled AVC (Average Variable Cost — but not variable in the short run). Optimum quantities and prices may be ascertained by finding a quantity such that P$_b$ and P$_r$ are equidistant from AVC, the former above and the latter below. This point occurs where S = 12, P$_b$ = $.80, and P$_r$ = $.20. This shows that fees should be $.80 per month for business subscribers and $.20 per month for residential subscribers; with this rate structure 12 switching units will provide proper capacity both day and night and the revenue requirement of $1,200/month will be exactly satisfied by collecting $960 from businesses plus $240 from residences.

This price structure appears to be a case of price discrimination, but in reality it is not.[14] The prices are exactly equal to their respective marginal costs, but because the costs are not easily identifiable some further explanation may be needed to make it clear that the fees are indeed equal to the respective costs of service. Consider the short-run marginal cost of adding another subscriber. The cost will be catastrophic system failure (or less dramatically, major inconvenience to phone users) unless an existing subscriber's service is simultaneously terminated. If the latter action occurs, the decrease in social utility and the revenue loss to the company is 20¢ or 80¢ (depending on whether the new subscriber is residential or business). Note that this reasoning applies only if the exchange is used to capacity (otherwise the opportunity cost of adding a subscriber is zero) and that capacity will be fully utilized only when the fees are 80¢ for business subscribers and 20¢ for residential subscribers. The long run marginal cost is $1.00 per pair of subscribers (one business, one residential)

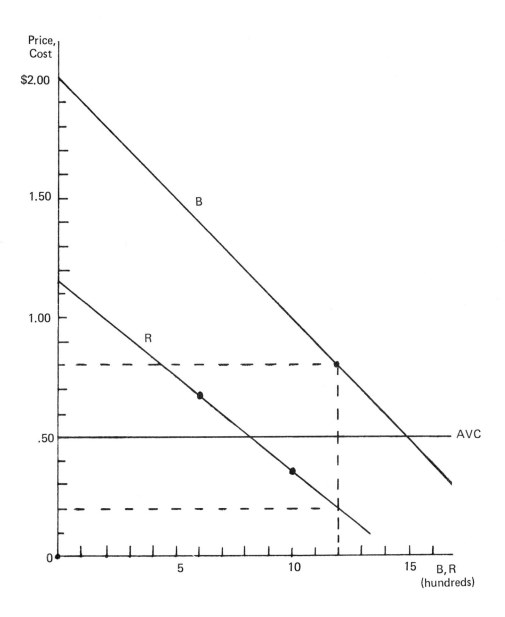

Figure 8.5. Allocation of Joint Cost by Demand Intensities

and the 80-20 split is needed in order to match the number of residential subscribers with the number of business subscribers. Accordingly, this price structure can be labeled value-of-service pricing because it does take into account the intensities and sensitivities of business and residential demand for exchange service, but it is not a case of price discrimination. Note also that the joint cost cannot rationally be allocated according to relative use because the aim of differential pricing is to equalize use.

Unfortunately even the simplest example becomes substantially more complex if we introduce decreasing long-run costs in place of the convenient constant-cost assumption used above; i.e., that the monthly cost of an exchange was simply $100 per switching unit. Let us now take a small step toward realism by assuming that the cost is $100 per switching unit plus a lump sum which is unaffected by the size of the exchange; e.g., the salary of a custodian. This does not affect long-run *marginal* cost, but it makes long-run *average* cost exceed long-run marginal cost. Consequently the optimal price structure (P = LRMC) no longer generates sufficient revenue to cover costs, and we are forced to consider a price structure which will be quasi-optimal or second best.

The logic of quasi-optimal pricing was developed rigorously by F.P. Ramsey in 1927, and has been discussed more recently by Baumol and Bradford.[15] Here the formula will be presented and illustrated, with no attempt to describe the reasoning behind it. It is important to appreciate, however, that the formula reflects maximization of social welfare, not maximization of profit. The measure of social welfare is the money value of consumers' utility minus the money cost of production. It thus assumes that a dollar has equal value to rich and poor alike, and the only question of income distribution it handles is the one of how to generate the revenue required with the least possible harm to consumers as a group.

In the case at hand (where cross-elasticities are zero) the Ramsey rule for quasi-optimal price structure is

$$\frac{P_b - MC_b}{P_r - MC_r} = \frac{MR_b - MC_b}{MR_r - MC_r} \qquad (8.2)$$

where P is price, MC is marginal cost, and MR is marginal revenue. (Marginal revenue is the first derivative of revenue; when the demand function is linear MR has twice the slope of the demand function.) The demand functions displayed in Figure 8.5 are $P_b = \$2 - \$.001B$ and $P_r = \$1.16 - .008R$, where B and R are the number of business and residential subscribers respectively. Substituting the numerical values for the terms MR and MC, we obtain

$$\frac{P_b - .80}{P_r - .20} = \frac{2 - 2\,(.001B) - .80}{1.16 - 2\,(.0008R) - .20} \tag{8.3}$$

Examination of this equation will reveal that the relative magnitudes of the margin between price and marginal cost vary with the vertical intercept of the demand function (i.e., \$2 for business demand, \$1.16 for residential demand) and with the slope of the demand function (i.e., $-.001$ for business and $-.0008$ for residences). If one thinks of the intercept as a measure of the intensity of demand, and the slope as a measure of the sensitivity of demand, we may say that the margin between price and marginal cost should vary positively with intensity and negatively with sensitivity. A commonly used concept in economics which combines these two aspects (in inverse form) is the price-elasticity of demand, and this concept may be used to express the Ramsey rule as

$$\frac{P_b - MC_b}{P_r - MC_r} = \frac{E_r}{E_b} \tag{8.4}$$

where E_r and E_b are the price-elasticities of demand for residential and business subscriptions. However, since elasticity varies along a given demand curve one cannot speak of *the* elasticity of demand function, and confusion is likely to result if one refers to the rule as prescribing a lower markup in the more elastic market. (Column (g) in Table 8.1 is included to illustrate this point.) Consequently,

most readers are advised to think in terms of intensity and sensitivity (height and slope) of demand rather than in terms of
elasticities.

Table 8.1 displays two quasi-optimal or "second best" price structures together with the optimal structure and the monopoly structure. Column (c) shows the net revenue; i.e., total revenue minus
the cost of switching units (@ $100 month). It is this net revenue
which is available to cover lump-sum fixed costs such as the expense of a custodian, amortization of design fees, or (in a more
complex situation) differences between average cost and marginal
cost due to decreasing marginal cost. The optimal or first-best
solution provides zero net revenue, and hence is not feasible. Two
second-best solutions are shown, generating net revenue of $360
and $576, to cover two possible levels of fixed cost. Finally, for
comparative purposes, the monopoly solution is shown which
generates net revenue of $648 which is the maximum possible.

Examination of the table will reveal why many discussions of
price discrimination generate more heat than light. Row (1),
the optimal solution, displays a situation which most people
would regard as "discriminatory" simply on the basis of the
differential prices shown in column (a), or with more sophistication, on the basis of the relationships to average total cost as shown
in column (e). But according to economic theory the situation is
really non-discriminatory, with prices for both types of service
equalling their respective marginal costs as shown in column (f).
The rate structures shown in the other rows may legitimately be
described as discriminatory, but it is not clear which group is
favored. A comparison between the fees and average total cost
(column (e)) suggests that there is discrimination against business
subscribers, but from the viewpoint of economic theory average
total cost is a meaningless concept in this context. The relevant
concept is marginal cost, and comparison of the ratios of fees to
marginal costs in column (f) reveals that the rate structures shown
in rows 2, 3, 4 involve discrimination *in favor* of business subscribers even though their rates are higher. One of the points of
this example is that no normative significance should be attached
to the term "discrimination" in the context of public utility

Table 8.1. Examples of Value-of-Service Pricing

	(a)		(b)	(c)	(d)	(e)		(f)		(g)	
	Monthly Fee		# of sub- scribers in each class (also variable cost in dollars)	Revenue minus variable cost	Average total cost	Fee/ATC		Fee/MC		Elasticity	
Rate Structure	Bus.	Res.				Bus.	Res.	Bus.	Res.	Bus.	Res.
Optimal	$ .80	$ .20	1200	$ 0	$ .50	1.60	.40	1.00	1.00	.67	.19
Quasi-optimal	1.00	.36	1000	360	.68	1.47	.53	1.25	1.80	1.00	.45
Quasi-optimal	1.20	.52	800	576	.86	1.40	.60	1.50	2.60	1.50	.81
Monopoly	1.40	.68	600	648	1.04	1.35	.65	1.75	3.40	2.33	1.42

economics; one must make an effort to avoid the common language connotation of wrongdoing. In the language of public utility economics the semantic difficulty is overcome by use of a double negative — an economist would testify that the rate structures displayed are "not unduly or unreasonably discriminatory." But such language fails to convey the optimal properties of the rate structures shown, and it should be emphasized that they maximize the sum of the utilities of business and residential subscribers, subject to the constraint that total revenue cover the cost of service. On the other hand, it must be stressed that the term "value-of-service pricing" also applies to many price structures which are neither optimal nor quasi-optimal and which may exhibit price discrimination in the pejorative sense of the term.

The major points which should be evident from this illustration of ideal value-of-service pricing are:

1. The ideal method of allocating a joint cost is on the basis of relative intensities and sensitivities of demand; the resulting price structure may appear to be discriminatory, but actually it is a sophisticated form of marginal-cost pricing.

2. The pricing dilemma created by decreasing costs may be resolved by using value-of-service pricing; the resulting price structure will be discriminatory but if it follows Ramsey's rule for quasi-optimality the discrimination is socially beneficial.

3. To ascertain whether a rate structure exhibits price discrimination and to determine whether the discrimination is socially beneficial or harmful requires information concerning the cost and demand functions for the services in question; gross errors result if judgments are based only on a comparison of the prices or rates.

In order to provide clarity, this illustration unrealistically assumed that the cost and demand functions were perfectly known, and that there was a hard and fast distinction between the two categories of subscribers. In reality the relevant functions are largely unknown and the distinctions between categories are more-or-less arbitrary for reasons to be discussed in the following section.

8.5 HISTORICAL SKETCH OF TELEPHONE RATE STRUCTURE

The example discussed in the preceding section was unrealistic in excluding the possibility of metering calls and in using extremely simple cost functions, but in a way the combination provides a fair representation of the circumstances which prevailed prior to the development of electronic switching. In reality the cost of providing telephone service is a complex function. Costs vary not only with the number of calls, but also with their distance, duration, and timing (i.e., the hour of the day/week/year, because lines and switching facilities are more congested at some times than others). Prior to 1921 calls were usually handled by operators, and it was possible to record such information for each call but it would have been inordinately expensive to do so. Most telephone companies in the U.S. made a judgment that detailed record-keeping was worth the effort only for "long-distance" calls, and since there is no independent standard which defines how long is "long", the truth of the matter is that a call distance was defined as "long" if the administrators of a system judged that it was worth the effort to record time, distance, and duration, and to charge accordingly. (In Britain the early custom was to classify calls according to whether they could be switched within a single exchange or whether they required use of an interexchange trunk-line, and therefore to use the term "trunk call" for non-local calls. However, with expansion of the system it became necessary to use trunklines for some local service, so that not all trunkline calls are "trunk calls.")

The introduction of dial systems after World War I altered the situation; mechanized message registers could cheaply record the *number* of local calls, but without operator handling it became more expensive to record the other aspects. Again it was a matter for judgment as to whether recordkeeping was worth the effort, and judgments varied widely. A few telephone companies in the U.S. set charges for local calls in accord with an elaborate metering system, a few offered pay-as-you-call coin-box service for residential subscribers, many adopted a rate structure which used a simple count of the number of monthly calls in one way or another, and many continued to charge a flat monthly rate regardless

of the number of local calls.[16] Outside of North America most
telephone companies counted local calls and billed accordingly.
The prevalence of flat rates in North America has been attributed
to the fact that the Bell System used this method in the early
years to compete with independent companies, but it must also
be noted that a billing system based on simple message counts
involves a significant increase in billing expense to obtain a rate
structure which is only slightly less crude than the flat rate method,
and that the flat rate method does involve higher charges for business service than for residential service on the reasonable assumption that the former is used more intensively.

By 1914 85% of local telephone service in the U.S. was provided
by subsidiary companies of American Telephone & Telegraph, and
almost all long-distance service was provided by AT&T's
Long Lines Department. At that same time in the U.K. telephone
and telegraph service were consolidated as a monopoly operated
by the British Post Office, and while historical details vary there
was a strong impetus toward a telephone monopoly in all nations
arising from the basic characteristic of *switching.* (A number of
technical terms are introduced in this section; when they first
appear they are emphasized and defined explicitly or implicitly.
It is important to bear in mind the meanings of these terms through
the discussion below.) Switching refers to the process of interconnecting subscribers through an *exchange*, and to provide the
possibility of interconnecting every possible pair of subscribers
in a nation it must be possible to link every subscriber directly or
indirectly with a common exchange. Consequently there is a compelling reason to establish a single network (possibly through
cooperation among separate companies if not through a single
corporate entity) for *switched services*. Facility for interconnection is less important for *private lines*; i.e., semipermanent links
for continuous communications by heavy users such as press wire
services, stock brokers, banks, etc., usually for telegraphic or teletypewriter service in the early years, and not to be confused with
single-party telephone service for ordinary businesses or residences.
However, economic and esthetic considerations virtually dictated
that private line service be supplied jointly with switched services
in order to avoid wasteful and unsightly proliferation of poles,

wires, and rights-of-way. (Telegraph signals could even be transmitted simultaneously with phone conversations on the same wire pair.) Consequently the provision of private line service was taken over by telephone and telegraph companies: in the U.S. by the Bell System, Western Union Telegraph Co., and to a small extent, by independent telephone companies; in most other nations by the government-operated telephone and telegraph service. Bearing in mind that in the U.S. each of the independent telephone companies had a telephone monopoly in the area it served and that competition between AT&T and Western Union was tepid, it is a fair summary to say that there was no competition in the market for switched services and very little competition in the market for private line services either in the U.S. or in other nations; telephone service was regarded as a natural monopoly.

Terminal equipment such as telephone instruments, subscriber-used switchboards, teletypewriters and the like might possibly have been purchased from independent manufacturers, but telephone companies prohibited the interconnection of such "foreign" equipment with the telephone network on the grounds that such interconnections would jeopardize the quality of the system; hence the market for terminal equipment was also insulated from competition.

Because they were insulated from competition telephone companies around the world were not compelled to set the price of each particular service equal to its cost, so there was little motivation to ascertain the costs of particular services. Even had the motivation existed, the task would have exceeded the capabilities of the available methods of gathering and analyzing the necessary data. Consequently we lack the information that would be needed to say to what extent the price structures reflected cost-of-service pricing for the period before 1960, but some discussion of the history of telephone rate structure is needed to understand the current issues.

The vocabulary used in discussions of telephone rate structure in the U.S. has been largely shaped by the Bell System. Telephone service is classified into two broad categories, *basic* and *non-basic*. The former refers to local exchange service for residential and business subscribers and the latter encompasses everything else:

message toll service (ordinary long-distance calls), Wide Area Telephone Service (WATS), private line service, special terminal equipment such as private branch exchanges (PBX), key telephone systems (KTS), teletypewriters, and conveniences for the ordinary subscriber such as extra-long cords, extension phones, and deluxe instruments.

According to representatives of the Bell System, their rate structure historically has aimed at obtaining maximum contributions to overhead and earnings from non-basic services so that basic service might be provided at the lowest rates consistent with the revenue requirements of the Bell Companies. [17] They term this policy the "residual basis" of setting rates for basic exchange service: rates for non-basic service allegedly were set to obtain the maximum *contribution* (i.e., revenue minus estimated cost), then rates for basic service were set so as to make aggregate revenues exceed aggregate costs by a margin sufficient to provide the permitted rate of return on aggregate investment. To the extent that non-basic services provide revenues in excess of their true costs they contribute support to the basic services; economists refer to such support as *cross-subsidization* but within the telephone industry the custom has been to say that basic service is *benefitted*. (The terms have come to be used as approximately synonymous, but since the telephone companies originally defined the latter in terms of their accounting methods it is possible that a service could be "benefited" in the eyes of the telephone company without being "cross-subsidized" in the eyes of an economist.) Cross-subsidization is accomplished through price discrimination, and to ascertain its presence, direction, and magnitude involves the same problem discussed in the previous section, namely, the accurate identification of the costs of the various services. With reference to Table 8.1, cross-subsidization is absent from the price structure depicted in row one and present in the others, but without the information relating fee to marginal cost (column (e)) one would not know this. (The concept of cross-subsidization is really more complex than implied here; it is not quite correct to infer its presence from the presence of price discrimination, but the identification of either one does involve the problem of accurate cost estimation.) [18]

Any meaningful discussion of rate structure must therefore deal with questions of cost structure, so we cannot avoid treating this extremely difficult and contentious subject. One must bear in mind that rate setting in the U.S. is an adversary process in which representatives of the regulated firm do their best to further the firm's objectives and the regulators attempt to satisfy their constituents. It would be erroneous to assume that AT&T has no concern for the welfare of the average American (if only because AT&T's long-run profitability depends upon public opinion) but it would be imprudent to accept at face value statistical information provided by AT&T. The regulators are supposed to represent the long-run public interest, but they may be influenced by myopic public opinion, by well-meaning but misguided public interest lobbies, by special interest pressure groups, and/or they may simply lack the resources needed to reach sound decisions. One therefore should not blindly trust either party in the controversy between AT&T and the FCC described below, but the complexity of the issues also makes it imprudent to trust a disinterested bystander whose fulltime job is something other than telephone rate analysis.

Readers who reside in nations where telephone service is nationalized may wish to skim much of what follows, but some attention is worthwhile to appreciate the hassles involved in the American approach. More important is the point that the American adversary system does generate useful information which is available at zero cost to other nations, whether or not it is worth its cost to the U.S.

Only a tiny portion of the cost-structure debate can be sketched here; we will briefly treat issues associated with rates for long-distance service and rates for special terminal equipment. Recall that these are both classified as non-basic services which, according to telephone company representatives, have historically been priced above cost so as to generate a financial contribution to support the provision of basic exchange service.

The most important source of benefit to basic service allegedly has come from AT&T's Long Lines Department, generated by message toll service (MTS), lease of private lines, and (since 1960) from Wide Area Telephone Service (WATS). The logic and method of transferring funds is unique to the telephone industry, and some discussion of the technical detail is appropriate not only to elaborate the immediate point but also to illustrate the enormous difficulty of dealing with joint cost in telephone service.

Consider the various items of equipment between a subscriber and the telephone exchange (in technical terms, the station equipment, the drop and block, and the subscriber loop) which provides *access* to the exchange. This access equipment is used for both toll and local service, so its cost must somehow be allocated between the two. But use of the equipment imposes no perceptible wear and tear, and for the usual case of single-party line there is no problem with congestion. Consequently the cost of providing access is a lump-sum investment cost which is unaffected by the number of calls, or their duration, or their distance (i.e., it is not *traffic sensitive*). Consequently it is impossible to allocate access cost by the method illustrated in Figure 8.5 for exchange cost (which is traffic sensitive); instead access cost is like the lump-sum expense of the custodian in the example above. If the demand functions for local and long-distance service were known it would theoretically be possible to apportion access expense in accord with the Ramsey formula for quasi-optimal pricing (as in rows 2, 3, 4 of Table 8.1). In reality the requisite knowledge is lacking, and the formula for apportionment has been determined more or less arbitrarily in negotiations among AT&T, the FCC, and State Public Utility Commissioners. It is widely believed that the current formula results in cross-subsidization from interstate MTS and WATS to local exchange service and (with less consensus) to intrastate toll service.[19]

Because jurisdiction over telephone rates in the U.S. is divided between federal and state governments, the most visible apportionment of access costs is between interstate and intrastate service, and the corresponding transfer of revenue appears in the form of *separations payments* from AT&T's Long Lines Department to local telephone companies (independents as well as Bell subsidiaries). The amount transferred to a local company is calculated in proportion to the company's investment in access equipment, and the fraction of this cost which is recovered through separations payments has risen from 5% in 1951 to 19% in 1974. During this period technological advances were reducing other costs of long-distance service, and it seems that much of this cost reduction was not reflected in lower rates for interstate MTS but instead the additional revenue was used to hold down rates for local exchange service by increasing the fraction of access cost allocated to interstate service. [20]

In 1974 separations payments amounted to $3 billion; if such payments were eliminated MTS and WATS rates could be reduced by about 29%, but rates for basic exchange service would have to rise by about 24% ($3/$12.6, where the denominator is basic exchange revenue for 1974 in billions). In the face of the probability that such a change would offend a large number of telephone subscribers who are more sensitive about rates for local service than for long distance, it is understandable that telephone companies and regulators would resist a move in this direction. However, technological and legal developments have forced them to confront the possibility.

In World War II the U.S. Army constructed a radio microwave communications network in North Africa, and in the late 1940's AT&T used microwave to provide carriage of network television which required a bandwidth of 600 voice-grade circuits and would have strained the capacity of the cable network. In 1950 the FCC licensed TV networks to build microwave relay stations in remote areas which lacked common carrier facilities but warned that this permission was temporary pending provision of service by the telephone common carriers. Developmental licenses were also granted to railroads, pipelines and lumber camps in remote locations. [21] In 1959 after three years of investigation (one of the important questions being whether the available spectrum could accommodate numerous users) the FCC decided to make available frequencies for private microwave for business use. [22] The FCC stated that such licenses no longer require a demonstration of exceptional circumstances or lack of common carrier facilities, but that applications would not be approved if there was a reasonable likelihood that adverse economic effects would result. Expansion of private microwave was limited, however, because the FCC did not permit separate organizations to share a system and the telephone companies did not permit interconnection of private microwave facilities to the telephone network.

The FCC relaxed its prohibition against shared private microwave in 1966 and took another step toward increased competition in telephone communication in 1969 when it approved the application of Microwave Communications, Inc. to offer common-carrier service between Chicago and St. Louis and nine intermediate points.[23] Up to five subscribers could share a single channel for the carriage of voice, facsimile, and data; rates were significantly lower than those charged by the existing common carriers in part because MCI offered a channel whose bandwidth was only 2 Hz, instead of the customary 4 Hz. Moreover, the FCC recognized that MCI's success would depend on its ability to link its microwave system with its subscribers through local distribution lines leased from telephone companies, and stated that such interconnections would be ordered in the absence of a showing that they were not technically feasible. Over the next few years a series of judicial rulings required the telephone companies to provide interconnections with microwave common carriers thus enabling the latter to compete more effectively with AT&T's Long Lines Department in the market for private line interstate service. By 1975 the new specialized carriers had gross revenues of $35 million, compared to telephone company gross revenues of $1,613 million from private line service. This was only a small percentage of the market (2%), but the economic effect of the new competition was greater than might be inferred from this statistic because AT&T responded to the competitive pressure by lowering its own rates for bulk services.

The telephone common carriers had opposed authorization of microwave common carriage on the grounds that such service could succeed only by practicing *creamskimming*; i.e., they charged that the new specialized carriers would provide service at lower rates between those points where traffic was heavy and which therefore could be served at low cost, but would leave the low-density, high-cost markets to be served only by the telephone common carriers (mainly AT&T's Long Lines Department). But since the FCC had not been persuaded to exclude competition, AT&T responded to meet it by lowering its rates for heavy users of Long Lines service. In 1960 it introduced Wide Area Telephone

Service and TELPAK, which was a bulk lease service offering packages of 12, 24, 60, or 240 voice-grade circuits at rates 51% to 85% lower than existing rates for private lines.

Motorola (a manufacturer of microwave equipment) and Western Union argued before the FCC that the TELPAK rates were unduly discriminatory, that they were set below the cost of the services in order to drive out competition in the market for bulk communications, and that AT&T was using profit from its monopoly markets (basic exchange service and MTS) to cross-subsidize TELPAK. After investigation the FCC ruled in 1964 that 12 and 24 circuit packages were unduly discriminatory but allowed the rates to stand for 60 and 240 circuit packages pending additional cost analysis. An investigation of comparative rates of return on Bell System's interstate services (commonly referred to as the Seven-Way Cost Study) showed that the rate of return on investment in 1963-4 was 10% for MTS and WATS, 4.7% for voice-grade private line, 2.9% for teletypewriter exchange service, 1.4% for telegraph-grade private line, and 1.3% for TELPAK.[24] It thus appeared that at that time AT&T's rate structure benefited those who were heaviest users (i.e., subscribers to TELPAK and private line service) in comparison to moderately heavy users (WATS) or light users (MTS).

AT&T has countered that the Seven-Way Cost Study erred in apportioning costs on a fully distributed cost (FDC) basis, and argued that a more appropriate method would use long-run incremental cost (LRIC). The dispute over the proper method of cost allocation has persisted; in 1976 the FCC rejected AT&T's proposed LRIC method and ruled that future rate filings must be justified on a fully distributed cost basis. The Commission also ruled that existing TELPAK rates were unjustly discriminatory and that some other private lines rates were unlawfully low, indicative of cross-subsidization.[25]

Discussion of the relative merits of LRIC and FDC accounting methods is beyond my scope; I will remark only that sound LRIC methods are superior for the purpose of managerial decisions, but that in an adversary proceeding a regulatory agency might opt for fully distributed historical costs because they are more readily verified by outside audit. For present purposes, the essence of the FCC's findings is that according to FCC accounting methods the evidence did not show that rates were structured in such a way as to cross-subsidize basic exchange service by means of contributions from private line service or TELPAK. The extent of cross-subsidization to basic exchange service from WATS and MTS continues to be a debatable question, but the FCC has made it clear that they do not intend to allow the new common carriers to compete for these services; competition will be encouraged only in private line service and in the judgment of the FCC this will have only a small impact on WATS and MTS revenues. (Note, however, that FCC decisions are subject to judicial review so it is possible that a court might order more competition than intended by the FCC.) The gist of AT&T's contention is that LRIC accounting methods show that their rates for bulk services are more than compensatory, that the FCC ruling prevents AT&T from competing effectively with the new carriers in the market for bulk services, and that because the new intercity carriers escape the separations burden they will be able to capture an increasing portion of intercity traffic thus diverting revenue from AT&T's WATS and MTS. This erosion of the contribution from intercity service will put upward pressure on rates for basic local service, unless a change is made so as to require private line service (whether provided by AT&T or by the new special carriers) to support basic exchange service in a manner similar to that which now applies to WATS and MTS. [26]

Similar developments occurred in the market for special terminal equipment. In a seemingly trivial legal case in 1956 a U.S. Circuit Court of Appeals overturned an FCC decision and ruled that the Bell Companies could not prevent a subscriber from attaching a Hush-A-Phone to his telephone; this device was a purely mechanical cuplike attachment which snapped onto the handset. [27] In 1968 the Commission (following guidelines set by the Court in the Hush-A-Phone case) ruled that the Bell Companies could not prohibit a subscriber from using his telephone in conjunction with Carterfone, a device which acoustically coupled the handset to a two-way radio thus permitting conversation between ordinary telephones and units of private mobile radio units. The effect of this decision was to end the Bell System's longstanding and adamant prohibition of "foreign attachments" to their network, and it became permissible for business customers to provide and attach to the phone system specialized terminal equipment such as computer terminals, facsimile machines, and teletypewriters if a protective interface device was used to avoid possible harm to the telephone network. The telephone companies thus became subject to competition in what is now known as the *interconnect* market. In 1978 an FCC ruling made it feasible for subscribers to purchase from independent suppliers such items as private branch exchanges (PBX), key telephone systems (KTS), answering devices, and have them interconnected directly to the telephone network provided that they have been certified for quality by the FCC.

Representatives of the telephone companies and of the State Utilities Commissioners (NARUC) contend that competition in the interconnect market will adversely affect local exchange rates in a twofold manner. First, as long as the telephone companies had a monopoly in this market they were able to lease special terminal equipment at rates which were high enough to generate net revenue which was used to support basic service; in the face of competition these rates must be lowered and/or the market will be lost, and in either case the contribution from these business services will be lost. Second, to the extent that terminal equipment is no longer owned by the local telephone companies, their share of separations payments will be diminished accordingly (recall the discussion above of the formula by which revenues from Bell's Long Lines Department are distributed to local telephone companies in proportion to their investment in access equipment), and basic exchange rates will have to rise to offset the loss of contribution from MTS and WATS. AT&T claims that if all of the services other than residential basic exchange were repriced to eliminate their contribution to access cost the average monthly residential exchange rate would have to be increased from $9.00 to $16.15, an increase of 79%.[28] In response, the FCC has stated:

> . . . we find the studies of contribution loss submitted
> by the telephone industry without merit. Among other
> things, the studies submitted generally utilize inappro-
> priate cost and allocation methodologies, thereby failing

to show whether there is, in fact, any direct contribution to basic local telephone service including residential exchange to be lost due to interconnect service. Bell's failure to allocate common and joint costs among its various services including business vertical services such as PBX and KTS (thereby understating vertical service costs and overstating contribution, if any), and its assumption that all such costs are to be borne by the local telephone ratepayer, is perhaps the best example of an inappropriate cost and allocation methodology ...

Indeed, it is likely that terminal equipment is a recipient of subsidy from basic local service rather than a donor. Under such circumstances, loss of terminal equipment business to interconnect competition could possibly result in rate reductions for local telephone service users rather than rate increases.[29]

While it is not clear to what extent the structure of telephone rates in the U.S. has utilized price discrimination or cross-subsidization, it is evident that technological and legal changes have substantially reduced the ability of telephone companies to offset losses on some services by overpricing others. Consequently they are now forced to move toward a rate structure which more rigorously reflects the cost of service.

8.6 TOWARD USAGE-SENSITIVE PRICING

The events described in the preceding section have sharply limited the ability of the American telephone system to collect revenues in excess of costs on some services in order to support deficits on other services. Additional pressures toward changing the rate structure have been generated by the combination of inflation, consumerism, and regulatory lag. Inflation forces telephone companies to seek frequent and substantial rate increases from regulatory agencies, and intervenors representing various constituencies have used the attendant hearings to question many aspects of rate structure which were formerly unquestioned. As we have observed in the preceding section, U.S. regulatory proceedings move slowly in any case; when disputes prolong the process while wages and material costs are rising rapidly, the upward adjustment of telephone rates will fall behind the rise in costs.[30] Consequently the telephone companies are being forced to restructure their rates so as to reflect the cost of service.

The FCC has made it clear that they favor continuance of customary procedures for rate averaging within various classes of service, so there is no expectation that cost-of-service pricing will be pushed to the extreme. (E.g., while the actual cost of an MTS call between Fargo and Pocatello exceeds the cost of an MTS call between Denver and Chicago, there is no expectation that the former will be priced higher.) The FCC also indicates that it intends to preserve AT&T's monopoly on interstate message toll service and WATS, so the possibility of transferring revenue from these to support basic exchange service will continue albeit possibly with some noticeable diminution.[31] Thus the term marginal-cost pricing would be too strong to use as a description of the rate structure which is envisioned; what is happening is a movement in the general direction of marginal-cost pricing but in view of various reservations it is appropriate and customary to describe this as a movement toward usage-sensitive pricing. The reforms to be described are in various stages ranging from study to implementation in various localities, and since practices are changing rapidly no attempt will be made to specify which particular reforms have been adopted in which specific locales.

Generally speaking, the reforms of telephone rate structure involve "unbundling" of rates — a move away from the practice of charging a lump-sum monthly fee which covered many different elements of service, toward a more elaborate system of separate prices for different aspects of service. For example, local directory assistance was provided without extra charge; 80% of the requests for information came from 20% of the subscribers, but the cost of this service was imposed uniformly on all subscribers adding about 30¢ per month to the average bill.[32] Many telephone companies now impose a separate charge for local directory assistance; long-distance directory assistance continues to be unpriced and is probably used heavily by businesses (it is said that a person's credit rating may be estimated according to whether or not he/she is a telephone subscriber). Service connection charges (hook-up or installation fees) were not differentiated adequately according to specific conditions, and on the average were less than the cost. Many companies now calculate such charges by summing the costs of five separate tasks.[33]

With the development and installation of electronic switching systems it is becoming feasible to measure not just the number of local calls but also their duration, timing, and approximate distance. It is to be expected that telephone rate structures will make increasing use of this information. To the extent that call charges are imposed it becomes possible to reduce the monthly rental. Some telephone companies now offer the option of a special rate which is advantageous for subscribers who make few outgoing calls — in Denver for example, an "access" charge of $4 per month with a charge of 7¢ per outgoing call. (In this case the "access" charge is intended to cover the expense of station equipment, inside wiring, drop and loop, but one may expect further unbundling to separate the charge for station equipment from the charge for pure access to the network, so as to facilitate competition in the interconnect market while allowing telephone companies to recover their expense for their unique service.)

Telephone company officials, regulatory agencies, and independent economists agree on the desirability of restructuring telephone rates so that they will more closely reflect actual costs of service, but there are limits set by public opinion. As an extreme example, note that a cost is imposed on the telephone system when an attempt to call succeeds only in reaching a busy signal, but the public would probably rebel against any proposal to charge for such attempts. Less dramatically, public resistance to change in the customary rate structure is to be expected. In order to achieve reform without too much disturbance it seems probable that telephone companies and regulators will gradually introduce usage-sensitive pricing schemes as options to the traditional unbundled rate. Allowing subscribers to choose among various pricing systems is not only politically convenient, it is also a theoretically justifiable method of overcoming the pricing dilemma in decreasing cost activities.[34]

As a byproduct of rate unbundling there will be a substantial increase in information about demand functions and cost functions for telephone service. Only when consumers are forced to pay for a product or service does one obtain reliable information about their demand for it, and the greater the variety of pricing schemes the greater the number of valid data points. Similarly, rate unbundling forces the suppliers of telephone service to estimate more accurately the costs of various services, and the introduction of competition into some of the telephone markets increases assurance that the true costs will be revealed. As we have seen, much of the controversy over telephone rates has persisted because analysts have had to conjecture about the relevant demand and cost functions; the increase in information which will emerge from rate reform in the U.S. will be of great value to rate makers in other nations whether their telephone companies are priviately or publicly owned, and will also shed light on the more fundamental question concerning the extent to which telephone service is indeed a natural monopoly.

In summary, current reform of the telephone rate structure aims at placing the burden of costs on those who cause the burden in order to improve efficiency and equity. Within the limits of practicality and the revenue requirement, the objective is to encourage use when marginal utility exceeds marginal cost and to discourage use when marginal utility falls short of marginal cost. Certain physical characteristics of telephone service together with the social, political, and judicial characteristics of American regulatory procedures make the process of reform difficult and slow, but the movement toward usage-sensitive pricing is well under way.

References

[1]   Baumol, W.J. and D.F. Bradford, "Optimal Departures from Marginal Cost Pricing, " *American Economic Review*, Vol. 60, June 1970, pages 265-83.

[2] Office of Telecommunications Policy, Executive Office of the President, *Competition in Telecommunications — The Telephone Industry Bill* (Washington, D.C.: June, 1976) page 17.

[3] Horn, C.E., "Factors Affecting Telephone Pricing," in J.T. Wenders (ed.), *Pricing in Regulated Industries: Theory and Applications* (Denver: The Mountain States Tel. & Tel. Co., 1977), page 6.

[4] Squire, L., "Some Aspects of Optimal Pricing for Telecommunications," *Bell Journal of Economics and Management Science,* Vol. 4, Aug. 1973, pages 515-25.

[5] Littlechild, S.C., "The Role of Consumption Externalities in the Pricing of Telephone Service," in J.T. Wenders (ed.), *op. cit.,* pages 44-5.

[6] Clemens, Eli W., *Economics and Public Utilities* (New York: Appleton-Century-Crofts, Inc., 1950), page 328.

[7] Mandanis, G.P. *et al., Domestic Communications and Public Policy* (Systems Applications Inc., 1973) as reported in J.H. Alleman, *The Pricing of Local Telephone Service* (Washington, D.C.: U.S. Dept. of Commerce, Office of Telecommunications, April 1977), page 114.

[8] An example appears in Kahn, Alfred E., *The Economics of Regulation* (New York: John Wiley and Sons, Inc., 1971), Vol. I, page 125.

[9] Hotelling, H., "The General Welfare in Relation to Problems of Taxation and of Railway and Utility Rates," *Econometrica,* Vol. 6, July 1938, pages 242-69.

[10] Dupuit, J., "On the Measurement of the Utility of Public Works," English translation by R.H. Barback in *International Economic Papers*, Vol. 2, page 89; original in *Annales des Ponts et Chaussées*, 2d series, Vol. 8, 1844.

[11] Hadley, Arthur T., *Railroad Transportation* (New York: G.P. Putnam's Sons, 1886), page 76.

[12] Samuelson, Paul A., *Foundations of Economic Analysis* (Cambridge: Harvard University Press, 1955), page 242.

[13] Lipinski, Jan, "The Correct Relation between Prices of Producer Goods and Wage Costs in a Socialist Economy," in D.C. Hague (ed.), *Price Formation in Various Economies* (New York: St. Martin's Press, 1967), page 124.

[14] Kahn, *op. cit.*, pages 89-109; Baumol, William J., *Economic Theory and Operations Analysis* (Englewood Cliffs, N.J.: Prentice-Hall, Inc., 1977) pages 173-5.

[15] Baumol and Bradford, *op. cit.*

[16] Alleman, *op. cit.*, pages 11-12.

[17] Testimony of Frank J. Alessio, New Mexico S.C.C. Docket No. 673, Hearing on Remand, Mountain Bell Exhibit B, *In the Matter of Rates and Charges of the Mountain States Telephone and Telegraph Co.*; Corman, W.F., "The Pricing of Telephone Service," *Telephone Engineer and Management,* Oct. 1, 1971, pages 93-110.

[18] Faulhaber, G.R., "Cross-Subsidization: Pricing in Public Enterprises," *American Economic Review*, Vol. 65, 1975, pages 966-77.

[19] Corman, *op. cit.*; Littlechild, S.C. and J.J. Rousseau, "Pricing Policy of a U.S. Telephone Company," *Journal of Public Economics*, Vol. 4, 1975, pages 35-56. For a contrary view, see Richard Gabel, *Development of Separations Principles in the Telephone Industry* (East Lansing: Institute of Public Utilities, Michigan State University, 1967).

[20] Horn, *op. cit.*, pages 19-26.

[21] Irwin, M.R., "The Communications Industry," in Walter Adams (ed.), *The Structure of American Industry* (New York: The Macmillan Co., 1971), pages 391-3; Kahn, *op. cit.*, Vol. II, pages 129-36.

[22] *Allocation of Frequencies in the Bands Above 890 Mc*, 27 FCC 359 (1959), 29 FCC 825 (1960).

[23] *In Re: Applications of Microwave Communications, Inc.*, 18 FCC 953 (1969).

[24] Reported in Irwin, *op. cit.*, page 394.

[25] *In Re: American Telephone and Telegraph Co. Long Lines Department*, Docket No. 18128, FCC Report No. 12343, Sept. 23, 1976.

[26] Horn, *op. cit.*, page 24.

[27] Kahn, *op, cit.*, Vol. II, pages 140-5.

[28] American Telephone & Telegraph Co., *Abstract of 1975 Residential Cost Study* (New York: 1976).

[29] *In Re: Economic Implications and Interrelationships Arising From Policies and Practices Relating to Customer Interconnection, Jurisdictional Separations and Rate Structures*, First Report, FCC Docket No. 20003, Sept. 23, 1976, pages 123-5.

[30] Horn, *op. cit.*, pages 14-15, 32-34.

[31] *In Re: Economic Implications . . .* , page 160.

[32] Alleman, *op. cit.*, page 103.

[33] Horn, *op. cit.*, page 14.

[34] Faulhaber, G.R. and J.C. Panzar, *Optimal Two Part Tariffs and Self Selection* (Holmdel, N.J.: Bell Laboratories, 1976).

Chapter 9

INTERNATIONAL ASPECTS OF
TELECOMMUNICATIONS OPERATIONS

A. Terrence Easton
Department of Telecommunication Management
Golden Gate University
San Francisco, California

9.1   HISTORY OF INTERNATIONAL COMMUNICATIONS

The history of modern international communication is the history of the modern European State. Today, in the late 1970's it is possible for a subscriber in North America to directly dial over 60 countries using his telephone set and be connected with over one million telex machines in 100 countries via his automatic telex machine.

Statistical data compiled by ATT Long Lines Department shows that as of January 1, 1976 there were 379,524,000 telephone sets in operation world wide, with the United States possessing 155,170,288 telephones accounting for approximately 68 telephones per 100 people, and 40% of the world's telephones. 99.8% of these telephones were connected to automatic dial exchanges. On the average, telecommunications growth world-wide is expanding at 15% per year, with the heavily industrialized North American community growing at the rate of 8%. (See Figure 9.1)

The revolution in international telecommunications has dramatically reduced the psychological size of the globe, and has enabled the less developed countries to be plugged into the industrialized world and its technological capabilities. For the past 125 years, telecommunicaitons has been the leading-edge tool of the industrialists.

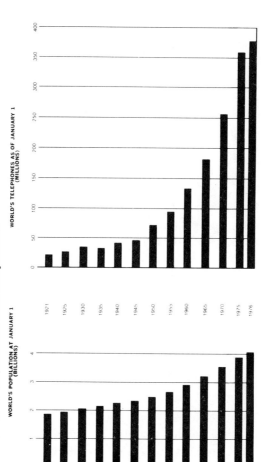

TOTAL NUMBER OF TELEPHONES IN SERVICE

1975	1974	1973	1972	1971	1966	CONTINENT
155,883,000	149,432,000	142,102,000	134,939,000	129,518,000	100,779,000	NORTH AMERICA
4,223,000	3,787,000	3,410,000	3,069,000	2,776,000	1,641,000	MIDDLE AMERICA
8,106,000	7,287,000	6,776,000	6,419,000	6,137,000	4,242,000	SOUTH AMERICA
124,103,000	115,022,000	106,173,000	97,986,000	90,301,000	62,432,000	EUROPE
4,291,000	3,985,000	3,733,000	3,531,000	3,342,000	2,474,000	AFRICA
54,479,000	49,270,000	44,244,000	39,216,000	34,704,000	19,261,000	? ASIA
7,322,000	6,818,000	6,463,000	6,139,000	5,879,000	4,271,000	OCEANIA
358,407,000	335,601,000	312,901,000	291,299,000	272,657,000	195,100,000	WORLD

Growth of population and telephones

WORLD'S POPULATION AT JANUARY 1 (BILLIONS)

WORLD'S TELEPHONES AS OF JANUARY 1 (MILLIONS)

Figure 9.1 Worldwide Telephone Expansion

The history of telegraphy pre-dates the modern era. The Romans had developed sophisticated visual communications systems 100 years before the birth of Christ. Recent historians have suggested that the French Revolution survived and was spared her surrounding enemies' wrath by the lack of communications among England, Spain, Germany and Italy. Within France, Claude Chappe had constructed a network of visual telegraphs which provided rapid dissemination of intelligence. By the 1850's (when the electric telegraph overtook the archaic system) over 550 towers were in operation using complicated semiphore arms and signaling techniques, spanning over 5,000 kilometers. Telegraph Hill in San Francisco is living testimony to the visual telegraph systems in operations in the United States in the last century.

Invented in the late 18th Century, the development of the modern electric telegraph did not proceed until after 1838, the year that the British scientist Charles Wheatstone constructed a commercial telegraph link between West Dreyton and Paddington, England along 13 miles of railroad right-of-way. By May 24, 1844, a commercial telegraph circuit between Baltimore and Washington was erected, based upon Samuel B. Morse's 1835 telegraph sender and key design, and the Western Union Telegraph Company was formed to literally carry the Union to the West.

Electronic communications connecting cities and countries together was a reality. By the 1880's even Lower Egypt had been criss-crossed with strands of telegrapher's wire. (See Figure 9.2) Trading and development companies sprang up in Europe, Africa, Asia and South America. The first underseas cable connecting two countries together was laid in September 1851 between France and England. By July 17, 1866, a successful transatlantic cable was cut into operation between the U.S. and Britain. On January 7, 1927, the first commercial radio-telephone service spanning the North Atlantic was inaugurated between New York City and London. The race was on.

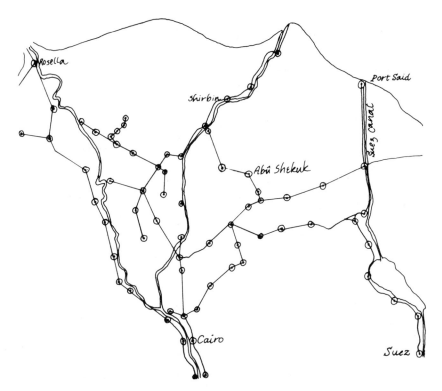

Figure 9.2   Telegraph Circuits in Egypt as of 1882

9.2 INTERNATIONAL ORGANIZATIONS

9.2.1 The International Telecommunications Union (ITU)

As nation after nation began to construct telegraph and telephone facilities, and agressive competition from the cable companies produced incompatible systems, it was apparent that an international engineering and standards organization was needed. Thus the International Telecommunications Union (ITU) was founded in 1863 to promote cooperation and systems compatibility. Nine decades later the ITU became a founding agency of the United Nations, and today comprises over 150 member nations. (See Figure 9.3)

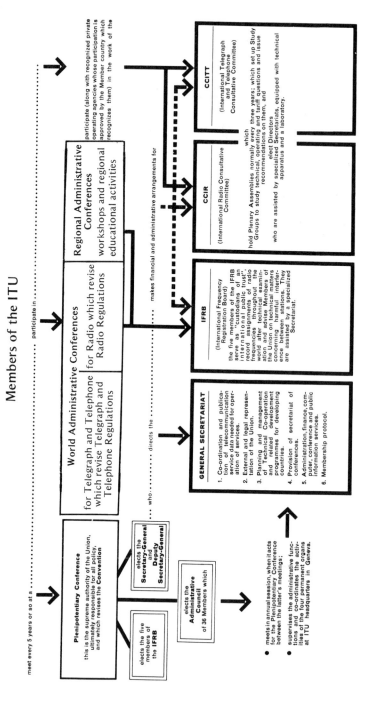

Figure 9.3   The International Telecommunications Union (ITU)

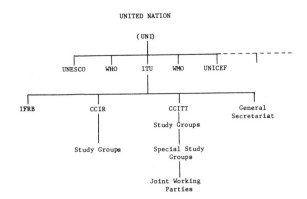

Figure 9.4   The CCITT within the ITU

9.2.2   Consultative Committee for International
Telegraph and Telephone (CCITT)

In Paris during its 1925 conference, the ITU concluded that the expanding nature of international communications would require a standing committee to regularly meet as an adjunct to the periodic ITU conferences. The decision was made to establish the "Comité Consultatif Internationale des Communications Téléphoniques à Grande Distance (CCIF) to study . . . the . . . standards regulating technical and operating questions for international and long distance telephony." A similar committee was also established to coordinate the international telegraph activities. This committee was named Consultation Committee for International Telegraph. The two standing organizations were merged in 1956 to form the present CCITT, an international body comprised of the national government bodies responsible for the regulation and provision of telecommunications services. Further details are given in Chapter 1.

The CCITT's other partners within the International Telecommunications Union include the CCIR (Radio Committee), the IFRB (Frequency Registration Board), and the General Secretariat. (See Figure 9.4). Within the CCITT, specialized study groups and joint working parties hold technical and administration conferences to develop operating practices, pricing and policy arrangements, and technical specifications. There are 14 joint working parties and 19 study groups presently meeting world-wide. (See Figure 9.5). Further details are given in Chapter 1.

DESIGNATION	STUDY GROUPS	CHAIRMAN
I	Telegraph operation and quality of service	Mr. K. Freiburghaus (Switzerland)
II	Telephone operation and quality of service	Mr. J. Biot (Belgium)
III	General tariff principles	Mr. M. Kojima (Japan)
IV	Transmission maintenance of international lines, circuits and chains of circuits; maintenance of automatic and semi-automatic networks	Mr. J. Kiil (Denmark)
V	Protection against dangers and disturbances of electromagnetic origin	Mr. G. Gratta (Italy)
VI	Protection and specifications of cable sheaths & Poles	Mr. J. Pritchett (UK)
VII	New net works for data transmission	Mr. V.C. MacDonald (Canada)
VIII	Telegraph and terminal equipment, local connecting lines	Mr. W. Staudinger (Germany, Fed.Rep. of)
IX	Telegraph transmission quality; specification of equipment and rules for the maintenance of telegraph channels	Mr. R. Brown (Australia)
X	Telegraph switching	Mr. E.E. Daniels (UK)
XI	Telephone switching and signaling	Mr. J. Ryan (USA)
XII	Telephone transmission performance and local telephone networks	Mr. F.T. Andrews (USA)
XIV	Facsimile telegraph transmission and equipment	Mr. M. Blanc (France)
XV	Transmission systems	Mr. D. Gagliardi (Italy)
XVI	Telephone circuits	Mr. S. Munday (UK)
XVII	Data transmission	Mr. V.N. Vaughan (USA)
XVIII	Digital networks	Mr. T. Irmer (Germany, Fed,Rep.of)
CABD	Circuit noise and availability	Mr. A.P. Bolle (Netherlands)
CMTT	Television and sound transmission	Mr. Y. Angel (France)
CMV	Definitions and symbols	Mr. R. Villeneuve (France)

DESIGNATION	WORKING COMMITTEES	CHAIRMAN
World plan	Worldwide telecommunication plan	Mr. M. Ghazal (Lebanon)
Plan Africa	Telecommunication plan for Africa	Mr. L. Dia (Senegal)
Plan Latin America	Telecommunication plan for Latin America	Mr. A.C. Nunez-a (Mexico)
Plan Asia	Telecommunication plan for Asia	Mr. A. Zaidan (Saudi Ariabia)
Plan Europe	Telecommunication plan for Europe and the Mediterranean Basin	Mr. L. Terol Miller (Spain)
GM/LTG	XI, XIV, XV, XVII, CMBD	Mr. L. Guillet (France)
GM/SMM	I, II, III	Mr. K.J.M. Jaspers (Netherlands)
GM/UMI	III	Mr. D. Tudge (UK)
GAS 3	Economic and technical aspects of the choice of transmission systems	Mr. G. Wallenstein (USA)
GAS 5	Economic conditions and telecommunication development	Mr. H. Longequeue (France)
GAS 6	Economic and technical aspects of the choice of switching systems	Mr. L. Ackzell (Sweden)
GR TAF	Tariffs (Africa)	Mr. A.D. Aithnard (Togo)
GR TAL	Tariffs (Latin America)	Mr. A.C.G. Ribas(Brazil)
GR TAS	Tariffs (Asia and Oceania)	Mr. F. Magallon (Philippines)
GR TEUREM	Tariffs (Europe and the Mediterranean Basin)	Mr. W. Jost (Switzerland)

FIGURE #5

Figure 9.5   CCITT Standing Committees

The Global System of Communications Satellites and Earth Stations

Earth Stations Served by Atlantic Ocean Satellites

Algeria: Lakhdaria 1
Angola: Cacuaco
Argentina: Balcarce 1 and 2
Barbados: Barbados
Balgium: Lessive
Brazil: Tangua 1 and 2,
  Cuiaba,* Manaus,*
  Boa Vista*
Cameroon: Zamengoe
Canada: Mill Village 1 and 2
Chile: Longovilo
Colombia: Choconta
Dominican Republic: Cambita
Ecuador: Quito
Egypt: Cairo
France: Pleumeur-Bodou 2, 3
  and 4
  French Guiana, Trou-Biran
  Martinique, Trois Ilets
Gabon: Nkoltang
Germany: Raisting 2 and 3
Greece: Thermopylae 2
Haiti: J-C Duvalier
Iran: Asadabad 1
Iraq: Dujail 2
Israel: Emeq Ha'ela
Italy: Fucino 1
Ivory Coast: Abidjan
Jamaica: Prospect Pen
Jordan: Baqa 1
Liberia: Sinkor
Madagascar: Philibert
  Tsiranana
Mali: Sullymanbougou 1

Mexico: Tulancingo 1
Morocco: Sehouls
Mozambique: Boane
Netherlands: Burum
Nicaragua: Managua
Nigeria: Lanlate 2
  Benin,* Enugu,* Ibadan,*
  Jos,* Kaduna,* Lagos*
Norway: Eik,* Ekofisk*
Panama: Utibe
Peru: Lurin
Portugal: Sintra
Romania: Cheia
Saudi Arabia: Taif
Senegal: Gandoul
South Africa: Pretoria 1
Spain: Buitrago 1 and 3
  Grand Canary: Aguimes
Sudan: Umm Haraz
Sweden: Tanum†
Switzerland: Leuk
Trinidad and Tobago:
  Matura Point
United Kingdom: Goonhilly 2
  and 3, Ascension Island
United States: Andover, Me.,
  2 and 3; Etam, W. Va., 1
  and 2; Cayey, P.R.
U.S.S.R.: Moscow (hotline),
  Lvov
Venezuela: Camatagua
Yugoslavia: Jugoslavija
Zaire: Nsele

Earth Stations Served by Indian Ocean Satellites

Algeria: Adrar,* Bechar,*
  Djenet,* Ghardaia,*
  In-Salah,* Lakhdaria 2,*
  Ouargla,* Tamanrasset,*
  Timimoun,* Tindouf *
Australia: Ceduna
Bahrain: Ras Abu Jarjur
Bangladesh: Betbunia
China: Peking 2
China: Taipei 2
East Africa: Longonot‡
France: Pleumeur-Bodou 1
  and 5,* Reunion*
Germany: Raisting 1
Greece: Thermopylae 1
India: Vikram, Dehra Dun
Indonesia: Djatiluhur 1
Iran: Asadabad 2
Iraq: Dujail 1
Italy: Fucino 2
Japan: Yamaguchi
Kuwait: Umm Al-Aish 1
Lebanon: Arbaniyeh 1
Malawi: Kanjedza
Malaysia: Kuantan 1
Mauritius: Cassis
Nigeria: Lanlate 1
Oman: Al Hajar
Pakistan: Deh Mandro
Philippines: Tanay 2
Qatar: Doha
Saudi Arabia: Riyadh 1,
  Abha,* Buraida*
Seychelles: Bon Espoir

Singapore: Sentosa 1
South Africa: Pretoria 2
Spain: Buitrago 2
Sri Lanka: Padukka
Thailand: Si Racha 2
United Arab Emirates: Dubai
United Kingdom: Goonhilly 1,
  Hong Kong 2
Yemen: Sanaa
Zambia: Mwembeshi

Earth Stations Served by Pacific Ocean Satellites

Australia: Carnarvon 2, Moree
Canada: Lake Cowichan
China: Peking 1, Shanghai
China: Taipei 1
Fiji Isands: Suva
France:
  Naw Caledonia (I'lle Nou)
Japan: Ibaraki 2 and 3
Korea: Kum San 1
Malaysia: Kuantan 3,* Kota
  Kinabalu*
Nauru, Rep. of: Nauru Island
New Zealand: Warkworth
Philippines: Tanay 1
Singapore: Sentosa 2
Thailand: Si Racha 1
United Kingdom: Hong Kong 1
United States:
  Brewster, Wash.;
  Jamesburg, Cal.; Paumalu,
  Hawaii, 1 and 2;
  Pulantat, Guam

* Domestic
† Tanum earth station is a joint undertaking of Denmark, Finland,
  Norway and Sweden
‡ Serves Kenya, Uganda and Tanzania

Earth stations in this section include those active in the global
system as of December 31, 1976. Where there is more than one
antenna at an earth station, the number of the antenna operating in
the region appears.

Figure 9.6   The INTELSAT System

9.2.3   International Telecommunications Satellite
Organization (INTELSAT)

By 1960, President John F. Kennedy had determined that a
global satellite communications network spanning oceans and
continents was feasible and desirable. By act of Congress, a private American corporation, the U.S. Communications Satellite
Corporation, was established to construct and operate a new global public communications satellite system. An international regulatory agency, INTELSAT, was established in 1964 to oversee
the operations of the multi-national network with the initial support of the eleven founding country members. Today, INTELSAT
has over 90 members, with the U.S. holding the largest investment
share (37.5%); COMSAT, the New York Stock Exchange Corporation responsible for technical and administrative operations, designs the INTELSAT-series satellites (which are built by Hughes
Aircraft and launched by NASA) and owns and operates a number of earth stations. Most overseas earth stations are owned and
operated by the local PTT (Postal Telepnone and Telegraph) organizations. (See Figure 9.6). Authorization for COMSAT activities within the United States is governed by the Communications
Satellite Act, but the Federal Communications Commission (FCC)
has the authority to issue earth station licenses and direct COMSAT
to activate or suspend certain channel links. Thus, through FCC
dictate, international communications to and from the United
States are routed about equally between COMSAT facilities and
the modern trans-oceanic cables owned and operated by the international communications consortia.

The INTELSAT network consists of multi-transponder satellites
positioned over the Atlantic, Pacific, and Indian Oceans, and dozens of earth stations scattered throughout the world. The system
provides for direct connection between any two operating members and can handle voice, data, and television transmissions. Satellite communications poses some interesting technical problems,
however. Because the satellites are positioned at geo-stationary
synchronous-altitude orbits approximately 22,300 miles above
the equator, the "turn around" time for a signal to transit through
the satellite to a distant station and back is approximately 700
milliseconds. This means that certain precautions must be taken
in data communications network design to minimize the problems

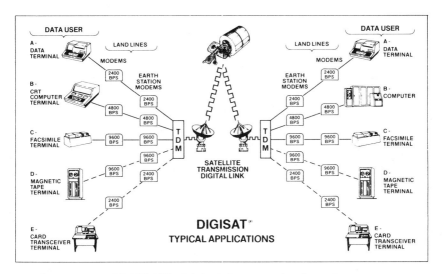

Figure 9.7   COMSAT's Digital Communications System

caused by such transmission delays; in voice communications, careful attention must be placed on the design of echo supressors to minimize the difficulties of talking over such channels. The satellite carriers (including AMSAT and other domestic carriers) have devised ingenious techniques to minimize these problems. AMSAT (American Satellite Corporation) has developed a "delay compensation unit" to reduce the impact of data delays. ATT uses a "split return" concept to cut delays in half by using a satellite link for one direction (say east-to-west communications), and a land line circuit for the opposite (west-to-east) direction. But these techniques need careful circuit adjustment to be successful. Ultimately the problems of voice echo will be eliminated with the retrofit conversion to all-digital transmissions. Then annoying echoes can be actually eliminated through new "echo canceler" equipment instead of merely being blocked via today's "echo suppresser" systems which were designed for land-line distances, not satellite hops. Recent advances by COMSAT have enabled an all-digital data communications system to be operated through the existing system. (See Figure 9.7)

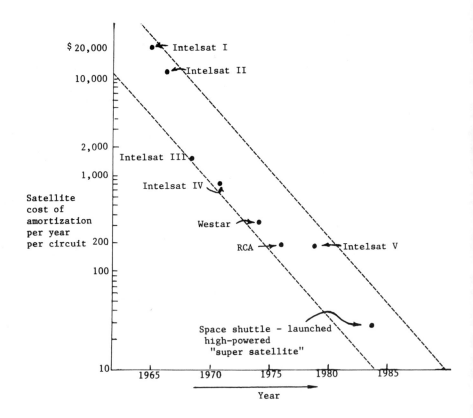

Figure 9.8   Satellite Communications Costs Collapsing

The reduction in cost of global communications circuits and metered-use rates has primarily been made possible through the development of the INTELSAT network. New INTELSAT V Satellites planned for launch during the 1980's should reduce the amortized per circuit cost to well under $100. Of course, as there is little or no relationship between actual cost and selling price in the international communications business, and earth station costs have not been included in this figure, it is unlikely that the end user will see such dramatic savings; but circuit costs in the 1980's should continue to decrease significantly, following the downward trend of the past decade. (See Figure 9.8)

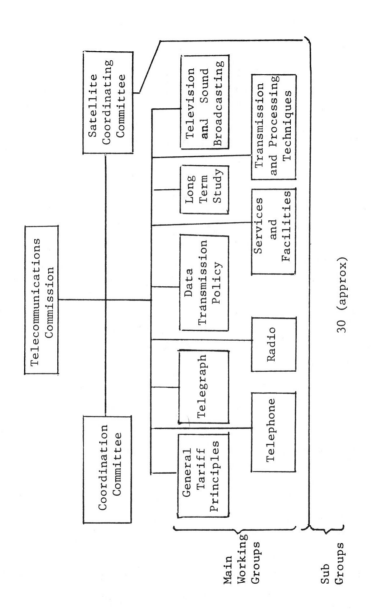

Figure 9.9   The CEPT and its Structure

9.2.4   European Conference of Postal and Telecommunications Administrations (CEPT)

In 1959, a number of European PTT administrations created the CEPT, a European version of the ITU. Consisting of 26 PTTs, the CEPT was encouraged by the success of the European Economic Community ("Common Market") in resolving intra-European differences and formulating joint administrative and technical programs. While the ITU meets every four years, the CEPT holds its plenary assembly every two years to discuss services, tariffs, techniques and policy. Unlike the ITU, there is no permanent CEPT headquarters; each member PTT assigns an internal delegation to coordinate the activities of its country's communications function. The CEPT has been successful at creating a unified European standard in many areas, and is considered a more militant organization than the ITU. (See Figure 9.9) The development of the Europacket network is a direct consequence of the CEPT "Working Committee on European Data Communications." However, neither the CEPT nor ITU have the force of regulatory authority internationally that the FCC maintains domestically in the United States. Indeed, because of its powerful position within the marketplace, the U.S. Federal Communications Commission significantly influences telecommunications policy world-wide through the application of its jurisdictional authority over American common carriers operating from the United States.

9.3   U.S. INTERNATIONAL COMMUNICATIONS FACILITIES

Communications between the U.S. mainland and international (overseas) points currently is provided within the U.S. by six common carrier organizations.: ATT Long Lines Department, RCA Global Communications Inc., ITT World Communications, Western Union International Inc., TRT Telecommunications Corporation, and The French Telegraph Cable Company.

By definition of the 1934 Communications Act, Canada and Mexico are considered domestic points, and Alaska and Hawaii are defined to be international locations. Communications to and from Canadian points is also provided by the Western Union Telegraph

Company (not related to Western Union International) which interconnects with the CN/CP Telecommunications Group in Canada and also with the Mexican PTT, TELMEX in the South.

The U.S. Communications Satellite Corporation (COMSAT) does not sell its international satellite communications services directly to the end-user but rather acts as a "common carrier's carrier" — selling its satellite circuits to the other U.S. common carriers. RCA Alascom Inc., affiliated with RCA Global Communications, provides toll telephone service from the "lower 48" to Alaska using the separate RCA Satcom satellite system.

The U.S. specialized common carriers, notably TELENET and TYMNET extend their U.S. facilities overseas through contractual arrangements with the IRC's, the international record carriers: ITT, RCA, WUI, TRT, and FTCC. Similar arrangements made with ATT (acting in its role as domestic carrier of U.S. communications to the Canada border) have extended Telenet and Tymnet network access throughout Canada as well.

9.3.1   ATT Long Lines

The Long Lines Department is an ATT division responsible for the construction and operation of the Bell System's interstate "backbone" microwave and coaxial cable network in addition to the Message Telephone Service (MTS) toll network of switching systems which interconnect the Bell System and independent telephone operating companies throughout the United States. The Long Lines Department owns a number of cable-laying ships and is the principal developer of underseas ocean cables in the world. International telephone service is provided to U.S. companies by the Long Lines Department, connecting over 250 countries and territories via satellite, cable, and radio telephone circuits. ATT Long Lines is authorized to provide public telephone service from the U.S. to overseas points. It maintains a partial or total ownership in several dozen undersea cable systems and the Communications Satellite Corporation. (COMSAT)

ATT is prohibited from providing "record carrier" services such as telex, telegraph, and message switching traffic; these services are provided by the international record carriers. Recently, ATT tariff 263 was modified to allow international Dataphone service to be offered. (Previously a U.S. international Dataphone user was officially restricted to using the IRC's archaic manual Datel service). In return, continued near-monopoly protection of the lucrative "leased line" market was assured the IRC's by the FCC. (Some debate, however, is still raging in these matters).

9.3.2 RCA Global Communications Inc.

RCA Global Communications Inc. (RCA GLOBCOM) is the largest International Record Carrier (IRC) operating in the U.S. As an IRC, RCA is authorized to provide leased telegraph and Alternate Voice Data (AVD) circuits, telegram and telex service, facsimile, data and television transmission facilities, and message switching services.

In 1977, RCA GLOBCOM realized operating revenues of over 140 million dollars. The subsidiary of RCA Corporation was formed in 1919 with the acquisition of the British-owned Marconi Wireless Telegraph Company of America. The company maintains active sales offices and operating centers in over 30 countries. Approximately 40% of the record communications market has been captured by RCA, which operates out of five international "gateway" cities: New York, San Francisco, Washington D.C., Miami, and New Orleans. The company also provides service in Honolulu and several U.S. Territories. The U.S. MARISAT Marine Satellite Communications System, built and operated by COMSAT, interconnects to the U.S. domestic telex network through RCA switching facilities (ITT & WUI also provide interconnection to the MARISAT system).

9.3.3 ITT World Communications

The second largest IRC in the U.S., ITT WORLDCOM had earnings in excess of 120 million dollars in 1977. Originally formed as All America Cable and Radios Inc. in 1878, through mergers with the Commercial Cable Company (England), Mackay Radio and Telegraph Company and Globe Wireless Ltd., the company has extended its network ownership interests world-wide. ITT WORLDCOM is operated as a division of the ITT International Communications Operations (ICO) group along with 16 other ITT foreign subsidiaries, primarily operating in Central and South America. ITT holds approximately 35% of the IRC market, with approximately one half of its revenues derived from international telex service. ITT operates out of the five "gateway" cities in the continental United States, and provides service in Hawaii and San Juan as well. Unlike RCA, which maintains telex switching machines in each of its major cities, ITT "hubs" all of its international telex traffic from the distant cities (such as Honolulu) back to New York City where the traffic is then routed overseas via a centralized computer switching facility. Like RCA, ITT maintains ownership in COMSAT, underseas cables terminating in U.S. territories, and its own marine radio stations.

9.3.4   Western Union International Inc.

The third largest IRC (capturing approximately 20% of the market), Western Union International is a separate company listed on the American Stock Exchange.

Originally operated as a division of the Western Union Telegraph Co. (WUT), WUI was divested by act of Congress in 1963 when the Western Union Telegraph Company acquired the Postal Telegraph Comapny of the U.S, thus giving it monopoly rights to U.S. domestic telex and telegraph traffic.

WUI revenues exceeded 90 million dollars in 1977 with approximately half of its income derived from international telex service, followed by leased-line and cablegram revenues (similar in proportion to the other IRC's). WUI also operates in the same international "gateway" cities of its larger competitors and maintains ownership in COMSAT and cable facilities. WUI has introduced several innovative services including International Digital Data Service (IDDS), a measured-use leased-line digital facility, and owns a number of non-regulated communications companies including the nationwide Able-One telephone answering service. Although WUI suffers from the inability to connect its telex subscribers to several important countries, including Japan, WUI is in an excellent competitive position with tariffs set at prices identical to the other IRC's and a record of excellent service and maintenance.

9.3.5   TRT Telecommunications

The fourth International Record Carrier, TRT had operating revenues exceeding 15 million dollars in 1977. Founded in 1913 as a subsidiary of United Fruit, the company operated under the name of Tropical Radiotelegraph and was instrumental in extending regular communications between the mainland U.S. and Central American countries. The company was renamed about the time its parent changed its name to United Brands, and in 1972 the FCC authorized the company to compete directly with the three major IRC's in all aspects of record communications and within the five "gateway" cities. (Previously the company had been restricted to Miami and New Orleans operations.)

An aggressive company which has maintained the fastest growth rate (TRT telex traffic quadrupled in the period from 1972 thru 1976), the company has been instrumental in introducing innovative services and lowering the price of certain telex rates, notably the U.S.-to-Britain telex charges. In addition to maintaining U.S. mainland carrier operations, TRT operates telex networks in Panama, the Panama Canal Zone, and Honduras and offers the "full house" of services identical to those of the larger IRC's.

9.3.6   French Telegraph Cable Company (FTCC)

An operating subsidiary of the French PTT, the FTCC has been granted a limited operating authorization by the FCC to maintain leased-line and telex services from the New York City "gateway". Although FTCC has less than 5% of the IRC market share, it is a spirited little company which provides high-speed telex switching capabilities through its French parent to all major countries. The company also owns certain U.S.-to-North Africa cable facilities, and a growing number of New York user companies are discovering the FTCC telex circuits are often less congested than the large IRC facilities.

Like its larger competitors, FTCC is completely interconnected with the Western Union Telex and TWX networks. It maintains a switching center for carrying U.S. message and Data Telephone Service (Datel) traffic overseas, and can provide leased-lines via its New York City gateway to most countries. The large French business community headquartered in New York City appears to do business almost exclusively with FTCC.

9.4 INTERNATIONAL COMMUNICATIONS SERVICES

A broad variety of international communications services are offered by ATT and the IRC's along with the specialized Common Carriers (Telenet and Tymnet) operating in conjunction with the IRC's. Within the IRC group there is little price competition, but service competition is severe. Each record carrier continually is developing new features to extend the value of its service and make it more attractive to the potential customer. Consequently all the record carriers offer computerized telex exchanges which allow a user to "camp on" a call for automatic retry if the circuit is busy; all maintain generous credit arrangements in most instances. All facilities are interconnected with the applicable domestic carriers including Western Union Telegraph Corporation and ATT Long Lines.

9.4.1   International Direct Distance Dialing

Introduced to New York City in March 1969, IDDD has grown from a single city access (London) to a 50 country subscriber-dialable service which can be accessed from almost 20% of the telephones in the United States.

To accomplish the cut over from the International Operating Center's (IOC) manually-placed calls to an automatic system, ATT had to wait for the introduction of the Electronic Switching Systems (ESS) and Traffic Service Position System (TSPS) computerized city operated centers. Because international telephone numbers vary in length and often exceed eleven digits (the maximum capacity of a crossbar register), only the newer computer exchanges could handle the dual-stage outpulsing scheme that was developed.

To introduce the international direct dialing service, ATT had to conform to the CCITT world/country zone plan. In this plan there are nine global regions, each containing one or more countries (See Figure 9.10). Large countries are assigned single digits

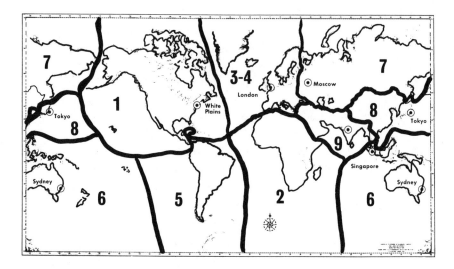

*Figure 9.10 Global Telephone Network with World Numbering
Zones and World Major Switching Offices*

(The U.S. is country code "1", Russia is country code "7"); medium-sized countries are assigned two digits (The United Kingdom is country code "44", France is country code "33"); and smaller countries which have fewer telephones are assigned three digit codes (Israel is "972.")

An international access code was assigned ("011") to enable the U.S. subscriber to access the overseas toll network, and a series of international transit centers were established along the lines of the CCITT recommendation to allow for through-country routing (See Figure 9.11).

COUNTRY AND ROUTING CODES FOR SOME CITIES

ANDORRA — Country Code 33

City	Routing Code
All points	078

AUSTRALIA — Country Code 61

City	Routing Code
Adelaide	8
Brisbane	7
Hobart	02
Melbourne	3
Perth	9
Sydney	2

AUSTRIA — Country Code 43

City	Routing Code
Graz	316
Innsbruck	5222
Klagenfurt	4222
Linz	7222
Salzburg	6222
Vienna	222

BELGIUM — Country Code 32

City	Routing Code
Antwerp	31
Brussels	2
Charleroi	71
Ghent	91
Liege	41
Malines	15

BRAZIL — Country Code 55

City	Routing Code
Belo Horizonte	31
Brasilia	61
Porto Alegre	512
Recife	81
Rio de Janeiro	21
Sao Paulo	11

CHILE — Country Code 56

City	Routing Code
Concepcion	42
San Bernardo	2
Santiago	2
Talcahuano	42
Valparaiso	31
Vina del Mar	31

CHINA, REP. OF — Country Code 86

City	Routing Code
Hualien	38
Kaohsiung	7
Pingtung	8
Taichung	42
Tainan	62
Taipei	2

COSTA RICA — Country Code 506

CYPRUS — Country Code 357

City	Routing Code
Famagusta	31
Kyrenia	81
Larnaca	41
Limassol	51
Nicosia	21
Paphos	61

DENMARK — Country Code 45

City	Routing Code
Aalborg	8
Aarhus	6
Copenhagen	1 or 2
Esbjerg	5
Odense	9
Randers	6

ECUADOR — Country Code 593

City	Routing Code
Ambato	2
Cuenca	4
Guayaquil	4
Machala	5
Manta	4
Quito	2

EL SALVADOR — Country Code 503

FIJI — Country Code 679

FINLAND — Country Code 358

City	Routing Code
Espoo-Esbo	15
Helsinki	0
Jyvaskyla	41
Kuopio	71
Lahti	18
Oulu	81
Pori	39
Tampere	31
Turku	21
Vantaa	14

FRANCE — Country Code 33

City	Routing Code
Aix-En-Provence	42
Bordeaux	56
Le Havre	35
Lyon	78
Marseille	91
Nice	93
Paris	1
Rouen	35
Toulouse	61
Tours	47

GERMANY, FED. REP. OF — Country Code 49

City	Routing Code
Berlin	30
Bremen	421
Cologne	221
Dusseldorf	211
Essen	201
Frankfurt	611
Hamburg	40
Munich	89
Nuremberg	911
Stuttgart	711

GREECE — Country Code 30

City	Routing Code
Athens	1
Corinth	741
Eretfos	1
Iraklion	81
Kavala	51
Larissa	41
Piraeus	1
Rhodes	241
Salonica	31
Volos	421

GUAM — Country Code 671

GUATEMALA — Country Code 502

City	Routing Code
Amatitlan	33
Antigua	32
Guatemala City	2
Quetzaltenango	61
Villa Nueva	31

HONG KONG — Country Code 852

City	Routing Code
Castle Peak	12
Hong Kong	5
Kowloon	3
Kwai Chung	12
Lantau	5
Ma Wan	12
Peng Chau	5
Sha Tin	12
Tai Po	12
Tsun Wan	12

IRELAND, REP. OF — Country Code 353

City	Routing Code
Cork	21
Drogheda	41
Dublin	1
Dundalk	42
Galway	91
Kilkenny	56
Sligo	71
Tralee	66
Waterford	51
Wexford	53

ISRAEL — Country Code 972

City	Routing Code
Ashkelon	51
Bat Iam	3
Beer Sheva	57
Hadera	63
Haifa	4
Jerusalem	2
Nazareth	65
Netanya	53
Rehovot	8
Tel Aviv	3

ITALY — Country Code 39

City	Routing Code
Bari	80
Bologna	51
Florence	55
Genoa	10
Milan	2
Naples	81
Palermo	91
Rome	6
Turin	11
Venice	41

JAPAN — Country Code 81

City	Routing Code
Gifu	582
Hiroshima	822
Kanazawa	762
Kobe	78
Kyoto	75
Nagoya	52
Osaka	6
Sapporo	11
Tokyo	3
Yokohama	45

KOREA — Country Code 82

City	Routing Code
Inchon	32
Kwangju	62
Masan	51
Pusan	72
Seoul	2
Taegu	82

KUWAIT — Country Code 965 (All points)

LIECHTENSTEIN — Country Code 41

City	Routing Code
All points	75

LUXEMBOURG — Country Code 352 (All points)

MONACO — Country Code 33

City	Routing Code
All points	93

NETHERLANDS — Country Code 31

City	Routing Code
Amsterdam	20
Eindhoven	40
Haarlem	23
Rotterdam	10
The Hague	70
Utrecht	30

NEW ZEALAND — Country Code 64

City	Routing Code
Auckland	9
Christchurch	3
Dunedin	24
Hamilton	71
Palmerston North	63
Wellington	4

NORWAY — Country Code 47

City	Routing Code
Bergen	5
Drammen	3
Oslo	2
Stavanger	4
Trondheim	7

PAPUA NEW GUINEA — Country Code 675

PERU — Country Code 51

City	Routing Code
Arequipa	54
Chiclayo	74
Cuzco	84
Lima	14
Piura	74
Trujillo	44

PHILIPPINES — Country Code 63

City	Routing Code
Angeles	542
Bacolod	34
Cebu	32
Davao	35
Iloilo	33
Manila	2

PORTUGAL — Country Code 351

City	Routing Code
Barreiro	19
Braga	23
Coimbra	39
Lisbon	19
Oporto	29
Setubal	15

VATICAN CITY — Country Code 39

City	Routing Code
All points	6

SAN MARINO — Country Code 39

City	Routing Code
All points	541

SINGAPORE — Country Code 65 (All points)

SOUTH AFRICA — Country Code 27

City	Routing Code
Bloemfontein	51
Cape Town	21
Durban	31
East London	431
Johannesburg	11
Pietermaritzburg	331
Port Elizabeth	41
Pretoria	12
Uitenhage	422

SPAIN — Country Code 34

City	Routing Code
Barcelona	3
Bilbao	4
Cadiz	56
Granada	58
Las Palmas (Canary Is.)	28
Leon	87
Madrid	1
Malaga	52
Palma de Mallorca	71
Pamplona	48
Santander	42
Seville	54
Valencia	6

SWEDEN — Country Code 46

City	Routing Code
Boras	33
Eskilstuna	16
Goteborg	31
Helsingborg	42
Karlstad	54
Linkoping	13
Lund	46
Malmo	40
Norrkoping	11
Stockholm	8
Sundsvall	60
Uppsala	18
Vasteras	21

SWITZERLAND — Country Code 41

City	Routing Code
Baden	56
Basel	61
Berne	31
Fribourg	37
Geneva	22
Lausanne	21
Lucerne	41
Lugano	91
Montreux	21
Neuchatel	38
St. Gallen	71
Winterthur	52
Zurich	1

THAILAND — Country Code 66

City	Routing Code
All points	2

UNITED ARAB EMIRATES — Country Code 971

Emirate	Routing Code
ABU DHABI — All points	2
AJMAN — All points	7
DUBAI — All points	4
FUJAIRAH — All points	70
RAS-AL-KHAIMAH — All points	77
SHARJAH — All points	6
UMM-AL-QUWAIN — All points	6
Fala, Amaka	9
Khawani	9

UNITED KINGDOM — Country Code 44

City	Routing Code
Belfast, N. Ire.	232
Birmingham, Eng.	21
Bournemouth, Eng.	202
Bristol, Eng.	272
Cardiff, Wales	222
Coventry, Eng.	203
Edinburgh, Scot.	31
Glasgow, Scot.	41
Liverpool, Eng.	51
London, Eng.	1
Manchester, Eng.	61
Nottingham, Eng.	602
Sheffield, Eng.	742
Southampton, Eng.	703

VENEZUELA — Country Code 58

City	Routing Code
Barcelona	81
Cabimas	64
Caracas	2
Coro	68
Maiquetia	31
Maracaibo	61
Merida	74
Puerto Cabello	42
Punto Fijo	69
San Cristobal	76
San Juan De Los Morros	46
Valencia	41
Valle De La Pascua	35

*Routing codes not necessary

**Military bases cannot be dialed directly

For Routing codes of cities not listed dial "0" (Operator).

Figure 9.11   International Direct Distance Dialing (IDDD)

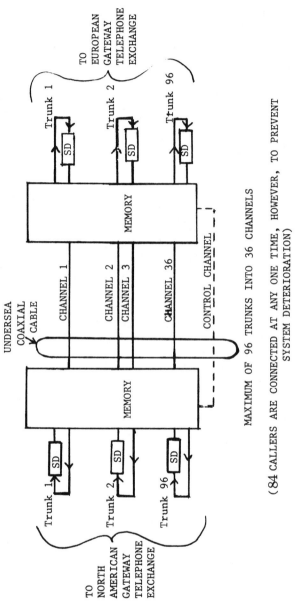

MAXIMUM OF 96 TRUNKS INTO 36 CHANNELS

(84 CALLERS ARE CONNECTED AT ANY ONE TIME, HOWEVER, TO PREVENT
SYSTEM DETERIORATION)

SD   Speech Detectors

Figure 9.12   The TASI System (Time Assignment Speech Interpolation)

IDDD is one of the fastest growing ATT services, exceeded perhaps only by WATS, and its universal acceptance has threatened certain IRC offerings including the lucrative telex service. For many years it was not permitted for a U.S. subscriber to transmit data through his telephone to an overseas point; this was the province of the IRC's Datel service. (The U.S. monopoly protection of IRC interests granted the record traffic-like Datel service to them.)

This prohibition extended to facsimile transmission as well, although many users regularly violated this prohibition. In fact the Datel service in most countries is simply a Dataphone service "piggy-backed" onto the domestic and international toll telephone service, just as ATT's Dataphone service is within the United States.

Aside from the political restrictions, however, there were a number of technical operating restrictions which often constrained the use of the ordinary ATT Dataphone for international data communications transmissions. First, on overseas cable circuits, ATT had introduced TASI (time assignment speech interpolation) a method of squeezing 84 conversations onto 36 circuits by taking advantage of idle time during conversations to provide more talking paths. (See Figure 9.12) TASI could play havoc with bursts of data, the beginning parts of which could potentially be chopped off by TASI in each transmission. Second, ATT's Data Sets did not correspond to CCITT specifications, and their communicating frequencies were different.

With the introduction of North American-supplied V23 international modems and modification of TASI systems to handle data communications, these technical restrictions have been virtually eliminated. Thus today it is possible to successfully transmit data over the IDDD network without problems in most instances.

ATT provides a number of pricing discounts for IDDD calls; a three minute direct dial call to London is $3.60 from the U.S. The same call placed manually by a toll operator costs $5.40, for a savings of 33% for the subscriber dialed call. IDDD rates are presently set at three minute minimums and subsequent one minute increments, but this should change within the next several years as ATT lowers its billing minimums to one minute and 30 seconds, respectively.

New tariffs filed by ATT to pass along the recent COMSAT re-
funds should lower IDDD calling costs to all countries, reducing
costs for a subscriber-dialed call by as much as 25%. (Ironically,
the deep discount given to IDDD calls to Britain may be removed,
causing a 3 minute IDDD call to London to cost approximately
$4.50 under the new tariffs. The net response to the new dis-
counts — and a special one-year "across the board" 15% international calling discount — will be to greatly stimulate the new
IDDD traffic from North America outward.)

Internationally, IDDD (often known abroad as ISD — International Subscriber Dialing) is available in many countries with
Britain currently leading the interconnection race with the
availability of subscriber dialing to over 70 countries. In most
countries, IDDD calls are billed in much shorter increments.
In Britain, a one minute call to Japan is charged approximately
$1.60 and billed in 2.4 second increments. When comparing
data transmissions via IDDD dataphone to conventional telex
services, one can easily realize a 48-to-1 compression of time
(50 bits per second over telex versus 2400 bits per second over
IDDD) and a lower overall rate ($1.20 per IDDD minute to the
United Kingdom versus $2.00 per telex minute). Thus North
American computer and terminal manufacturers such as Digital Equipment Corporation, Hewlett-Packard, and Wang offer
high-speed message systems which capitalize on this fact.

9.4.2   Datel (Data Telephone Service)

Datel service, the international public network dial-up dataphone
service, is offered by the IRC's to a growing list of countries (See
Figure 9.13). Datel rates vary from approximately $2.00 to $4.00
per minute depending upon the country, and are usually billed in
one minute increments. There are three ways of accessing the Datel
system: leased-line from subscriber location to IRC New York Datel
operating center, Western Union Broadband dial-up access, and ATT
telephone network dial-up access. In all instances the call is manually
established by the IRC Datel "gateway" city operator who must in-
tervene to establish the call before the data sets are connected to the
circuit. Because of this archaic operation, Datel service has not expanded in the US nearly as rapidly as it has in Western Europe.

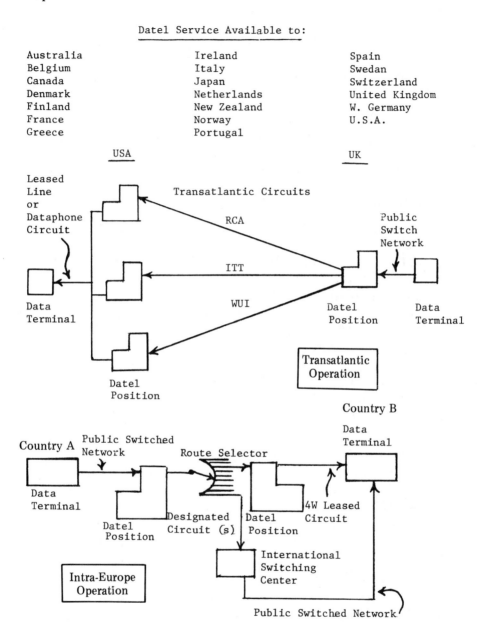

Datel Service Available to:

Australia	Ireland	Spain
Belgium	Italy	Swedan
Canada	Japan	Switzerland
Denmark	Netherlands	United Kingdom
Finland	New Zealand	W. Germany
France	Norway	U.S.A.
Greece	Portugal	

Figure 9.13   DATEL (Data Telephone) System

It is likely that existing Datel services will be modified and new, more competitive features such as automatic retry-on-busy, will be offered as ATT's IDDD Dataphone service bites into the Datel market.

9.4.3   International Telex

The IRC's maintain the monopoly rights to provide telex service to and from the United States. Unlike most other countries, in which a single unified agency provides both record carrier and telephone services, the United States maintains this division based upon historic operating precedents.

Within the five "gateway" cities, the major IRC's are authorized to install their own international telex machines on the premises of their customers in direct competition with the domestic Western Union Telegraph Company's own telex machines. (FTCC is also authorized to provide its own telex machines in New York City only.) All the IRC's are also interconnected with the Western Union Telegraph Telex and TWX networks to enable a user in the hinterlands (such as Chicago or Los Angeles) to send or receive an overseas telex message via his domestic Western Union Telex or TWX machine. Thus, in the five "gateway" cities, it is possible to obtain international telex service from both the Western Union Telegraph Company and it's four IRC competitors, with fierce competition underway.

The provision of these international telex machines is restricted to international traffic only; it is not possible to send or receive any domestic U.S. traffic through these machines, although it is possible to transmit telex messages to Hawaii and Mexico over them. Canada is considered a domestic point and is interconnected directly to the Western Union Telegraph Company's Telex and TWX networks through the CN/CP telex network and TCTS Canadian TWX network. Mexico is also connected directly to the Western Union Telegraph Company's facilities, and it is cheaper to send a message to Mexico via WUT than through the IRC's.

Automatic subscriber dialing to over 100 countries is available, and most connections are established either through direct New York/San Francisco/Miami circuits or by way of the London,

Rome, or Hong Kong transit centers. Most direct-dialed telex calls are billed in one minute increments with a "grace period" varying from 6 to 18 seconds of connection time before the call begins billing. This allows for wrong number calls to be aborted without billing consequence to the user.

Within the five "gateway" cities, the IRC's, unlike Western Union Telegraph, are allowed to supply their telex machines and circuits on a "no cost" basis to the customer if his telex traffic exceeds a certain yearly dollar amount. Various computer interfaces allow telex networks to be directly accessed by data processing centers using automatic calling units (ACU's) and data sets.

9.4.4   Leased-Line Service

The IRC's are authorized to provide a variety of slow speed teletype and voice-grade "alternate voice data" (AVD) leased-lines from the United States to other countries. There are an estimated 2,000 such circuits operating from the United States, and their use is growing rapidly.

There are four kinds of leased-line offerings: AVD (Voice) circuits, full speed channels (operating at 66 words per minute rates), half speed circuits (33 words per minute), and quarter speed circuits (16 words per minute). International leased-line rates are based upon recommendations developed by the CCITT, and are related to the replacement costs for dial-up toll telephone rates between the two countries concerned. An AVD circuit is charged at a monthly rate which is equivalent to approximately 9,000 minutes of toll telephone charges. Thus at approximately $1.20 per minute for a US-UK telephone call, an AVD circuit should sell for approximately $10,800. In fact, some countries discount their circuits to obtain additional business; other countries charge a premium as the market will bear. Table 9.1 presents a typical AVD leased-line cost between the United States and European overseas locations. Note that the leased-line charges are levied at both ends: each country's PTT or operating carrier is responsible for billing its end of the circuit to the customer within its jurisdiction. Full-speed circuits are billed at approximately 40% of the AVD rate, and quarter-speed circuits are billed at approximately 25% of the equivalent voice circuit costs.

Table 9.1 *European Monthly Terminal Rates for Leased Channels with the U.S.A.*

Country	Local Currency	AVD With M-102 Cond. Local Currency	AVD With M-102 Cond. US DLRS $	AVD W/O M-102 Cond. Local Currency	AVD W/O M-102 Cond. US DLRS $	Full Speed Local Currency	Full Speed US DLRS $	Half Speed Local Currency	Half Speed US DLRS $	Quarter Speed Local Currency	Quarter Speed US DLRS $	
Austria	G. FR	18,600	7,330	18,000	7,094	4,500	1,773	4,000	1,576	1,576	2,550	
Belgium	B. FR	216,310	5,363	206,500	5,120	73,600	1,825	57,400	1,423	34,200	850	
Denmark	D. FR		5,912		5,675		2,246		1,655		1,005	
Finland	MK		5,321		5,108		2,022		1,490		904	
France	F. FR	24,960	5,200	24,960	5,200	9,600	2,000	6,480	1,350	3,840	800	
Germany	DM	15,780	6,582	15,149	6,319	4,734	1,974	4,208	1,753	2,683	1,119	(C)
Greece	DR	203,280	5,912	197,400	5,675	65,760	2,246	43,830	1,655	26,310	985	
Italy	L	6,311,250	7,537	6,113,250	7,301	1,771,770	2,116	1,303,170	1,557	798,270	953	
Luxembourg	G. FR	15,000	5,912	14,400	5,675	4,500	1,773	4,000	1,576	2,550	1,005	
Netherlands	G	16,440	5,912	15,785	5,675	4,935	1,773	4,385	1,576	2,895	1,040	
Norway	N. KR	6,682	5,912	6,414	5,675	2,539	2,246	1,871	1,655	1,136	1,004	
Portugal	P$		6,800		6,600		2,200		1,467		880	
Spain	PTS		6,800		6,600		2,200		1,467		880	
Sweden	S. KR		5,321		5,108		2,022		1,490		904	
Switzerland	S. FR	18,000	7,349	18,000	7,349	6,000	2,450	4,000	1,633	2,400	980	(C)
Turkey	T. L		6,800		6,600		2,200		1,467		880	
United Kingdom	£	2,479	4,215	2,396	4,073	833	1,417	625	1,063	375	638	(C)

Notes:

1. The rate for 100 wpm leased channel service is the rate for full speed service plus 10 percent.

2. Where not shown, monthly terminal rates in local currency will be provided at a later date.

3. The United Kingdom Terminal rates are to be paid in the United Kingdom in Pounds Sterling. The U.S. dollar amounts shown are estimated and should be used with caution.

4. 8% VAT must be added to the London terminal rates.

5. 8% Gov't. Tax must be added to the Greece terminal rates. (C)

6. The Switzerland terminal rates are to be paid in Switzerland in Swiss Francs. The U.S. dollar amounts shown are estimated and should be used with caution.

It is possible to arrange the AVD circuits in a number of ways, and to connect them both to the subscriber's domestic telephone facilities and computer equipment. In most instances, however, the customer is restricted from interconnecting the circuit to the distant country's toll telephone network and thus circumventing the international toll telephone charges that would otherwise be applicable. This restriciton does not apply to the U.S. terminations, of course, as the Carterfone Interconnect Supreme Court Decision allows any network circuit to be connected to a customer-owned equipment and arranged by the customer as he sees fit. The Carterfone Supreme Court ruling supersedes, within the USA, the CCITT recommendation to prohibit such interconnection.

A variety of circuit conditioning arrangements and corresponding data sets are supplied by the IRC's and PTT's, and a "speech plus" facility supported by most common carriers enables a simultaneous voice conversation as well as up to four low-speed teletypewriter conversations to be transmitted over the same AVD channels to be divided into multiple low speed teletypewriter circuits totaling 24 to as high as 192 (using computer concentrator equipment). Teletypewriter and derived-teletypewriter speed channels may be interconnected to the customer's message switching system; these message switching computer facilities can, in turn, be provided by the IRC's/PTT's under monthly or yearly lease.

9.4.5   Message Switching Service.

Within the United States, the IRC's all offer advanced computerized message switching facilities which may be rented on a per-port connection on a month-to-month basis. Internationally, large PTT's including the BPO in the UK, Itelcable (Italy), Cable and Wirless (Hong Kong and Bahrain), and the French PTT offer competitive computerized switching systems which are often available in a heavily discounted package.

RCA's shared computerized message switching service is known as AIRCOM; ITT calls its system ARX (Automatic Re-transmission Exchange); and WUI's facility is known as TELUS.

Recent TELUS additions have expanded the message switching service to allow access to the domestic US Telex and TWX networks on a "real time" basis from an overseas location via a leased-line or derived leased-line channel. It is possible also to access the European Telex network via a leased-line extension to the BPO's message switching exchange center in London.

In addition to providing classic store-and-forward services, the message switching systems offered by the IRC's enable the interconnection of leased-lines to Telex and TWX facilities, and the speed and code conversion of Telex (operating at 66 words per minute, five level CCITT alphabet II "Baudot" codes) to TWX (100 words per minute eight-level ASCII).

9.5 PTT SERVICES OVERSEAS

Internationally, most administrations offer a range of telephone, telegraph, and data communications services which vary in sophistication in direct proportion to their status as industrialized, developing, or third world country. As most telecommunications facilities are owned by the government agency charged with communications responsibility (usually the Post Office), a more integrated offering of voice and data facilities is usually possible in the industrialized nations. The British Post Office (BPO) provides for a variety of inland and external telecommunications services including most of those facilities available within the United States from ATT and the IRC's.

With the advent of the INTELSAT satellite network, it became possible for even the least developed country to provide international subscriber-dialed telex and telephone service. Because of the heavy investment in plant equipment that is required of a modern telecommunications facility, most developing countries have concentrated on providing public exchange access services first. These include toll telephone and telex systems. Specialized business services such as message switching systems, derived channels and "speech plus" arrangements data terminals and computer dialing equipment are often not available in developing countries because of its low priority on their capital expenditure lists. Even the industrialized European countries tend to encourage the consumption of dial-up facilities before leased-line or specialized business services are used.

Since a switched service can in theory be loaded much more
heavily and efficiently than a private channel (which has been
removed from the public "pool" usage), it is reasonable to assume that public policy encourages shared facilities whenever possible. This appears to be confirmed by the rather high price for
an international leased-line when compared against a typical
domestic facility of similar distance. To illustrate, a voice circuit from New York to San Francisco costs approximately
$1,000 per month via domestic satellite common carrier. A
similar circuit carried over the INTELSAT IV Pacific satellite
from San Francisco to Japan costs 21 times more.

Because of the PTT encouragement of the use of public switched
facilities, the development of packet switching networks in Europe
and elsewhere is proceeding at a rapid pace. Euronet, the Europacket network, will be fully operational by 1980; similar networks exist in Scandinavia, and domestically within Britain, Germany, Spain and other European countries.

Some PTT's (such as TELMEX in Mexico) require that the customer purchase stock and/or bonds in its communications organization as a method of capitalizing its plant development. Other
organizations such as the German Bundespost require that a minimum time commitment be given upon establishment of certain
expensive services such as international leased-lines. Still other
PTT's (such as the British Post Office) require the customer to
purchase his own terminal and switching equipment such as certain modems, PBX's, and teletypewriters. In many instances, the
sophistication of the service to be provided is directly proportional to the distance from the international "gateway" city switching
center. This is especially true in developing countries which may
often possess excellent global communications facilities at the
"gateway" earth station site, but due to inadequate local loop
distribution facilities cannot provide these services farther than
a few kilometers away. In many instances, the relocation of a
business office of only a few hundred meters may mean the difference between obtaining service within two weeks versus two
years.

Because of the encouragement of switched dial-up services, most industrialized PTT's have been eager to interconnect to the U.S. Telenet and Tymnet specialized common carrier networks which have been extended abroad through the auspices of the U.S. IRC's— notably RCA and ITT. These two packet-like switched networks are now accessible in over a dozen countries, thus allowing a user on a portable terminal in London, for example, to be connected to his computer data base in Los Angeles by dialing a local London telephone number. Access rates are low to moderate, with the typical country charging its customer between $10 to $20 per hour in connect time and from $0.25 to $1.00 for each thousand characters transmitted or received.

In Canada, the provision for both overseas voice and international telex/data Communications is the responsibility of Teleglobe Canada, a Crown Corporation established by act of Parliment for this purpose in the 1950's. Domestically, national telex service is provided by the CN/CP telecommunications organization formed by the combination of the two national railroads' communications networks. All international telex service is routed through to the domestic Telex (and TWX) systems for subscriber access. Public telephone service is provided by the Trans-Canada Telephone System (TCTS) consortium of operating companies with Bell Canada (no relation to ATT) and BC Telephone (owned by General Telephone) providing the majority of subscriber telephones. The telephone companies also provide TWX service in Canada in competition with the CN/CP telex network, and there is no interconnection between these two competing facilities, unlike the U.S. arrangement. Thus a U.S. Telex machine can dial a Canadian TWX machine, but a Canadian Telex machine cannot! Customer owned equipment as is known in the United States is not currently permitted in Canada, although certain ancillary devices such as answering machines are allowed to be connected to the public network.

Domestically, Canada offers a variety of services identical to those found within the United States (including WATS lines which, unfortunately, do not connect across the border) and a variety of other services such as switched facsimile Telco-provided packet and digital data networks, and advanced computer

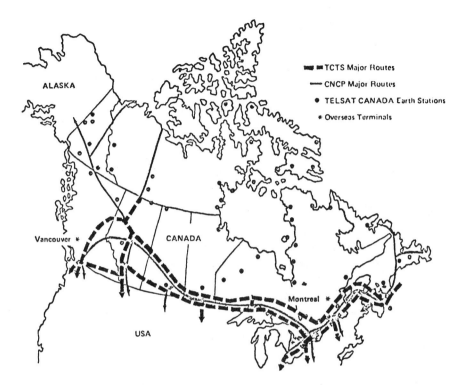

Figure 9.14   Canadian Telecommunications Trunk Routes

communications services not currently provided for within the
United States. As Canada is considered to be a domestic point
under the communications act of 1934, the interconnection of
ATT and Western Union facilities has enabled the free exchange
of U.S. leased-line and dial-up services to be expanded without
hindrance to Canada. Thus it is likely that a call from Los Angeles
to New York may will be routed by way of Regina, Canada during
certain times of the day.

Teleglobe Canada provides message switching service in Montreal
and the other "gateway" cities of Toronto and Vancouver. Through
the introduction of a radial discount pricing arrangement Teleglobe
Canada is now in a position to compete favorably with the U.S. IRC
from its official bank rate, it is possible to price a facility from
to route their calls by way of the northern-bound system instead of
the San Francisco or New York IRC "gateways. (See Figure 9.14)

As all PTT leased-line and international dial-up charges are determined based upon bilateral agreements with the countries concerned, pricing schemes for international service vary extraordinarily from country to country. Thus it is possible to telex from London to Hong Kong for less money than from San Francisco to Hong Kong. In fact, because each country computes the exchange rate at a different figure, which in turn is often different from its official bank rate, it is possible to price a facility from both sides and obtain two different figures! This factor should always be considered in designing any multi-country network.

As most PTT's are as voracious in competing with each other abroad as they are in preventing competition internally, tremendous flexibility can be had by the international organization which takes its time to optimize its world-wide communications network on a service-to-price basis. Many instances exist in which the routing of traffic (either dial-up or leased-line) by way of a strategic third country may dramatically reduce the costs as compared with communicating directly from country A to country B. The Cable and Wireless group is noted globally for its flexibility in providing radial discounts for multi-leased-line networks, and it is considered to be a private operating agency in Hong Kong even though it is owned by the British Treasury. Although the official refile of third party traffic by an unrelated party is prohibited by CCITT recommendation, a number of organizations recently have established successful businesses to perform just such a service. The proliferation of time sharing services which tacitly allow "storage and retrieval" mailbox facilities cannot but continue to undermine the PTT monopoly position and its potential revenue base.

9.6 MANAGEMENT TECHNIQUES, STRATEGIES, PROBLEMS AND PITFALLS

While the typical international communications carrier is faced with classic expansion problems such as outdated plant facilities, lack of capital, bureaucratic inefficiencies, and political direcitves, the international user is faced with a sudden explosion of service opportunities. The "21 flavor" problem of having too many choices may pose to the user a problem as potentially serious as the historically recent lack of options available only a decade ago. The acquisition of third and fourth generation technology by the PTT's has often created a bizarre two-state configuration: highly advanced services operated side-by-side or in conjunction with archaic 1920's equipment.

It is the responsibility of the communications manager to sort between these services and PTT capabilities, but too often reliance is laid upon the home country's PTT or IRC. In many instances, staff members of these common carriers do not have the necessary familiarity with their correspondent's equipment and service capabilities, and thus are unable to develop a sophisticated network which will truly optimize the customer's needs. It is, of course, unreasonable to expect that a sole-source supplier will not tend to optimize his own profits before minimizing his customer's costs. This is as true of any PTT as it is for any large computer manufacturer. Until recently, there has been precious little in the way of quantified and accurate information available on the subject of global communications services, offerings, and individual PTT capabilities and pricing arrangements.

The US organization has a significant advantage because the IRC's (notably RCA and ITT) maintain liason offices in most countries and are thus able to provide technical and coordination support in conjunction with the local PTT. Nonetheless, it is still easy to overlook certain potential problems or not realize certain cost savings.

For example, it is almost as expensive to lease a multi-drop data communications network connecting New York, London, Paris, Zurich, and Rome as it is to configure a far more reliable network consisting of separate circuits hubbed to a London switching center. Few communications managers are aware that communications PTT personnel are not available in Switzerland over the weekends, and that the Swiss PTT is totally automated at these times. Routing such a multi-drop network by way of Switzerland could create a weekend service problem with the Rome terminal if a failure should occur in the multi-country circuit within the Swiss borders. Such an arrangement would likely be unacceptable for use by a multi-national airline, although such a configuration is quite typical within the United States.

A British-based bank with direct teletype circuit to its Nassau
facility would most likely order the facility through the BPO
in London. It would not be aware that the BPO does not have
direct circuits to the Bahamas, and that such an arrangement
would be routed, most likely by way of the Cable and Wireless
operations in Jamaica, and back-hauled through a Jamaica-Batelco
circuit to Nassau. By routing a circuit instead to New York, interconnecting this channel to a Western Union Telegraph circuit
to Ft. Lauderdale, Florida for futher interconnection to a TRT
Telecommunications circuit from Florida to Nassau (via the 1380-
pair coaxial cable), several thousand dollars can be saved monthly.
Moreover, if the London Bank has a New York subsidiary as well,
it would be possible to combine communications to create a three-
point network as desired. In point of fact, the coordination problem has not increased dramatically although three separately —
billed circuits are involved.

As there is no international equivalent of ATT Long Lines, coordination and administration of the maintenance function often falls upon the shoulders of the international user organization itself. To familiarize the communications manager with all
aspects of the far-flung network operations, it is often recommended that a personal "fly-by" trip be taken while the network
is being built. Meeting with PTT personnel and understanding
a country's culture and technical capabilities first hand is a priceless
experience which cannot be substituted by reading tariffs from a
distance. Several international consultancies have been established
in recent years specifically to assist in the global design and management of international corporate communications networks.

9.7 FUTURE TRENDS

As INTELSAT moves toward its fifth generation satellite network
and packet switching networks proliferate, significant new international services are just beyond the horizon. Packet switching networks can be configured to provide "datagram" service (similar to
telegram per-word billing), virtual leased-line arrangements (which
appear to look like today's typical point-to-point private line circuits), and virtual line switch configurations (commercially equivalent to today's Telex and voice dial-up switched services). Integration of these separate networks into a common facility seems likely in the 1990 time period.

Digital voice terminals which can simultaneously squeeze four voice conversations onto a 9,600 bit per second voice channel (supplied by companies such as E-Systems, ICS, and Time and Space Processing) will proliferate in an increasingly digital world.

The implementation of the Space Shuttle program will enable sixth and seventh generation "super satellites" to be fabricated. These systems will allow relatively inexpensive video communications and multi-national point-to-point handheld communicator conversations to take place. The ability to manufacture sophisticated electronic circuitry through "printing press" integrated circuit technology will enable solid-state high fidelity telephone subsets to be manufactured economically. (Such sets are already under trial in Scandinavia and Canada). An overall broadening of interconnection policies within the industrialized countries will continue to allow flexible expansion of electronic mail and multi-national word processing and data transmissions.

At our present rate of North American and global expansion, it is likely that by the year 2001 there will be 120 phones for every 100 people within the United Staes; world-wide almost a billion telephone sets will be in operation. In that year, the telephone will celebrate its 125th birthday.

BIBLIOGRAPHY

Tood, Keith T., *A Capsule History of the Bell System,* American Telephone & Telegraph Company, New York, New York, 1975.

A Half Century of Cable Service to the Three Americas, All America Cables, Inc., New York, New York, 1928.

Brooks, John, *Telephone — The First Hundred Years,* Harper & Row, San Francisco, 1976.

Business Telecommunications, The Chameleon Press, London, 1977.

CCITT Sixth Plenary Assembly, ITU — Orange Book, Geneva, 1977.

Clarke, Arthur C., *Voice Across the Sea,* Harper & Row, Inc., San Francisco, 1974.

Communications Equipment and Systems, Institution of Electrical Engineers, London, 1976.

Easton, A.T., *Executive Telecommunications Management Handbook,* ICM Publishing, San Francisco, 1977.

Field, Henry M., *The Story of the Atlantic Telegraph,* Charles Scribner's Sons, New York, 1892.

"1977 Handbook & Buyers Guide," *Telecommunications Magazine,* Dedham, MA., July, 1977.

Handbook of Data Communications, NCC Publications, Manchester, England, 1975.

Martin, James, *Telecommunications and the Computer,* 2nd ed., Prentice-Hall, Inc., Englewood Cliffs, 1976.

Notes on Distance Dialing 1968, American Telephone and Telegraph Company, New York, 1968.

Renton, R. N., *The International Telex Service,* Pitman Publishing, Bath, Great Britain, 1974.

Telephone Engineer & Management Directory, Harcourt-Brace-Jovanovich, Duluth, MN, 1977.

Telephony's Directory, 82nd, Telephone Publishing Corp., Chicago, 1977-78.

The World's Telephones, American Telephone and Telegraph Company, New York, 1976.

Chapter 10

TELECOMMUNICATIONS SYSTEMS

S.W. Maley
Department of Electrical Engineering
University of Colorado
Boulder, Colorado

10.0 Introduction

10.1 Trends in Telecommunications

10.2 Communication Signals

    10.2.1 Representation of Information by Signals
    10.2.2 Frequency Domain Representation of Signals
    10.2.3 Sampled Signals
    10.2.4 Digital Representation of Analog Signals
    10.2.5 Modulation
    10.2.6 Effects of Noise on a Communication Signal
    10.2.7 Error Control in Digital Systems
    10.2.8 Coding for Information Security

10.3 Quantitative Characterization of Telecommunication Signals and Systems

    10.3.1 Information Theory
    10.3.2 Entropy
    10.3.3 Channel Capacity
    10.3.4 Fundamental Theorem of Information Theory

10.0 INTRODUCTION

Telecommunications, in its many forms, is in a state of rapid expansion throughout the world. The capabilities of existing systems and the character of current expansion are governed by several factors: the influence of governmental regulatory agencies, the economics of the telecommunications industry and the technology of telecommunications. This chapter is concerned specifically with the influence of technology on the current and future trends in expansion of telecommunications throughout the world.

Following a general discussion of current trends in the evolution of telecommunications, foundations are laid for a more detailed study of its technological aspects first by a reasonably broad study of communication signals and the manner in which they represent information, then by a study of some aspects of information theory which provides a means of numerical measure of information in signals, of information handling capabilities of telecommunication systems and of coding procedures for matching a signal to a system.

Telecommunication systems are then categorized, and the various categories are analyzed with the objective of providing insight into current and future trends in system design. In this regard, point-to-point switched systems are treated first, followed by a brief discussion of broadcast systems.

10.1 TRENDS IN TELECOMMUNICATIONS

Many types of telecommunication systems are in use throughout the world today, and the variety is becoming steadily greater. It may be said with little fear of contradiction that communication capability in almost all types of systems is in a state of rapid expansion. The underlying reasons are discussed in detail in other chapters of this volume. Very briefly one could cite the improved standard of living in many parts of the world. Further, there is a strong desire to improve relations and cooperation among people on all levels, personal as well as the various levels of government. The potential contribution of telecommunications toward these ends is complex, but certainly improved communication can only be beneficial over an extended period of time.

In recent times the demand for telecommunications capability has, on a world-wide basis, always exceeded the supply; despite the rapid expansion there seems no possibility of satisfying that demand in the foreseeable future. The reasons are simple. Telecommunications systems are characterized by a long life and by high initial capital requirements. Indeed, at the present time, new telecommunications capability is consuming a substantial proportion of the investment capital available throughout the world. The rate of expansion of telecommunications capability is thus limited only by the availability of resources that can be currently devoted to new systems.

Examples of increasing demand for telecommunications are commonplace. Consider nonbusiness telephone usage. The number of telephones in homes is increasing rapidly in the lesser developed countries. In the more highly developed countries, the rate of increase is much less. However, in such countries, the proportion of telephone calls that are made to distant cities is increasing; that is, the average distance over which calls are made is increasing. This greater proportion of toll-calls requires significant expansion in intercity communication links. It may therefore be said that expansion is occurring throughout the world to accommodate non-business telephone demand although the character of the expansion takes different forms in different countries.

The other major category of communications will be called business communications, although it will include government, military and educational communications as well as all other types that cannot be considered as personal. This is a very broad category of communications and most aspects of it are increasing rapidly. Business, in the broad sense used here, is changing in character in the sense that it is being conducted over greater distances than has been true in the past. Even small businesses now are increasingly finding it necessary to communicate with other businesses in distant cities. This increase in business communications is taking several forms: voice, data and facsimile. Perhaps the most dramatic increase is in data communications which may be described as communication between computers. This has occurred as a consequence of the ongoing electronics revolution which has so drastically reduced costs of electronic data processing systems. These systems reduce costs of record keeping and information handling to such a degree that a business cannot remain competitive without them. Their use requires communication links between the computers or terminals at the various locations at which business operates. This linking of computer terminals, called data communications, is one of the most rapidly growing types of telecommunications.[5]

The increasing use of telecommunications seems destined to continue at least into the near future, but there is another significant factor that may influence future trends in telecommunications expansion. That factor relates to the cost of energy. Modern business involves a great amount of travel, with the consequent consumption of large amounts of energy, for the purpose of permitting face-to-face discussion. The impending shortage of low cost energy has dictated that alternatives be found for at least part of such travel. Some travel could be avoided by using the telephone in place of face-to-face conversation; undoubtedly this has occurred, to some extent, during the past several years while fuel costs and thus travel expenses have increased. It is apparent that the telephone does not provide adequate communication capability to substitute for all face-to-face communication, but a significant improvement such as could be provided by a versatile video telephone could reduce the required amount of travel significantly. A video signal requires vastly greater communication capacity than a voice signal, at least one hundred times as much; so the cost of video telephone service is significantly greater than for voice telephone. Despite this fact, video telephone communication offers the possibility of great savings in cost as well as energy and time. Such possibilities are emphasized here because of the role they may play in future expansion trends for telecommunication systems.

The foregoing discussion makes it clear that telecommunication systems are being expanded as rapidly as is possible with the resources available. A question may be asked as to the role played by research and development. This is easily answered; technological progress determines the amount and quality of telecommunications capability that can be achieved in return for the investments of capital that can be made. The potential technological benefits over the course of the next few years appear to be highly significant; their nature will be the subject of the remaining sections of this chapter.

In conclusion it may be said that a high rate of expansion is occurring throughout the world in most types of telecommunications. The rate of expansion is limited, not by demand, but rather by the world-wide resources available to devote to that purpose. The intensifying shortage and consequent higher cost of energy appears destined to influence trends toward more telecommunications and less travel. Technological development will greatly increase the telecommunication capability achievable for the available investment capital. This latter point will be examined in detail in the following sections of this chapter.

10.2 COMMUNICATION SIGNALS

10.2.1 Representation of Information by Signals

The function of a telecommunication system is the transfer of information from one location to another. The information transferred is in the form of so-called communication signals. Such signals need to be introduced before embarking upon any further discussion of systems. There are various kinds of signals representing various kinds of information sources. For example a voice signal, which often takes the form of an electrical voltage varying in time, is obtained from an actual voice (by means of a microphone) and it represents the voice-in the sense that it can be converted by means of a loudspeaker into an understandable replica of the voice from which it was obtained. Similarly a video signal is obtained from a picture and it can be converted back into a replica of that picture. A digital or data signal, on the other hand, represents a sequence of symbols, usually alphabetic or numeric characters. It is usually obtained from a computer or from a device such as an electric typewriter (more specifically, a typewriter having an electrical output). In the latter case, depression of one of the keys not only prints an impression on the paper but it also generates an electrical signal having a graph of amplitude versus time which is characteristic of the particular key that was depressed. Each key has a different graph. The electrical signal, then, is the communication signal representing the symbol labeling the key that produced it. Just as in the case of voice and video signals, a digital communication signal can be converted back into the symbol that it represents (by means

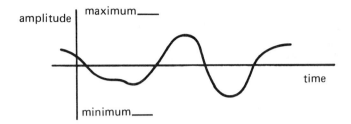

Figure 10.1 Typical Analog Signal

of typewriter with electrical control capability or by means of a teletype console or other similar device). In the case of a digital signal generated by a computer, the same sort of data signal is generated but it occurs as a consequence of control signals in the computer rather than a keystroke as in the case of the electric typewriter.

Analog Signals. The signals discussed above are of two distinct types. Voice and video signals are of the first type, and a typical such signal is sketched in Figure 10.1. The amplitude varies with time in a manner determined by the information source that generated the signal. The amplitude can take any value between two extreme values labeled maximum and minimum amplitude which are determined by the device that converts the information into the signal. Such a signal, that may take any amplitude within a range of values, is named an analog signal. It is apparent that within a specified length of time there can be an infinite number of different analog signals.

Digital Signals. The other type of signal is the digital (or data) signal; a typical example is sketched in Figure 10.2. In this case the signal represents a sequence of decimal digits. Each digit is represented by a pulse having an amplitude equal to the digit it represents. The digits represented by the various pulses are given in parentheses below the pulses. This is only one of the many possible ways to represent decimal digits by pulses. In another version, the pulses could be widened to completely fill in the interpulse spaces between the pulses. In another version both positive and negative pulses could be used. In still another version, pulses of only two possible amplitudes could be used but with four such pulses for each decimal digit. An example of this type

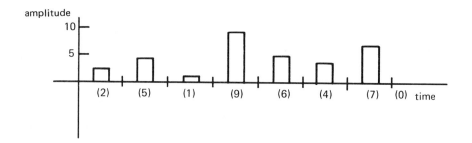

Figure 10.2 Typical Digital Signal

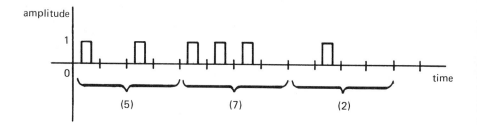

Figure 10.3 Typical Binary Digital Signal

of signal is sketched in Figure 10.3. The first four pulses have amplitudes 1,0,1, & 0 and, as indicated by the decimal numbers in
parentheses, represent the digit 5. The second block of four pulses
represents the digit 7. The rule relating the decimal digit to the
block of four binary (two-level) pulses is, in this case, based upon
conversion of the decimal digit to a four digit binary number and
the representation of that four digit binary number by four binary
pulses, least significant digit first. The significant difference between analog signals and the data signals sketched in Figures 10.2
and 10.3 is the fact that each of the pulses in the data signals can
have only a finite number of amplitudes, ten in Figure 10.2 and
two in Figure 10.3. This is in contrast to the analog signal which
at any time can have any of an infinite number of amplitudes.
This limitation to a finite number of possible pulse amplitudes is
a consequence of the fact that the signal represents a sequence of
symbols selected from a finite set (or alphabet), and it is the reason such signals are often referred to as digital signals. (Digital in
this usage is intended to imply a finite symbol set rather than im-

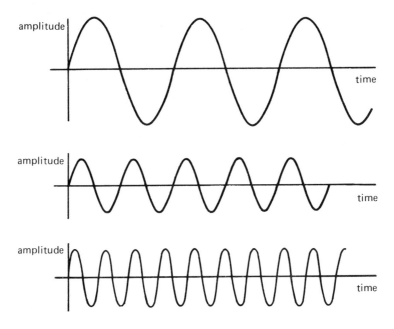

Figure 10.4 Sinusoidal Signals of 3 Different Frequencies

plying a strictly numerical symbol set). The finiteness of the pulse set, in the case of digital signals, has far reaching consequences in the design of communication systems to handle such signals. This will be discussed further in later sections of this chapter.

10.2.2 Frequency Domain Representation of Signals

One of the characteristics of both the types of signals discussed above is the range of amplitudes that they may assume. In the case of the signal sketched in Figure 10.1, the range extends from the value marked minimum amplitude to the value marked maximum amplitude. In the case of the signal of Figure 10.2 the range is discrete rather than continuous and extends from 0 to 9 in steps of one. Similarly, the signal of Figure 10.3 has a discrete range of zero and one. Another, and much more important, physical characteristic of all signals is the bandwidth. In order to explain this characteristic refer to the signals of Figure 10.4 in which there are graphs of three different signals. The shape of each is called sinusoidal because they are mathematical sine functions of time, but each has a different frequency. The frequency, by definition, is the number of times the cycle, consisting of one complete positive and one complete negative excursion of the amplitude, occurs in one second. Only a sinusoidal wave can be assigned a precise frequency because the definition requires the shape of the graph to be that of a sine function. It follows that signals such as those sketched in Figure 10.1, 10.2, and 10.3 cannot be assigned a precise frequency; however there is an alternative way to characterize these signals in terms of the concept of frequency. To do so it is first observed that any signal can be shown to be equal to the sum of a number of sinusoidal signals, different signals in the sum having different frequencies, different amplitudes and different phases. (A change in phase corresponds to shifting the signal to the right or left on the time scale.) The number of sinusoidal signals in the sum may be finite or infinite depending upon the signal that is to be expressed as the sum. Furthermore there is only one way that the sinusoidal signals in the sum can be selected. These facts can be verified experimentally or theoretically, but no attempt at such verification will be made here. The matter is more fully explored in the appendix. The representation of a signal in terms of the amplitudes, frequencies and phases of its sinusoidal components is called a frequency domain representation. Another frequency domain description expresses a signal in terms of the power and frequency of the various sinusoidal components; it is called the power spectrum and will be discussed further later. The sinusoidal signals in the sum all have different frequencies and the range of values of frequencies (that is, the difference between the highest and lowest frequencies) is called the bandwidth of the signal.

It is an extremely important characteristic of a signal because the cost of transmission of a signal is fundamentally dependent upon its bandwidth. Generally speaking, higher bandwidth signals cost more to transmit. Frequency and thus bandwidth are measured in cycles (of the sine wave) per second. This unit of measurement is often called the Hertz; one Hertz is one cycle per second. Often frequencies and bandwidth are so large that the kilohertz, megahertz and gighertz are more convenient (these are 10^3 Hertz, 10^6 Hertz and 10^9 Hertz, respectively).

Figure 10.5 Sampled Signal Obtained from the Signal of Figure 10.1

10.2.3 Sampled Signals

Now that the concept of bandwidth has been introduced, a
method for changing an analog signal to a digital signal can
be discussed. Suppose the signal of Figure 10.1 is passed through
a gate which passes the signal for only a short period of time at
periodic intervals. The resulting signal, called a sampled signal,
is sketched in Figure 10.5. A theoretical analysis shows that if
the number of samples per second is at least two times the bandwidth of the signal being sampled (the signal of Figure 10.1 in
this case), then the original signal can be exactly recovered from
the sampled signal by passing it through a low-pass filter.[3, 11]
(The output of the filter may differ from the original signal in
amplitude, but that can be corrected by an amplifier or an at-
tenuator). This is true regardless of the width of the samples.
There are several reasons for sampling; one is the fact that the
sampled signal has unused time intervals between pulses. These
intervals can be used for some other purpose. For example, the
pulses of another sampled signal could be inserted in those intervals. Then the transmission of the single composite signal effectively accomplishes the transmission of the two analog signals which were sampled and interleaved. At the receiving end
of the transmission system the samples for the two analog signals are separated and passed through individual filters to recover
both of the original signals. This combining of two signals into
one in such a manner that they can be recovered is called multiplexing,[3] and this particular technique is called time-division
multiplexing (TDM). Another method will be discussed later
In this example, only two signals were multiplexed, but in principle any number can be multiplexed. In practical systems oftentimes thousands of signals are multiplexed.

10.2.4 Digital Representation of Analog Signals

Another reason for sampling a signal is its use as a first step in the conversion of an analog signal to a digital signal. It may be noted that the tops of the pulses in Figure 10.5 take the shape of the signal being sampled. If the tops were all flattened at their average value, the signal recovered from the resulting distorted, sampled signal by passage through a filter would not be exactly the same as the original signal, but if the width of the samples is narrow compared to the sample spacing, the difference between the original signal and the recovered signal (which could be termed the error) can be made sufficiently small that the information carried by the original signal can be extracted from the approximate replica of it obtained by passing it through a filter. More specifically the error approaches zero as the width of the sampling pulses approaches zero. Thus the error caused by the flattening of the top of the pulses can be made arbitrarily small.

But why flatten the pulse tops if it causes errors however small? The reason is that the flattening makes all pulses have the same shape, and the advantage of this is that each pulse can be completely specified by its amplitude; that is, by a number. The number describing the pulse amplitude must be rounded off with the consequence that it does not exactly describe the pulse amplitude. The error can be made arbitrarily small by choice of the number of digits used in the number describing each sample amplitude. If the pulse amplitudes are changed to the values given by the rounded off numbers, the resulting signal is called a quantized version of the original signal. Each pulse in a quantized signal has one of only a finite number of amplitudes and is thus a digital signal. Therefore, if an analog signal is band limited (has finite bandwidth) it can be converted to a digital signal with arbitrarily small error; and, of course, the digital signal can be converted back to a close approximation to the analog signal. It can be concluded that an analog signal can, if desired, be converted to a digital signal and then transmitted over a digital communication system. The converse problem, that of converting a digital signal to an analog form is trivial because actually no conversion is needed; an analog communication system can handle a digital signal without any change if the frequencies present in the digital signal can be handled by the analog channel. If not, then the frequencies can be shifted by some sort of modulation. It may be now said that there are two fundamental types of signals, analog and digital. An analog signal can, if desired, be converted to a digital signal for transmission and then converted back to analog form. Conversely, a digital signal can, if desired, be transmitted without modification by an analog communication system. The relative advantages and disadvantages of the two types of signals and systems will be discussed in detail later.

10.2.5 Modulation

An analog signal obtained from an information source can be converted into any of a great variety of different but equivalent analog signals; each of course, carries exactly the same information. A similar statement can be made concerning digital signals. The process used to convert one form of signal into another is called modulation, and the inverse of the process is called demodulation. (Since no information is lost in the conversion to a different form, the inverse of the conversion, which recovers the original form is always possible.) The objective in changing a signal to a different but equivalent form (modulation) is the matching of the signal to a communication system. More specifically, a communication system is designed to handle signals of a specific form. If a signal does not have the proper form, its form must be changed by a suitably chosen modulation process. It can be said then that modulation is a reversible process used to match a signal to a communication system.

Pulse Modualtion.   One class of modulation processes consists of methods of changing the form of pulses in a pulse type of signal, and another class consists of methods changing the band of frequencies that make up the signal. The class concerned with the changing of pulse shapes will be discussed first.

Pulse Amplitude Modulation (PAM).   A diagram of several pulse type signals and the processes for conversion between types is shown in Figure 10.6. An analog signal is shown in the diagram to permit inclusion of the process of conversion from an analog signal to a digital signal by sampling.[3] The result is a pulsed

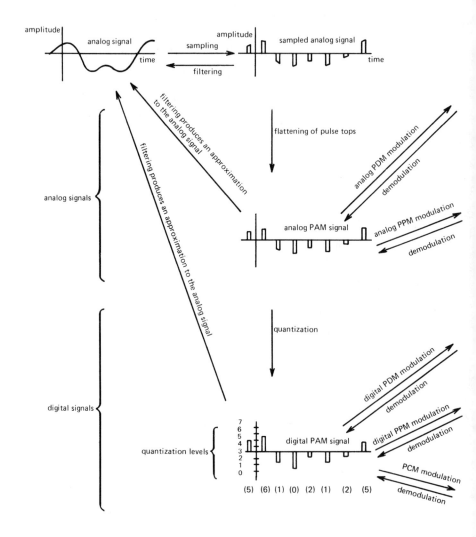

Figure 10.6 Various Forms Taken by Pulse Type Communication Signals and Processes for Conversion between Forms

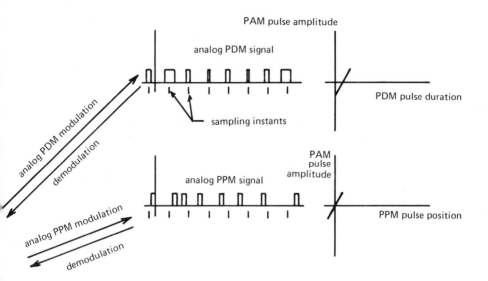

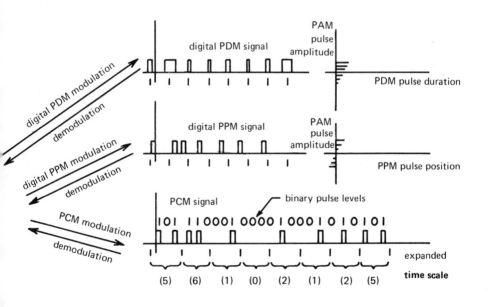

analog signal which has been discussed above. If the tops of pulses are flattened to their average value, the result is a pulse amplitude modulated (PAM) signal.[3] This is still an analog signal since the pulse heights can be any value, and it is distorted due to the flattening of the tops of the pulses. As mentioned above, this signal can be converted to an approximation of the original analog signal. The error in the approximation approaches zero as the pulse width in the sampling process approaches zero.

Pulse Duration Modulation (PDM). An analog PAM signal can be converted to other forms of pulse signals. One of these forms is called a pulse duration modulated (PDM) signal or a pulse width modulated (PWM) signal.[3] (The terminology PDM will be used here.) It is obtained from the PAM signal by replacing each PAM pulse by a PDM pulse having a duration which is a function of the amplitude of the PAM pulse it replaced. Of course, the PDM signal can be converted back to the PAM signal with no distortion or loss of information. All PDM pulses have the same amplitude. The relationship between the PAM pulse amplitude and the PDM pulse duration is under the control of the system designer. A possible relationship is given in the graph sketched near the PDM signal in Figure 10.6. There are, of course, some constraints on the relationship; for example, if the slope of the graph in Figure 10.6 were too small, the pulses could overlap. The control over the relationship between PAM pulse amplitude and PDM pulse width possessed by the system designer has important consequences. To facilitate a description of those consequences, assume that the relationship between PAM pulse amplitude and PDM pulse duration is linear. (It is not necessarily linear, but such an assumption simplifies the reasoning in relation to the following ideas and yet does not diminish their generality.) A decrease in the slope of the linear relation implies a greater range of values of pulse duration which in turn implies a greater bandwidth. However, a greater range of values of pulse duration also results in greater tolerance to degradation of the signal by noise. Thus a designer using PDM can make the signal more or less tolerant of noise by adjusting the bandwidth — the greater the bandwidth the greater the resistance to degradation by noise. Since greater bandwidth requires greater communication system capacity and consequent higher cost, a price must be paid for the aforementioned greater resistance to degradation by noise. The question as to the degree of resistance to noise the designer should employ depends upon its cost in terms of bandwidth (and thus also in terms of investment capital); this in turn is different for different systems. Each system design therefore requires a complete analysis of costs and benefits.

It has just been argued that a telecommunication system designer can diminish the effects of noise on signals at the expense of increased bandwidth if PDM is being employed. (As will be seen, the same is true for pulse position modulation, frequency modulation and phase modulation.) Another method for diminishing noise effects consists in increasing the signal power but keeping bandwidth constant. (If signal power is increased while noise power remains constant, it is easily reasoned that the signal will be less affected by the noise.) It may thus be said that if signal quality is to remain constant then bandwidth can be increased while signal power is decreased or bandwidth can be decreased while signal power is increased. It can be said that there is a *trade-off between bandwidth and signal power.* This trade-off has been introduced as a characteristic of pulse duration modulation, but it also occurs with a number of other types of modulation as well. The nature of the trade-off depends upon the type of modulation. The theoretical optimum trade-off can be found by studying the channel capacity theorem of information theory. This will be discussed in section 10.3.

Pulse Position Modulation (PPM).   Returning now to the various forms of pulse signals as diagrammed in Figure 10.6 another alternative form obtainable from the analog PAM signal is the pulse position modulated (PPM) signal.[3] In this signal there is one pulse for each of the PAM pulses, and all pulses are of the same amplitude and width. However their positions with respect to the PAM pulses they represent are shifted right or left on the time scale an amount dependent upon the amplitude of the PAM pulse they represent. (The PPM signal must of course lag behind the PAM signal by an interval sufficiently long to permit the necessary signal processing.) Needless to say, the inverse of the conversion process recovers the PAM signal exactly. The relationship between the pulse position and the pulse amplitude may have any form; a linear relation is sketched to the right of the

PPM signal in Figure 10.6. The communication system designer
has control over this relationship subject to certain constraints
just as in the case of PDM. If the slope of the line in the linear
relationship of Figure 10.6 is decreased (it must not be decreased
to an extent causing pulses to overlap), the bandwidth is increased
and at the same time the tolerance of the signal to degradation
by noise is increased. Thus with PPM the designer also has the capability of trade-off between signal power and bandwidth just as
was true with PDM.

Digital Pulse Modulation.   The analog PAM signal can be quantized to convert it to a digital PAM signal. In the quantization
process a finite number of discrete amplitude levels are selected
and each analog PAM pulse is changed in amplitude to the nearest
discrete level.[3] Then each pulse can be represented by a number having a prespecified number of digits. (The number of digits is the same for every pulse and is determined by the number
of quantization levels and the number system used.) The discrete
levels used in the quantization process may be equally spaced or
non-equally spaced. Sometimes closer spacing is used for small
amplitudes than for large to improve the quality of low level signal. The changing of the amplitude of a PAM pulse to one of the
discrete amplitude levels causes an irreversible loss of information;
the original amplitude cannot be recovered by any sort of inverse
operation. However if the discrete amplitude levels are spaced sufficiently close, the loss of information can be made negligibly
small, and the analog signal obtained from the digital PAM signal by filtering will differ negligibly from the original analog signal.

The digital PAM signal obtained from the analog PAM signal can
be completely described by a sequence of multidigit numbers,
each number representing the amplitude of one pulse. In a sense
the sequence of numbers may be described as a code for the signal; for this reason the signal is sometimes called a pulse code
modulated (PCM) signal. A PCM signal can take a variety of forms,
in fact a different form for every number system that can be used

to express the pulse amplitudes. An example may clarify this point. Suppose that the quantization in Figure 10.6 utilizes 10 levels, then each pulse can be described by one decimal digit and the signal could be coded into a sequence of single decimal digits. Now suppose each decimal digit is written as a four digit binary number. Then a digital PAM (or PCM) signal having pulses of either of two possible amplitudes (binary pulses) can be generated to correspond to this binary description of the signal. Such a signal is sketched in Figure 10.6; successive blocks of four pulses represent decimal digits. Binary PCM signals are widely used. It may be noted that less distortion of the signal would occur if the quantization had been into 16 discrete levels rather than 10. The 16 levels could still be converted into 4 binary pulses; this is the procedure that is customarily used. More generally, if binary pulses are to be used, the number of discrete amplitude levels is selected to be a power of 2; then the number of binary pulses needed to represent each PAM pulse is the logarithm (base 2) of the number of levels.

The digital PAM signal can be converted to a digital PDM or to a digital PPM signal in the same manner that the analog PAM signal was converted to the analog PDM or to the analog PPM signal. In the case of a digital PDM signal the pulse duration can be of any of a discrete set of values, one value for each of the possible amplitudes of the digital PAM pulse it represents. Similarly a discrete PPM signal has a pulse shifted by any of a discrete set of values, one for each possible amplitude of the digital PAM pulse it represents.

The various forms of pulse signals shown in Figure 10.6 have been discussed. Each of these signal forms is used in practical systems. The characteristics of different communication systems dictate the use of different forms of signals; the choice must be made by the system designer. The presentation of pulse signals in Figure 10.6 is by no means complete since there are many others; these are simply the more common.

Modulation Terminology.   Some discussion is needed concerning the terminology of the conversion between signal forms. The conversion from PAM signals to PDM signals is usually called pulse duration modulation or pulse width modulation with the adjectives analog or digital added, if necessary, to clarify the meaning. The conversion from PDM to PAM is called pulse duration demodulation (or pulse width demodulation). Similarly pulse position modulation and demodulation are used for the conversion between the PAM and PPM signals. The conversion from one PCM (or digital PAM) signal to another is called pulse code modulation (or digital PAM). The term modulation fundamentally means a reversible change in the form of the signal and demodulation refers to the reverse change which exactly (in principle) recovers the original signal. The conversion from an analog to a sampled analog signal is a reversible conversion but it is usually not called modulation, but rather is called sampling. The inverse conversion from the sampled analog signal to the analog PAM signal does not have a standardized name and often is considered as part of a conversion from an analog signal directly to an analog PAM signal and is then referred to as pulse amplitude modulation.

The discussion of the signals of Figure 10.6 may have implied that the various pulse signals are all alternative forms for analog signals; this is not true. In many systems the information source (a computer, for example) generates digital signals which can then be converted to any of the various alternative digital signal forms.[5]

Modulation for Frequency Shifting.   The other class of modulation processes involves changing the band of frequencies present in the signal. As explained above, any signal can be expressed as the sum of a number of sinusoidal signals each of different frequency, amplitude and phase. A graphical representation of the signal consists of a graph of the power (or the square of the amplitude) of the sinusoidal wave at the various frequencies. Examples are shown in Figure 10.7. In Figure 10.7(a) it is assumed the signal is a sum of a finite number of sinusoidal signals; the height of the vertical lines indicates the power of the individual sinusoidal signals. In Figure 10.7(b) it is assumed that the signal is a sum (or integral) of an infinite number of sinusoidal signals, one at every frequency for which the graph is non-zero. In either case,

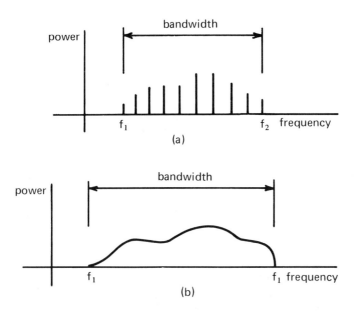

Figure 10.7 Frequency Domain Representations of Signals

the important characteristics of the signal are the upper, f_2, and lower f_1 frequencies present in the signal and the difference between them which (as mentioned above) is called the bandwidth (meaning the width on the frequency axis of the band of frequencies in the signal). Such graphs were introduced in section 10.2.2 and are called frequency domain descriptions of signals. More specifically those in Figure 10.7 are called power spectra. These are incomplete descriptions because they do not give the phase of the various sinusoidal components of the signal. The consequence of this is the fact that a power spectrum is not a description of a single signal but rather of an entire class of signals. The sinusoidal signals of power and frequency specified by the power spectrum may have any phase relationship among themselves. Since there are an infinity of ways the phases of the various sinusoidal signals can be chosen, there are an infinity of signals, all of which have exactly the same power spectrum. The incompleteness of the power spectrum as a description of signals may, at first thought, seem to be a disadvantage; but communication systems are not designed to handle precisely defined, prespecified signals but rather to handle broadly defined classes of signals. Furthermore these classes of

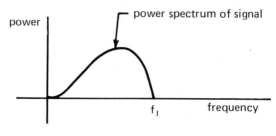

(a) of signal

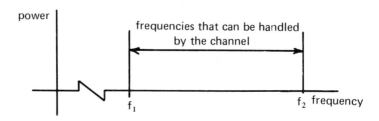

(b) of the Communication Link

Figure 10.8 Power Spectrum

signals generally have frequencies in the same band. Therefore the power spectrum is an ideal description of signals for use in the study of communication systems.

Another reason for the use of the power spectrum is the fact that communication links (the transmission facilities in a communication system) are customarily defined in terms of the band of frequencies that they will transmit. In the overall design of a communication system, the characteristics of the signal to be transmitted over the communication link must be of a nature that can be handled by the link, and since the link is usually defined in terms of the power spectrum that can be transmitted over it, it is clear that the most useful type of signal description is the power spectrum.

The problem of matching a signal to a link may be described by reference to Figure 10.8. The signal has the power spectrum shown at the top in Figure 10.8 (the signal to be shifted in frequency is often called the baseband signal), and the channel has the capability of transmitting signals in the frequency band from f_2 to f_3. The matching of power levels is trivial since the levels can be changed using amplifiers or attenuators. Nothing more will be said of this aspect of the matching problem.

Practical and economical communication links often are such that f_2 and f_3 are both very much larger than the highest frequency, f, present in the signal. In order to match the signal to the link, its frequency band must be shifted up into the region between f_2 and f_3. This can be done by any of several modulation processes. The definition of modulation in this case is the same as in the case of pulse modulation; it is a reversible transformation of the signal. The inverse transformation which recovers the original signal is called demodulation. There are three fundamental types of modulation whereby frequency bands may be shifted: amplitude modulation (AM), frequency modulation (FM) and phase modulation (PM).

Amplitude Modulation (AM).   There are three common types of amplitude modulation: double sideband, double sideband-suppressed carrier, and single sideband.[3,9] They are described in terms of manner in which they shift frequencies in Figure 10.9. Double sideband amplitude modulation shifts the power spectrum and its image with respect to the zero-frequency line up to the region between f_2 and f_3. (It is assumed that $f_3 - f_2 = 2f_1$ in this discussion.) In addition a so-called carrier is added to the signal at the center frequency, f_c. The presence of the carrier requires much more power thus making transmission less efficient than double sideband-suppressed carrier. However, it's presence permits use of less complex, and thus less expensive, receivers. The two halves of the double sideband amplitude modulated signals are called upper (right hand) and lower (left hand) sidebands. It is possible to shift only the upper sideband or only the lower sideband. These are called single sideband amplitude modulation. There are a variety of amplitude modulation techniques intermediate between double sideband and single sideband; these shift parts of both sidebands and are called vestigial sideband (VSB) amplitude modulation. The bandwidth of amplitude modulated signals ranges

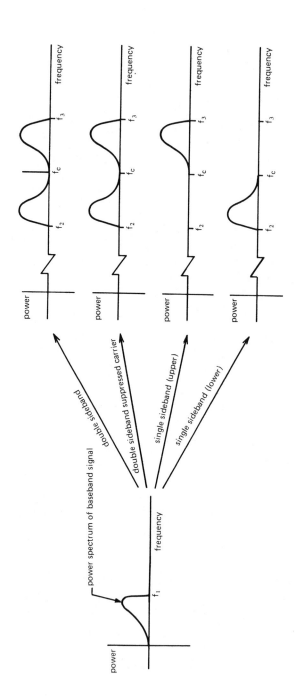

Figure 10.9 Frequency Shifting Characteristics of Various Types of Amplitude Modulation

from f_1 (the bandwidth of the modulating signal) for single sideband to $2f_1$ for double sideband with intermediate values for
vestigial sideband. Single sideband AM is widely used because
its narrow bandwidth permits more radio links to operate in the
same area within a specified range of frequencies. (To prevent
interference, radio links operate on non-overlapping frequency
bands.) This conservation of the spectrum is becoming increasingly important because of the steadily increasing use of radio
systems. The disadvantage of single sideband is the complexity
of equipment compared to that needed for double sideband.
Vestigial sideband has bandwidth and equipment complexity
characteristics intermediate between those of double sideband
and single sideband; it is used in commercial television broadcasting.

Frequency Modulation (FM). Frequency modulation [3,9] can
be described in terms of the diagrams in Figure 10.10. Here again
the band of frequencies making up the baseband signals is shifted
up into the region between f_2 and f_3. (In this discussion it is assumed that $f_3 - f_2$ is greater than $2f_1$.) This process, however,
does not preserve any of the characteristics of the shape of the
power spectrum of the baseband signal. Furthermore, the bandwidth, $f_3 - f_2$, of the modulated signal is usually somewhat greater than for amplitude modulation; it is approximately given by
$2mf_1$ where m is the modulation index. It is equal to, or greater
than, one and is often in the vicinity of 5 or so. Thus a typical
FM signal has several times the bandwidth of the equivalent AM
signal. In view of the electromagnetic spectrum crowding problem
mentioned above, it may be asked why FM is ever used. The answer lies in the fact that FM, like some of the pulse modulation
techniques mentioned previously, permits the system designer to
control the bandwidth (by choice of the modulation index, m)
and simultaneously the vulnerability of the system to degradation
by noise. As discussed before, if the quality of the signal is to be
kept constant (in terms of noise), the designer can increase the
bandwidth and decrease the transmitter power or alternatively
decrease the bandwidth and increase the transmitter power. Because of this tradeoff it is possible to transmit very high quality
signals using very low transmitter power and correspondingly high
bandwidth. For this reason FM is used in many applications in

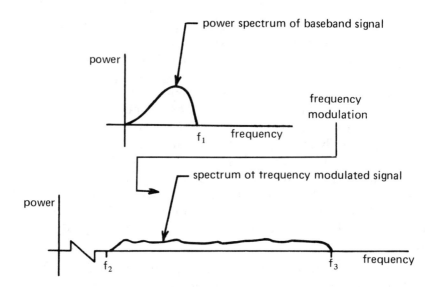

Figure 10.10   Frequency Shifting Characteristics of Frequency
                Modulation

which transmitter power is limited. These include such diverse uses
as transmission of biological information from birds in flight to recorders on the ground and transmission from earth satellites and
other space vehicles to the earth.

Phase Modulation (PM).   Phase modulation has characteristics
which are very similar to those of FM, and it has the same band-
width-transmitter power tradeoff characteristics.[3,9] For this
reason no further discussion of phase modulation is needed.

Digital Modulation Terminology.   AM, FM, and PM take particularly simple forms in cases of digital baseband signals. Because
of the simplicity of the processes, they are easily described by
simple phrases which are often used as names for the modulation
processes. Those phrases (names) are so widely used that they
need to be introduced at this time. In the case of digital baseband
signals, AM is called amplitude shift keying (ASK), FM is called
frequency shift keying (FSK), and PM is called phase shift keying
(PSK). A device, used at a digital terminal, that performs such
modulation on the outgoing signal and the corresponding demodulation on the incoming signal is commonly called a modem (*mo*du-
lator-*dem*odulator).

Multiplexing. The types of modulation that involve the shifting of frequency bands to facilitate the matching of a signal to a channel also provide the means for a second method of multiplexing, frequency division multiplexing. The first method, time division multiplexing (TDM) was introduced in the discussion of sampling and pulse amplitude modulation.

Frequency Division Multiplexing (FDM). Multiplexing in general means the combining of a number of signals into one, called the multiplexed signal, in such a way that the separate signals can be recovered from the multiplexed signal.[3] A multiplexing system is sketched in block diagram form in Figure 10.11(a). Frequency division multiplexing is described in terms of power spectra in Figure 10.11(b). Each of the individual input signals has its band of frequencies shifted to a different band. These are then added together to produce the multiplexed signal. The demultiplexing process consists of extracting the various bands of frequencies from the multiplexed signal by means of filters. The outputs of the various filters are then frequency shifted, using the same modulation process as used for multiplexing, to the proper band so as to exactly recover the individual signals that were originally multiplexed. The modulation process (FDM) used for the frequency shifting can be AM, FM, or PM; often single sideband AM is used (as is done here) because the narrow bandwidth permits more signals to be multiplexed into the frequency band available for use. It is apparent from the diagram that the multiplexed signal has a bandwidth equal to the sum of the bandwidths of the signals multiplexed.

Time Division Multiplexing (TDM). Time division multiplexing which was discussed previously is described diagramatically in Figure 10.11(c). The interleaving and the separation processes are indicated by arrows. This illustrates multiplexing for a sampled signal, but obviously any sort of pulse signal can be time division multiplexed. It can be shown that just as it is true for FDM, the minimum possible bandwidth of the multiplexed signal is the sum of the bandwidths of the signals originally sampled. Multiplexing permits many individual, relatively narrow bandwidth signals to be transmitted by a single wide bandwidth communication link. The only justification for multiplexing results from the fact that in many cases a single wide band link is less expensive than many narrow band links.

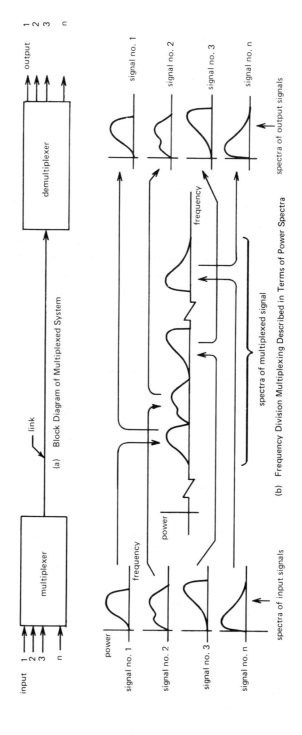

(a) Block Diagram of Multiplexed System

(b) Frequency Division Multiplexing Described in Terms of Power Spectra

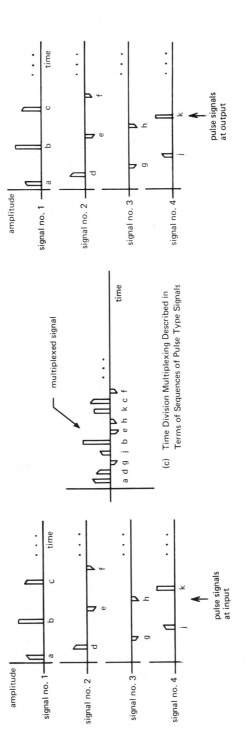

Figure 10.11   Multiplexing Techniques

10.2.6 Effects of Noise on a Communication Signal

Little has been said, in this discussion, of the effects of noise upon communication signals. Every device that generates, transmits, modulates, amplifies, multiplexes, etc. contributes noise to the signal. It may be said that the contamination of signals by noise is inevitable. The noise distorts the signal and thereby tends to obscure its information content. The degree to which information is obscured or lost depends upon the amount of noise power added in relation to the amount of signal power. It also depends upon the type of signal and the use to be made of it. All signals have some degree of noise tolerance; that is the information can be extracted from the signal despite the presence of noise. The noise tolerance is relatively low for video signals; in other words, the noise power must be very small compared with signal power to produce a high quality picture. In the case of audio signals, speech is intelligible even with substantial amounts of noise power, although music requires lower noise power levels for pleasant listening. Digital signals are highly tolerant of noise; indeed it is possible to transmit digital signals with low error rates in the presence of noise power of levels nearly of the order of magnitude of the signal power. As noise power levels increase, communication is still possible but with somewhat lower quality; video signals produce pictures somewhat distorted or with less detail, audio signals are still understandable but with greater difficulty due to the noise in the background, and digital signals have more errors in the symbols.

Signal-to-Noise Ratio (SNR): A Measure of Analog Signal Quality
In the analysis of telecommunication systems a quantitative measure of signal quality is needed. In the case of analog signals that measure is customarily taken to be the signal-to-noise ratio which is the signal power divided by the noise power. This measure will be abbreviated SNR. It is a number which is usually in the range of about 2 up to 100,000 or so, but in some cases it is outside these limits. Because of the large numbers that SNR often takes, it is customary to express them in decibels. To do so, the logarithm (base 10) is taken and the result is multiplied by 10. The range 2 to 100,000 then becomes 3 dB to 50 dB. (Decibel is abbreviated dB.)

Error Probabilities in Digital Signals: A Measure of Digital Signal Quality.   In the case of digital signals, their quality can be measured in terms of the SNR (signal-to-noise ratio) just as in the case of analog signals, but it is more meaningful to use a different measure, namely the probablity of error, BER (meaning bit error rate), of a signal. This measure usually lies in the range 10^{-2} to 10^{-8}, meaning error rates ranging from 1 digit in 100 to 1 in 100,000,000. Since SNR and BER are both measures of the quality of digital signals, it may be expected that they are related. This, of course, is true and the relationship depends upon the type of modulation used. The relationships are somewhat involved but approximations (good for signal-to-noise ratios of about 10 or higher) can be written concisely for ASK (amplitude shift keying), FSK (frequency shift keying and PSK (phase shift keying). The approximations are

$$\text{BER} \simeq \tfrac{1}{2}e^{-\alpha\,(\text{SNR})} \text{ for non-coherent systems,} \qquad (10.1)$$

and

$$\text{BER} \simeq \frac{\tfrac{1}{2}e^{-\alpha\,(\text{SNR})}}{\sqrt{\pi\alpha\,(\text{SNR})}} \text{ for coherent systems.} \qquad (10.2)$$

The parameter α is to be chosen ¼ for ASK, ½ for FSK, and 1 for PSK. Different expressions are needed for coherent and non-coherent systems. The distinction between such systems is discussed below.

Signal Quality Using Coherent Demodulation.   Coherent systems are those for which a reference signal is transmitted from the modulator to the demodulator to be used in the demodulation process. When this is done it is possible to reduce the noise power in the output of the demodulator to a greater extent than is possible for a non-coherent system (a system without such a reference signal). Thus a coherent system gives a higher quality signal (a signal of higher SNR) than a non-coherent system. However, there is a price to be paid since coherent systems are more complex and costly due to the necessity of transmitting the reference signal to the receiving end of the system.

A study of the above relations between BER and SNR shows that for a specified SNR, a coherent PSK system gives the lowest BER followed by non-coherent PSK, coherent FSK, non-coherent FSK, coherent ASK and non-coherent ASK in that order. The best choice of a system depends upon costs and other factors. All six of the systems are in wide use.

10.2.7 Error Control in Digital Systems

It is apparent that occasional errors in digital signals are inevitable because of the ever present noise that contaminates the signal. In some systems those occasional errors are so serious that substantial effort is justified in reducing their number. They can be reduced by reducing noise levels in the system, but there is a limit to the improvement that can be achieved by this method at a reasonable cost. However there is another method which is widely used; it involves the coding of the information in such a manner that errors can be detected and corrected. The codes used for this purpose are called error detecting and error correcting codes. [10] The concepts underlying such codes will be briefly discussed below. Error detecting codes will be considered first.

Error Detecting Codes. The simplest code for error detection makes use of a so-called single parity digit. To illustrate its use, assume the digital signal is coded into successive blocks of 5 binary digits. (This assumption does not limit the applicability of the results of this discussion because any digital signal can be recorded into blocks of any length and of any set of symbols.) Now assume several of the successive blocks are as shown in Figure 10.12. One additional binary digit is added to each block making the total block length 6 digits. The added digit is called a parity digit and it is chosen so as to make the total number of 1's among the 6 digits an even number. (An alternative to the code described here could be based on selection of a parity digit to make the total number of 1's an odd number.) At the receiving end of the communication system the number of ones in each block is counted; if it is an odd number, the block is in error and a signal is sent to the transmitting end of the system causing the block that was in error to be transmitted again. Then a parity check (a count of the number of 1's) is made again at the receiver. If the parity check is correct (an even number of

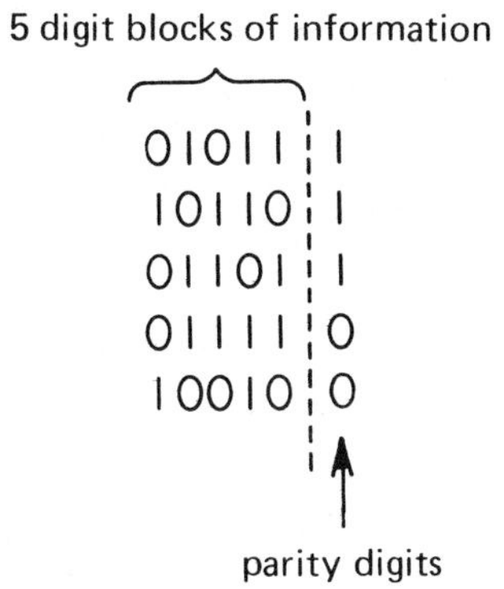

5 digit blocks of information

```
0 1011 | 1
1011 0 | 1
0 1101 | 1
0 1111 | 0
1 0010 | 0
            ↑
```

parity digits

Figure 10.12   A Code Utilizing Single Parity Digits

ones) the block is accepted as correct; if not, a signal is sent
causing another retransmission. This procedure is then repeated
until the parity check on the received block is correct. (In practice, more than one retransmission is rare.) It is apparent that
the use of the single parity digit will not detect all possible errors. It will detect all possibilities of a single error but it will not
detect any double errors. (If two errors occur in a block, the
number of 1's will still be an even number.) More generally, it
may be said that the use of a single parity digit in each block
permits detection of an odd number of errors but it will not
detect an even number of errors. Even though it will not detect
all possible errors this code is useful because in many telecommunication systems the great majority of errors occur not more
than one to a block; so in such a system the use of a single parity digit would detect the great majority of errors. In systems
in which the probability of two errors per block is substantial,
the single parity digit is not particularly successful in reduction
of the error rate; but there are other codes which make use of
more than one parity digit per block (each chosen to have even
parity on a selected set of digits within the block) which are capable of detecting double errors. Generally it may be said that
an error detecting code can be devised to detect any type of
error pattern; so codes are, in essence, designed to fulfill the
needs of a particular communication system. The example a-

bove illustrates that the use of an error correcting code can reduce the error rate, but only at a price, the price being the reduction of the rate at which the information is transmitted by
the channel. Without parity digits, all of the digits transmitted
carry information, but with parity digits, only 5/6 of the digits
carry information; so the information rate is reduced to 5/6 of
the value possible without parity digits. There would be an additional reduction in the information rate due to the necessity
of retransmitting a block when an error occurs. The system designer must decide whether the advantages gained by reduction
of the error rate are worth the cost in terms of the reduced information rate. The reduction in the information rate in the example above is deceptively large; often the length of the block
to which the parity digit is added is somewhat larger, which results in a smaller reduction in the information rate.

Error Correcting Codes.   The other type of code used for error
control is called an error correcting code.[10] Such codes will
be illustrated by a simple example. Consider information grouped
into blocks of 4 binary digits. Let 3 parity digits be added to each
block for a total block length of 7 digits. Let the information digits be labeled I_1, I_2, I_3, and I_4 and let the parity digits be labeled
P_1, P_2, and P_3. Each parity digit will be selected for even parity
(an even number of 1's) among itself and a selected subset of the
information digits; the rules for determining each parity digit are
given in the encoding-decoding table in Figure 10.13. The top row
of the table is labeled P_1. The P_1 is selected so that there will be an
even number of 1's among I_1, I_2, I_4, and P_1. P_2 is selected so there
will be an even number of 1's among I_1, I_3, I_4, and P_2; and finally
P_3 is selected so there will be an even number of 1's among I_2, I_3,
I_4, and P_3. An example may make this clear; assume the information digits are $I_1 = 1$, $I_2 = 0$, $I_3 = 0$ and $I_4 = 1$. P_1 is selected so
there will be an even number of 1's among I_1, I_2, I_4 and P_1 that is
among 1, 0, 1 and P_1; so P_1 must be 0. Similarly the rules dictate
that $P_2 = 0$ and $P_3 = 1$; so the code word, complete with its parity
digits, is 1001001. This then is transmitted instead of the block
1001, of information digits only. It is evident that the "cost" of
the use of this code is the reduction of the information rate to

	I_1	I_2	I_3	I_4	P_1	P_2	P_3
P_1	✔	✔		✔	✔		
P_2	✔		✔	✔		✔	
P_3		✔	✔	✔			✔

Figure 10.13 Encoding-Decoding Table For a Simple Error Correcting Code

4/7 of that which would be possible without the addition of parity digits.

To illustrate the use of this code, assume the above codeword, 1001001, is transmitted over the communication system. At the receiving end of the system the parity checks indicated by the encoding decoding table are made. More specifically, the number of 1's among I_1, I_2, I_4 and P_1 is determined. This is called parity check number 1, and if the number of 1's is even, the parity check is said to be correct; otherwise it is said to have failed. Similarly, parity checks number 2 and 3 are made according to the definitions given by the 2nd and 3rd rows of the encoding-decoding table. Suppose that an error occurs and the block is recieved as 1011001. At the reciever the first parity check is correct but the second and third both fail. Then it is assumed that only one error has occurred and reference to the encoding-decoding table indicates the location of the error. (For systems for which the probability of double errors is appreciable this code is unsuitable, but other codes are available for use.) If the error were in I_1 the table shows that I_1 enters into parity checks 1 and 2 but not 3; so the first two parity checks would fail and the third would be correct. This is not the result of the parity checks so the error cannot be in I_1. Each of the other possible positions is considered, and it is seen that the only position where an error could cause the first parity check to be correct and the other two to fail is I_3; therefore the error must be in digit I_3. I_3 is then changed producing 1001001, which was the block transmitted. This procedure is called error correction, and it gives the code its name: an error-correcting code.

Efficiency of Codes. The procedure for making the parity checks and performing the corrections is automatically done by computer-type circuits in the receiver. In this example the "cost" of the use of the error correcting code is the reduction in the information rate by the factor 4/7; alternatively, the efficiency of the code may be said to be 4/7. It may be said that, in general, error detection and error correction is accomplished by the placement of parity digits among the information carrying digits or, in other words, by the incorporation of redundancy into the message. (The parity digits are redundant because they carry no message-information.) This redundancy provides protection against errors by permitting error detection or correction in much the same manner that redundancy in natural languages makes them resistant to errors. The redundancy in error control codes or in natural languages reduces communication efficiency by necessitating longer sequences of symbols to carry the information. Codes can be devised with whatever degree of error control needed for a specific application. Generally the greater the probability of error, the greater the redundancy, the less the efficiency, and thus the greater the cost of implementation. Error control codes take a great variety of forms. Many have longer block length than those discussed; there are also codes that don't have a block structure, but in general codes can be devised to overcome errors with any sort of characteristics.

10.2.8 Coding for Information Security

Codes as discussed above are used strictly for error control purposes; that is, for protection of digital messages from distortion and thus errors due to noise. Another important use for codes is in preserving the security of information or, in other words, coding for secrecy.[8] At one time, this need existed primarily in military systems, but now it exists in all systems that transmit information, protected by privacy laws, concerning all citizens. Techniques for coding and decoding digital information for secrecy purposes have been standardized, and equipment for such purposes is widely available.

10.3 QUANTITATIVE CHARACTERIZATION OF TELECOM–MUNICATION SIGNALS AND SYSTEMS

Telecommunication systems exist to transmit information from one place to another. In order to describe and discuss such systems quantitatively, a method of measuring information must be devised. It is also necessary to devise a method of measuring the capacity of a telecommunication system to transmit information. These needs are met by modern information theory, the primary results of which will be discussed below.

10.3.1 Information Theory

Information theory will be discussed in terms of digital systems, although the ideas can be extended in a straightforward manner to analog systems. Digital information takes the form of a sequence of symbols, each symbol being selected from a symbol set of finite size. Assume a digital information source uses an alphabet consisting of symbols A,B,C,D, and E; then a message may take the form ACDBAABDECBE, another may be BCEADCA. The information source can be studied statistically to determine probabilities for every possible message. Strong arguments can be made for defining a measure of information that assigns higher information content to messages of lower probability. One may reason that a message of low probability is more informative because it is more unexpected and in that sense, contains more information than one that is expected. (The information is greater when the surprise of receiving the message is greater.) A great variety of other arguments support the basic idea that lower probability should correspond to greater information. This is the most fundamental concept in information theory; to implement it, assume the probabilities of occurrence of the various symbols are p(A), p(B), p(C), p(D) and p(E). Assume further that the probabilities of occurrence are independent of previous symbols in the sequence. The arguments can be modified for the case of non-independence yielding the same concepts but with greater mathematical complexity.

One possibility for the definition of information content is that of making it equal to the reciprocal of the probability. If this definition is used, the information content in symbol A would be $1/p(A)$ and that in symbol B would be $1/p(B)$. The two symbol message AB would have probability p(A)p(B) and information content 1/p(A)p(B) which is (1/p(A)) (1/p(B)), the product of the information content of A and that of B. The information content of a sequence of symbols would be the product of the information in the various symbols. Such a definition of information could be used but it would be awkward. An additive measure of information would be much more convenient than a multiplicative measure. A simple modification of the definition produces an additive measure of information; the information content, I, of a message will be defined as the logarithm of the reciprocal of the probability, p, of the message;[11] or

$$I = \log \frac{1}{p} \tag{10.3}$$

Thus the information content of the symbol A is $\log \dfrac{1}{p(A)}$ and

that of B is $\log \dfrac{1}{p(B)}$. Now the information content of the message AB is $\log \dfrac{1}{p(A)p(B)}$ which is $\log \dfrac{1}{p(A)} + \log \dfrac{1}{p(B)}$ or the sum of the information in A and B. Generalizing, it may be said that the information in a sequence of symbols is the sum of the information in the various symbols in the sequence. The logarithm in the formula for information may be for any base that is convenient, but almost all analytical investigations are conducted using the base 2, in which case, the unit of information is called the bit.

10.3.2 Entropy

Considering again a sequence of the symbols A,B,C,D, and E, the information content of the message would depend upon the number of the various symbols present in the message. If the message is reasonably long then the numbers of the various symbols can be closely estimated from their probabilities, and it is possible to calculate the average information content per symbol. A straightforward averaging procedure produces the result.

$$H = p(A)\log \frac{1}{p(A)} + p(B)\log \frac{1}{p(B)} + p(C)\log \frac{1}{p(C)} + p(D)\log \frac{1}{p(D)}$$

$$+ p(E)\log \frac{1}{p(E)} . \tag{10.4}$$

The average is called the entropy because of the close conceptual correspondence with thermodynamic entropy and has the units of bits per symbol. It is easily shown that to maximize the entropy (which will then maximize the efficiency of utilization of a communication link) the probabilities of the various symbols should all be equal. Let k be the number of symbols, then maximum entropy H_{max} occurs if the probabilities of all symbols are equal; in other words, $p = \frac{1}{k}$ for all symbols. Then the above formula gives $H_{max} = \log k$. Sometimes a recoding of information into a new set of symbols is done so as to increase the entropy and thus the efficiency of communication.

Methods of measuring the information, I, in a message (of any number of symbols) and for the average information per symbol, have been introduced. One more convenient quantity is needed to quantitatively describe messages, that is the average information rate, r. If the message source produces n symbols per second and the entropy is H, then the information rate (average) is

$$r = nH, \tag{10.5}$$

and it is measured in bits per second.

10.3.3 Channel Capacity

The second major use of information theory concepts, that of characterizing the information handling capabilities of a channel, will now be discussed. The discussion will be phrased in terms of a system handling a digital PAM signal but the result can be generalized to any system. A well known theorem relates the maximum possible rate, N of digital pulses that can be transmitted by a channel to the channel bandwidth, B. The relation is

$$N = 2B \tag{10.6}$$

Achievement of this rate requires sophisticated circuitry; usually simpler equipment and lower rates are used, but the determination of the maximum possible channel capacity must be based upon this relation.

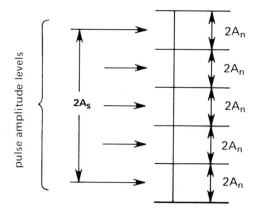

*Figure 10.14   Diagram Illustrating Selection of Pulse Amplitudes
so as to Prevent Errors Caused by Noise*

The amount of information in a pulse depends upon the number
of possible levels, K, the pulse can take. Each level can represent
a different symbol, so the maximum information represented by
a pulse is the maximum information per symbol for a set of K
symbols; this was previously found to be log K. The maximum
rate at which information can be transmitted over a channel is
called the channel capacity, C, and the expression for it is

$$C = N \log K \tag{10.7}$$

Another and more useful expression for C is obtained by expressing K in terms of the signal-to-noise ratio, SNR, at the receiving
end of the channel. The relationship is[2]

$$K = \sqrt{1 + SNR} \tag{10.8}$$

The derivation of this expression is somewhat involved but a simple discussion provides insight into concept upon which it is based.
Referring to Figure 10.14 which illustrates the selection of pulse
amplitude levels, assume that the noise amplitude has a maximum
value of A_n. Since the noise amplitude can be either positive or
negative, it can add or subtract from the amplitude of a pulse.
Therefore, to prevent noise from changing a pulse level to a value nearer that of another pulse level, the pulse levels must be sep-

arated by at least $2A_n$. Letting A_s be the maximum pulse amplitude, and taking into consideration the fact that pulses can be positive or negative, a simple relation can be found for the number, K, of possible pulse amplitudes. From the figure it is seen that

$$K = \frac{2A_s + 2A_n}{2A_n} = 1 + \frac{A_s}{A_n}. \tag{10.9}$$

The quantity A_s/A_n is the ratio of signal amplitude to noise amplitude. SNR is the signal to noise ratio in terms of power; it is the square of the signal to noise ratio in terms of amplitude; thus $SNR = (A_s/A_n)^2$, and

$$K = 1 + \sqrt{SNR} \tag{10.10}$$

An approximation to this relation is

$$K \simeq \sqrt{1 + SNR} \tag{10.11}$$

which is the one cited above as being the correct one. The simplified discussion above is based upon the assumption that a maximum value, A_n, can be assigned to noise amplitude. Actually this cannot be done and a much more involved derivation is needed, but the simplified discussion above gives meaningful insight into the significance of the relationship between K and SNR. This relation provides another formula for channel capacity,

$$C = B \log (1 + SNR) \tag{10.12}$$

10.3.4 Fundamental Theorem of Information Theory

Now that methods of measurement of information in signals and methods of measurement of information handling capability of channels have been introduced, it is possible to state what is sometimes called the fundamental theorem of information theory.[2] It is convenient to present the theorem in the diagram below. A key point concerning the fundamental theorem is that if the information rate, r, does not exceed the channel capacity, C, the information in the signal can be transmitted by the channel. The theorem does not say the signal can be transmitted, but rather that the information in the signal can be transmitted. The signal can be transmitted only if it is matched to the channel; the matching conditions are $n \leqslant N$ and $k \leqslant K$. If these are not both satisfied, the signal is not matched to the channel and cannot be transmitted. A signal that is not matched to a channel (assuming $r \leqslant C$) can always be recoded into a signal that does satisfy the matching conditions and can therefore be transmitted. The reader is referred to other sources for a discussion of recoding techniques.

Signal Description	**Channel Description**
pulses per second = n	pulses per second = N = 2B
number of levels = k	number of levels = K $=\sqrt{1 + SNR}$
entropy = H = $\sum_{i=1}^{k} p_i \log \frac{1}{p_i}$	channel capacity = C = N log K
average information rate = r = nH	or C = B log (1 + SNR)

Fundamental Theorem of Information Theory

If $r \leqslant C$, then the *information* in the signal can be transmitted by the channel with arbitrarily small probability of error.

Matching the Signal to the Channel

If $r \leqslant C$, the channel will transmit the information but not necessarily the signal. If

$n \leqslant N$ and $k \leqslant K$

then the channel will transmit the signal. (Some modulation may be needed for spectrum modification.) If these conditions are not both satisfied, it is always possible to recode into a signal which does satisfy both conditions.

10.4 TYPES OF TELECOMMUNICATION SYSTEMS

10.4.1 Point-to-Point vs. Broadcast Systems

Telecommunication systems can be categorized in several ways. Perhaps the most fundamental classification is into switched, point-to-point systems on the one hand and into broadcast systems on the other. A switched point-to-point system provides communication between any two (or sometimes anong a small number larger than two) of a group of terminals. This is accomplished by a communication link from each terminal to a switching center. The switching center (or centers) are capable of connecting together the communication links of any two of the terminals. The most familiar example of a switched point-to-point system is the worldwide telephone system in which the many switching centers in many different locations are connected together to effectively function as one large, world-wide switching center which is capable of connecting together any two of the terminals connected to any of the individual switching centers. The communication links between terminals (telephones) and switching centers are customarily cables strung on poles or buried underground, but in some cases they are radio links. Similarly, the communication links between switching centers can be cables or radio links. The switched point-to-point type of communication system basically provides for two-way communication between two persons or between two machines (computer or computer terminals) or between a person and a machine. Sometimes the communication involves several persons or machines; and in some cases such a system has very limited one-to-many or broadcast capability.

A broadcast type system, on the other hand, provides for one way communication from one terminal to many, although some broadcast type systems such as cable television can be made to have very limited point-to-point capability. A familiar example is a television broadcasting station. The uses, operation, regulation, technology and financing of broadcasting type of communication systems are so different from those of switched point-to-point systems that analysis of them requires a completely separate study.

10.4.2 Guided Wave vs. Unguided Wave (Radio) Systems

A second type of categorization of communication systems is based upon the type of communication link used. The two types may be referred to as the guided wave type such as cables or waveguides and the unguided wave type such as radio and microwave links. Each of the two types has unique characteristics which strongly influence the choice of the type of systems to use in various applications.

A guided wave system requires the use of a cable or waveguide preventing application in any communication system in which terminals are located in moving vehicles. Thus such a communication system requires use of an unguided or radio type of link. (Isolated exceptions such as guided wave communication, with vehicles on railroad tracks occur infrequently.)

Unguided wave, or radio type systems, radiate signals that propagate through the troposphere. The signals can be received by any radio receiver located in regions having sufficient signal power. For this reason two or more radio systems operating in the same region will interfere with each other if their signals are in the same frequency band. Interference can be avoided by operating each system in a different frequency band, an excellent solution to the interference problem unless there is a demand for more systems than there are frequency bands available for use. This is a problem, called the spectrum crowding problem, that is becoming common in the more heavily populated regions of the world. A solution is the use of guided wave communication links for new systems in those regions. Guided wave systems do not interfere with each other or with unguided wave systems. (This is not a completely accurate statement, but the interference is very small and can be made even smaller if necessary.) For this reason much of the future expansion of communication systems will involve guided wave links. This fact is of considerable importance in an economic sense both because guided wave systems are highly capital intensive and because new communication systems are already consuming a substantial proportion of all investment capital currently available.

Optical Fiber Communication Links. The high cost of guided wave communication links provides strong incentive to achieve as much communication capacity as possible on every link. Channel capacity is highly dependent upon the technology at the time of system design; at the present time it appears that much of the new guided wave communication capacity for the foreseeable future will utilize optical fiber communication links.[4] These links have broad bandwidth and are moderately priced. They have one highly undesirable feature, their dispersion, which causes severe signal distortion. The distortion can be completely overcome by devices called equalizers, but they are intricate and expensive. An alternative is a system in which all signals are digital, thus allowing the use of much simpler and less expensive equalizers; it may possibly (depending upon the design) make equalizers unnecessary.

10.4.3 Analog vs. Digital Systems

The third and last major categorization of communication systems is into those with analog communication links and those with digital communication links. This classification of systems is not nearly as clear cut as those previously discussed because an analog link can handle digital signals by the simple expedient of inserting a modem at each end of the link, and a digital link can handle analog signals by inserting samplers and filters. However, the distinction between the two types of links is important in understanding the current trends in communication system design. Most of the existing communication capacity utilizes analog links, but there are strong arguments favoring digital links for new designs. Digital links are more efficiently adaptable to transmission of data than are analog links. Since demand for data communication is expanding rapidly, and since the cost of digital communication equipment has steadily become more favorable compared to analog equipment (partly as a consequence of the drastically reduced costs of computer-type circuitry made possible by the development of integrated circuits), digital communication links are likely to be widely used in future communication systems.

A "Venn" type diagram in Figure 10.15 illustrates the 8 distinct types of communication systems resulting from the 3 types of categorization discussed in this section. Some of the uses and trends of the various types of systems are mentioned on the diagram.

10.5 POINT-TO-POINT TELECOMMUNICATION SYSTEMS

10.5.1 Guided Wave Communication Links

Point-to-point telecommunication systems, as the name implies, provide two-way communication between two locations or, in some specific cases such as a conference call on the telephone system, among several locations. The communication links used in such a system may be of a guided wave type or of an unguided wave (radio) type, and the signals handled by the links may be in

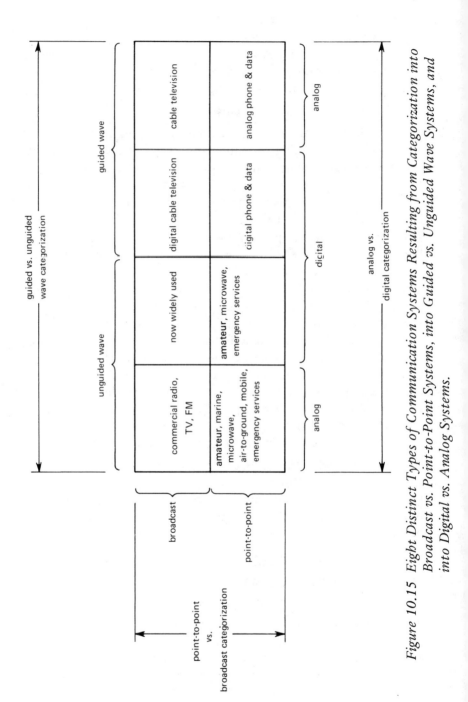

Figure 10.15 Eight Distinct Types of Communication Systems Resulting from Categorization into Broadcast vs. Point-to-Point Systems, into Guided vs. Unguided Wave Systems, and into Digital vs. Analog Systems.

either digital or analog form. The choices of the types of links and of the types of signals for use in various types of systems depend upon their characteristics. A study of those characteristics provides insight both into the choices that have been made in existing systems and into trends in system design likely to persist into the near future.

Guided wave communication systems are characterized by a physical structure which serves to guide the electromagnetic signals that carry information. This is in contrast to unguided wave communication systems in which electromagnetic energy is radiated into the atmosphere or into space in all directions from the radiating structure (antenna); the radiation may be much stronger in some directions than in others, but nevertheless it exists in all directions. The guiding structure in the case of guided wave systems confines the signal to the immediate vicinity of the structure so well that it is possible to operate several such systems side by side in close proximity without interference among them. In contrast, any radio systems operating closely spaced, side by side, will interfere with each other unless each operates on a different frequency band. (In principle, if not in practice, this interference probem can be overcome in a different way by a technique presented later in this section.) Since only a limited number of usable frequency bands exist, only a limited number of closely spaced, side by side radio systems can operate without interference. This frequency spectrum crowding problem has been a compelling reason for use of guided wave systems for uses requiring a very large number of communication links in a limited geographical area. There is essentially no upper limit to the amount of communication that can be achieved in a region with guided wave systems. This is a strong argument favoring guided wave systems for some applications, but there are also strong arguments favoring unguided wave systems. For example, they generally offer lower cost and simpler right of way problems. A choice as to the type of system for use in a specific application obviously requires a more thorough examination of the characteristics of the two types of systems. Further discussion of those characteristics is presented in the remainder of this section.

10.5.2 Repeater Spacing vs. Signal Power in Guided Wave Systems

Guided wave communication links make use of a number of types
of guiding structures including wire pairs, coaxial cables, waveguides and optical fibers. Many such links have bandwidths and
lengths such that no amplification is needed between the two ends
of the system. These are easily analyzed and designed, but to gain
insight into the problems involved in planning new systems of high
capacity and great length, a knowledge of the relationship between
channel bandwidth, repeater spacing, signal power and signal-to-
noise ratio is needed.

Consider a baseband guided wave communication system as shown
in Figure 10.16 consisting of n repeater sections, each of length d.
The total length of the system is D = nd. The baseband system is
slightly easier to analyze than a system handling modulated signals
but the results are the same, except for trivial modifications, for
two cases. In this discussion, no consideration will be given to the
character of the information carrying signal except that it has bandwidth B. Assume the output of the transmitter consists of the communication signal power, S_O', as well as thermal and device noise
power, $kB\delta (T + T_t)e^{2\alpha d}$, where $k = 1.38(10)^{-23}$ joules per degree
Kelvin, is Boltzmann's constant, T is the temperature of the sys-
tem in degrees Kelvin, T_T is the noise temperature of the transmitter, and δ is given by

$$\delta = \frac{h}{BkT}\int_{f_1}^{f_2} \frac{f df}{e^{\frac{hf}{kT}}-1} \qquad (10.13)$$

where $h = 6.62(10)^{-34}$ joule-sec. is Plank's constant, and f_1 and f_2
are the lower and upper limits in Hz of the frequency band of the
transmitted signal. (Note that $f_2 = f_1 + B$). At microwave frequencies $\delta \simeq 1$, and at optical frequencies δ is so small it can be considered to be zero. The factor $e^{2\alpha d}$ where α is the attenuation constant
of the waveguide in nepers per unit length, is included in the expression for noise in order to express all noise in terms of the power levels that exist at the repeater input. In addition, there are two
other noise terms in the output of the transmitter, N_{oT}, representing other noise (if any) not proportional to the bandwidth and, N_{bT},

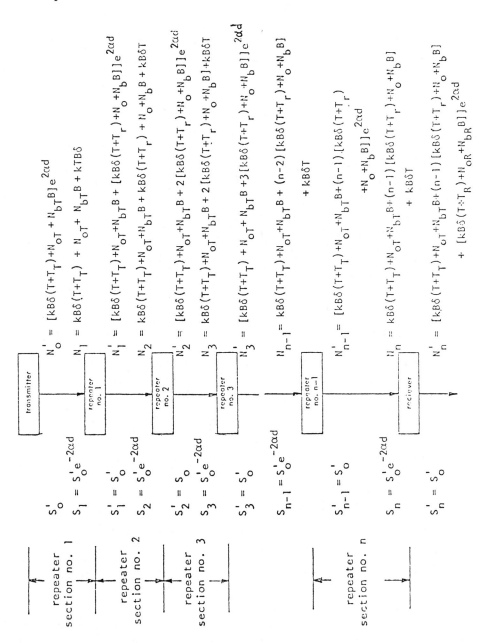

*Figure 10.16 Signal and Noise Levels in a Baseband Cable System
with n Repeater Sections*

representing other noise proportional to bandwidth. Thus the total noise power output from the transmitter is

$$N_0' = \left[kB\delta(T + T_T) + N_{oT} + N_{bT} B \right] e^{2\alpha d}. \tag{10.14}$$

The breakdown of noise into these 3 contributions is somewhat artificial and is done in an effort to make the resulting expressions applicable to a wide variety of systems. In applying the formulas to any particular system, the noise can be broken up into 3 terms in an unlimited number of different ways, any of them being satisfactory as far as this analysis is concerned.

The signal, S_1, and the noise, N_1, at the input to the first repeater are

$$S_1 = S_0' e^{-2\alpha d} \tag{10.15}$$

$$N_1 = kB\delta(T + T_T) + N_{0T} + N_{bT} B + kB\delta T \tag{10.16}$$

where $kB\delta T$ is the thermal noise from the communication link in repeater section number 1. It will be assumed that each repeater has a gain, $e^{2\alpha d}$, so as to exactly compensate for the attenuation in the preceding section of the communication link. Thus the signal S_1' and noise, N_1', at the output of repeater number one are

$$S_1' = S_0' \tag{10.17}$$

$$N_1' = \left[kB\delta(T + T_R) + N_{0T} + N_{bl} B \right] e^{2\alpha d} + \left[kB\delta(T + T_r) + (N_0 + N_b B) \right] e^{2\alpha d} \tag{10.18}$$

where T_r is the noise temperature of the repeater, and $(N_0 + N_b B)e^{2\alpha d}$ is other noise added by the first repeater.

As the signal progresses through the system, the noise keeps increasing, with each repeater section adding the same amount of noise to the total. The expressions for signal and noise are given at various places in the system in Figure 10.16. The receiver has a noise temperature T_R and adds other noise terms $(N_{0R} + N_{bR} B)e^{2\alpha d}$

It is convenient and reasonable to let

$$kB\delta(T + T_T) + kB\delta(T + T_R) = kB\delta(T + T_r) \tag{10.19}$$

In other words, it is assumed that the thermal and device noise of the transmitter plus that due to the receiver is equal to that due to one repeater. This is reasonable if a repeater consists of a receiver followed by a transmitter (as is true of some types). Even if it is not true, the assumption will introduce little error if there are a large number of repeater sections (since the transmitter and receiver contribute only a small fraction of the total noise). Using the same reasoning it will be assumed that

$$N_{0T} + N_{bT}B + N_{0R} + N_{bR}B = N_0 + N_bB \qquad (10.20)$$

As a consequence of these assumptions, the noise, N_n', at the output of the receiver can be written

$$N_n' = n\left[kB\delta(T + T_r) + N_0 + N_bB\right]e^{2\alpha d} \qquad (10.21)$$

This shows that each repeater section contributes thermal and device noise $kB\delta(T + T_r)e^{2\alpha d}$ and other additive noise $(N_0 + N_bB)e^{2\alpha d}$. These are additive from section to section because of their random nature. The signal-to-noise ratio, SNR, at the output of the receiver is

$$SNR = \frac{S_0'}{n\left[kB\delta(T + T_r) + N_0 + N_bB\right]e^{2\alpha d}} \qquad (10.22)$$

This relation gives some insight into the design of the system. At a first glance it would seem that n should be small to minimize the noise. However, since d = D/n, making n small makes d large thus offsetting the small n in the denominator. Substitution of d = D/n into the expression for signal-to-noise ration results in

$$SNR = \frac{S_0'}{n\left[kB\delta(T + T_r) + N_0 + N_bB\right]e^{\frac{2\alpha D}{n}}} \cdot \qquad (10.23)$$

Solving for S_0' gives

$$S_0' = SNR\ n\left[kB\delta(T + T_r) + N_0 + N_bB\right]e^{\frac{2\alpha D}{n}} \cdot \qquad (10.24)$$

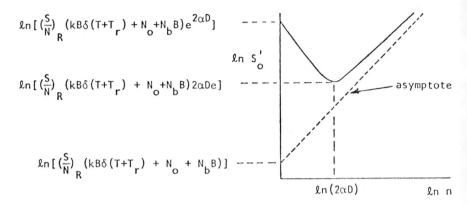

$\ell n \left[\left(\frac{S}{N} \right)_R (kB\delta(T+T_r) + N_o + N_b B) e^{2\alpha D} \right]$ ----

$\ell n \; S_o'$

$\ell n \left[\left(\frac{S}{N} \right)_R (kB\delta(T+T_r) + N_o + N_b B) 2\alpha D e \right]$ ---- ---- asymptote

$\ell n \left[\left(\frac{S}{N} \right)_R (kB\delta(T+T_r) + N_o + N_b B) \right]$ ----

$\ell n \, (2\alpha D)$ $\ell n \; n$

Figure 10.17   Repeater Output Power, S_o', as a Function of Number of Repeaters, n

In system design, the signal-to-noise ratio, the total length, D, and the bandwidth, B, may be considered to be specified. Furthermore, the state of the art in equipment manufacture usually approximately determines T_r, N_0, N_b and α. Then it remains for the designer to determine S_0' and n. S_0' as a function of n is approximately as shown in Figure 10.17.

Generally, n should be chosen near the minimum of the curve, the exact point being dictated by the economics of the system; that is, the cost of repeaters. n is usually selected at a point slightly to the left of the minimum of the curve.

The above analysis is concerned with output power for a single repeater as a function of repeater spacing. It is interesting to find an expression for the total signal power needed for the system; that is, the power per repeater times the number of repeater sections. This is given by

$$nS_0' = (SNR) \; n^2 \left[kB\delta(T + T_r) + N_0 + N_b B \right] e^{\frac{2\alpha D}{n}} \qquad (10.25)$$

which is sketched in Figure 10.18. It is seen that the minimum total power occurs with half as many repeaters as for the case of minimum power per repeater.

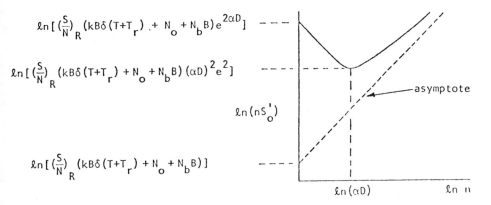

$$\ln[(\tfrac{S}{N})_R (kB\delta(T+T_r) + N_o + N_b B) e^{2\alpha D}]$$

$$\ln[(\tfrac{S}{N})_R (kB\delta(T+T_r) + N_o + N_b B)(\alpha D)^2 e^2]$$

$$\ln(nS_o')$$

$$\ln[(\tfrac{S}{N})_R (kB\delta(T+T_r) + N_o + N_b B)]$$

asymptote

$$\ln(\alpha D) \qquad \ln n$$

Figure 10.18   Total System Signal Power, nS_0', as a Function of Number of Repeaters, n

Signal Power vs. Length for Guided Wave Systems. One further observation may be made concerning the above expression for total signal power, nS_0' required for communication. The state of the art of manufacture of components at the time of system design essentially determines repeater spacing. Suppose it is fixed at the value d for all systems regardless of length. Then since d = D/n, n may be expressed as n = D/d and nS_0' may be expressed as

$$nS_0' = (SNR)\frac{D^2}{d^2}\left[kB\delta(T + T_r) + N_0 + N_b B\right] e^{2\alpha d} \qquad (10.26)$$

Since d is now considered constant, the total signal power required is seen to be proportional to the square of the total length of the communication link. Note that if some of the constants in the above expression change because of improvements in component design, only the constant of proportionality changes, and the total power required is still proportional to the square of the total length of the link. This relationship, total power proportional to the square of total system length, is the same relation that exists for unguided wave systems and thus seems to have the character of a universal law of communication systems.

Dispersion in Guided Wave Systems.   The above analysis produced relationships which permit a designer to determine the spacing of repeaters, knowing signal power levels, noise power levels and attenuation rates. The design problem is not quite as simple as suggested here for a number of reasons, one of which is the fact that for a high capacity communication link the attenuation rate, α, is different at different frequencies. This causes signal distortion which can be overcome by amplifiers which amplify more at some frequencies than others and which can thus be made to compensate for the variation in the attenuation constant.

A still more serious problem (or, if not more serious, at least more difficult to overcome) is the velocity dispersion problem which is more severe in guided wave systems than in unguided wave systems. All waveguiding structures have characteristics (dispersion) which result in different velocities of propagation for different frequencies. This in turn causes severe distortion in signals. The distortion can be corrected by use of specially designed amplifiers (called equalizers) just as was true of distortion caused by variation of attenuation with frequency, but the cost is substantial because fo the intricacy of the equalizer. The amount of distortion of the signal increases with the signal bandwidth, and it increases with the length of the communication link. The greater the distortion, the more complex and costly the equalizer. This dictates that, for cost effective systems, equalizers should be inserted at intervals such that distortion never becomes excessive. This requirement may, in some cases, result in an equalizer spacing less than the spacing calculated for repeaters on the basis of power levels. When this occurs, repeaters are combined with equalizers and the spacing appropriate for the equalizers is used. In many existing systems the equalizer spacing is several times the repeater spacing, and equalizers are needed only on a fraction of the repeaters.

Digital Signals in Guided Wave Systems.   The characteristics of digital signals are particularly significant with regard to dispersion-caused distortion because for digital signals the problem of correcting signal distortion is greatly simplified. A digital signal consists of a sequence of pulses, each pulse being one of a finite set. Dispersion causes pulse shapes to become distorted. The

proper shape could be restored by an equalizer; but this is unnecessary if the distortion is not too great to prevent assignments of the proper symbol to each pulse. For then, instead
of restoring the proper shape to the pulse, a new pulse of proper shape is generated by a device called a regenerative repeater.
Regenerative repeaters are much simpler and less costly than
equalizers, but their use may require closer spacing of repeaters
than would be required for a system using equalizers and ampli-
fying repeaters because the signal must be regenerated before
distortion becomes so great that pulses representing different
symbols cannot be distinguished. An alternative design could
involve greater spacing of the regenerative repeaters but with
partial equalization at the input of each repeater. This alternative is important because partial equalization can be much
less expensive than full equalization. Despite the possible closer
spacing, the cost advantages of regenerative repeaters are sufficiently great to give digital systems a significant advantage over
analog systems in long,high-capacity, guided wave communication systems.

Uses of Guided Wave Systems. Guided wave point-to-point
communication systems, or cable or wired systems, as they
are often called, are widely used throughout the world. They
are used where systems using radio links are impractical or impossible because of spectrum crowding, cost of equipment, lack
of line-of-sight propagation path because the system is in rough
terrain or inside a building, or because of a requirement for secrecy of the information carried by the signal. Telephone systems are the most important example of this type of system.

10.5.3 Switching Systems

Point-to-point communication systems, particularly of the guided
wave type, often involve more than two terminals with capabilities
for making a connection between any two of them. The capability
is provided by a switching center as sketched in Figure 10.19. Each
of the communication terminals is connected to the switching center by means of a waveguide (wire-pair, cable, optical fiber, etc.)
The switching center has the capability of connecting the waveguide of any terminal to the waveguide of any other terminal, thus
providing communication between the two terminals. In a typical

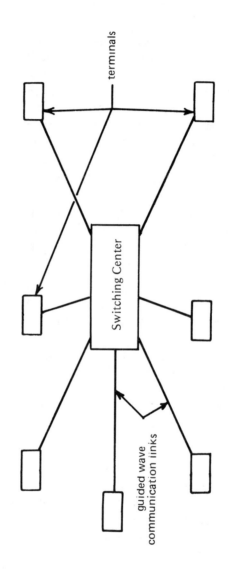

Figure 10.19 Block Diagram of a Switching Center and Several Terminals

system (for example, the telephone system) many such pairs are
connected and are communicating at any time. Continuous analog
signals (that is, analog signals not of the pulse type) require a permanent connection through the switching center for the duration
of the communication. Pulse signals, on the other hand, require
connection only during pulses. Of course, a permanent connection
will serve to switch pulsed signals; but the fact that it is not needed
is significant because it allows a type of switching, called time division switching, in which the switching equipment is shared by many
terminals each making use of it only during pulses, the pulses being
staggered in time for different terminals. The consequence of the
use of time division switching is a switching center of complexity,
and thus cost, roughly proportional to the number of terminals associated with the switching center. Furthermore the type of equipment used is the type used in computer systems which has under-
gone drastic reductions in price due to technological advancements
during the past few years with further reductions foreseen for the
future. This is to be contrasted with switching systems which make
permanent connections during the communication. These are called
space division switching systems, and they are characterized by a
complexity (and cost) proportional to the square of the number of
terminals serviced. Furthermore, the cost of this type of switching
equipment has not decreased recently as has been the case with
time division switching equipment.

It is apparent that time division switching systems have favorable
characteristics compared with space division switching systems
and for these reasons are likely to be widely used in new systems
during the next few years. This observation complements the previous observation concerning the favorable characteristics of digital guided wave communication links as compared with analog
guided wave links, and it thus strengthens the potential of digital
communications systems for the future. One further observation
is appropriate at this point. It relates to the versatility of digital
systems as compared with analog systems. Conventional analog
systems handle analog signals with ease and can be made to handle
digital signals with addition of costly equipment. On the other
hand, proposed digital systems are quite easily adaptable to a
great variety of digital and analog signals. In short, it can be said

that digital systems are more versatile than analog systems, a definite advantage in future systems because of the great variety of types of signals to be handled.

10.5.4 Communication Traffic Theory

The switching systems discussed above are described only in terms of their functions. Some further discussion of their characteristics is in order. Large switching systems (those having a large number of terminals) almost always have terminals that are in use for communication only a fraction of the time. Furthermore, the times that terminals are in use are randomly distributed, and the distributions are independent for the various terminals. An excellent example is the telephone system. At any time only a small fraction of all telephones are in use. The times at which a telephone is put into use are random, and there is little tendency for two or more terminals to initiate calls at the same time. The consequence of the random nature of the communications handled by the switching center is the possibility of using a switching system with only sufficient equipment to handle communication for a small fraction of the terminals at any time. If all of the equipment is in use and another terminal requests service, the service is unavailable and the communication is said to have been blocked (or delayed if the terminal waits for equipment to become available.) If the amount of equipment in the switching center is properly chosen, based upon the statistics of communications for the various terminals, the probability of blocking can be made arbitrarily small. The study of the relationship between the statistics of communications of the terminals (called the traffic), the amount of equipment in the system, and the quality of service (as measured by the probability of blocking or by the distribution of waiting times) is called traffic theory or queueing theory.[1] It basically solves the problem of determination of amount of traffic, amount of equipment, or quality of service if the other two of the three measures are known. Traffic theory is useful not only in the design of switching systems but also of all other parts of communication systems that are characterized by random usage.

10.5.5 Repeater Spacing vs. Signal Power for Unguided Wave (Radio) Systems

Unguided wave communication systems, or radio systems as they are commonly called, are capable of substituting for guided wave systems in many applications but their characteristics are such that they generally are complementary rather than competitive with guided wave systems. Their characteristics will now be discussed.

The relationship among signal power, SNR, bandwidth, antenna gain, repeater spacing, and total system length is developed in the same way as was done for guided wave systems.

The signal-to-noise ratio of an unguided wave system having n repeater sections will be analyzed and compared with the waveguide system discussed above. Such a system is shown in Figure 10.20.

The pre-detection signal-to-noise ratio at the receiver is

$$SNR = \frac{S_0'}{n\left[kB\delta(T + T_r) + N_0 + N_b B\right] \dfrac{(4\pi \frac{D}{n\lambda})^2}{G_t G_r L_c}} \tag{10.27}$$

after making assumptions similar to those made in the guided wave system. G_t and G_r are the power gains of the transmitting and receiving antennas respectively; L_c is the composite power efficiency of the transmitting and receiving antenna systems; and λ is the free space wavelength of the radiated signals.

Signal Power vs. Length for Unguided Wave Systems.   The above expression may be rearranged to give an expression for total signal power, nS_0' required for the n repeater sections.

$$nS_0' = (SNR)\left[kB\delta(T + T_r) + N_0 + N_b B\right] \frac{\left(4\pi \frac{D}{\lambda}\right)^2}{G_t G_r L_c} \tag{10.28}$$

It may be noted that the expression on the right is independent of n. Thus, in an unguided wave system, the total signal power required is proportional to the square of the total length, D, of

transmitter

$$N_o' = [kB\delta(T+T_T)+N_{oT}+N_{bT}B]\frac{(4\pi\frac{d}{\lambda})^2}{G_t G_r L_c}$$

$$N_1 = kB\delta(T+T_T)+N_{oT}+N_{bT}B+kTB\delta$$

repeater no. 1

$$N_1' = [kB\delta(T+T_T)+N_{oT}+N_{bT}B+[kB\delta(T+T_r)+N_o+N_bB]]\frac{(4\pi\frac{d}{\lambda})^2}{G_t G_r L_c}$$

$$N_2 = kB\delta(T+T_T)+N_{oT}+N_{bT}B+kB\delta(T+T_r)+N_o+N_bB+kB\delta T$$

repeater no. 2

$$N_2' = [kB\delta(T+T_T)+N_{oT}+N_{bT}B+2[kB\delta(T+T_r)+N_o+N_bB]]\frac{(4\pi\frac{d}{\lambda})^2}{G_t G_r L_c}$$

$$N_3 = kB\delta(T+T_T)+N_{oT}+N_{bT}B+2[kB\delta(T+T_r)+N_o+N_bB]+kB\delta T$$

repeater no. 3

$$N_3' = [kB\delta(T+T_T)+N_{oT}+N_{bT}B+3[kB\delta(T+T_r)+N_o+N_bB]]\frac{(4\pi\frac{d}{\lambda})^2}{G_t G_r L_c}$$

S_o'

repeater section no.1

$$S_1 = S_o'\frac{G_t G_r L_c}{(4\pi\frac{d}{\lambda})^2}$$

$$S_1' = S_o'$$

repeater section no.2

$$S_2 = S_o'\frac{G_t G_r L_c}{(4\pi\frac{d}{\lambda})^2}$$

$$S_2' = S_o'$$

repeater section no.3

$$S_3 = S_o'\frac{G_t G_r L_c}{(4\pi\frac{d}{\lambda})^2}$$

$$S_3' = S_o'$$

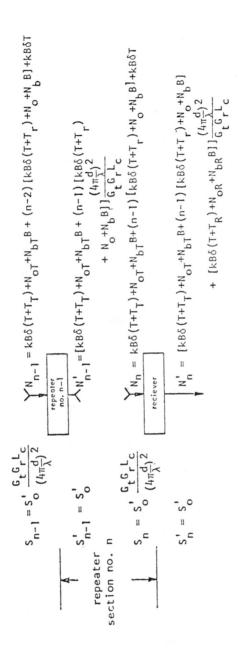

Figure 10.20 *Signal and Noise Levels in a System Using Radiation Through the Troposphere and having N Repeater Sections*

the system but is independent of the number of repeater sections. The spacing of repeaters in such a system is determined primarily by the requirement of line-of-sight propagation paths.

In the case of a waveguide system as shown in Figure 10.16, signal power requirements place distinct limits on repeater spacing. Once that spacing has been selected, however, then the total signal power, nS_0', required is proportional to n^2; but since for a fixed repeater spacing, d, n = D/d, the total signal power is proportional to the square of the total length of the system just as is true for the unguided wave system of Figure 10.20.

Satellite Communication Systems. Satellite communication systems are a special case of unguided wave systems in which there are two repeater sections (one repeater) each very long.[5,6] The absence of a large number of repeaters and the associated sites and towers reduces the cost of the system as compared with a very long terrestrial microwave system thereby giving satellite communication systems significant cost advantages compared with terrestial systems. In a geosynchronous satellite communication system (one in which the satellite remains stationary with respect to a location on the surface of the earth) the signal is transmitted from the earth up to the satellite 36,000 km above the center of the earth then back down to a point on the surface of the earth a few hundred or a few thousand km from the transmitter. The signal travels a distance of about 70,000 km to communicate a distance of only a few hundred or a few thousand km. Furthermore, the costs are the same regardless of the locations of the transmitter and the receiver. Thus, in a satellite system, communication costs are independent of distance making such systems very economical compared with terrestrial systems for communication over long distances.

10.5.6 Comparison of Repeater Spacing vs. System Length Relations for Guided and Unguided Wave Systems

It has been observed that the relationship among signal power, bandwidth, SNR and system length is basically the same for guided wave systems as for unguided wave systems. (Power is proportional to the square of total system length.) The only significant difference is the fact that in the case of a guided wave system, repeaters must be placed at reasonably closely controlled intervals to keep total signal power within reasonable limits, whereas for unguided wave systems the repeater spacing is unrelated to total signal power and is determined primarily by the economics of the construction of towers to provide for line-of-sight propagation paths and by meteorological characteristics of the troposphere. The choice between a guided wave system and an unguided wave system is thus not made on the basis of technical aspects related to communication signals but rather on the basis of such factors as right of way, construction and operating costs and spectrum crowding. Contruction of either type of system involves right-of-way problems, but the problems obviously are much greater in the case of guided wave systems since with unguided wave systems (radio or microwave systems) all that is required is space for a repeater every 30 to 60 km and access to that site.

The cost of a guided wave system is generally much greater than that of, say, a microwave system, but the right of way established for such a system permits a great amount of expansion of communication capacity at relatively low incremental cost. So for communication between locations that may have greatly expanded traffic in the future the guided wave system may be the best choice.

10.5.7 Operation of Several Side-by-Side Unguided Wave Systems on the Same Frequency Band

For communication routes with very large amounts of traffic, strong arguments can be made for guided wave systems on the basis of spectrum crowding. There is no practical limit to the number of side-by-side guided wave systems that can be operated on the same frequency band without mutual interference. However, in the case of radio systems, closely side-by-side systems are operated on different frequency bands. This is a practical requirement rather than an absolute requirement as will be discussed. Its effect though is to limit the number of side-by-side systems operating over a single communciation route because of the limitation of the frequency bands available for use. To see that it is possible though not generally practical to operate several systems side-by-side on the same frequency band refer to Figure 10.21. The spacing and radiation patterns are

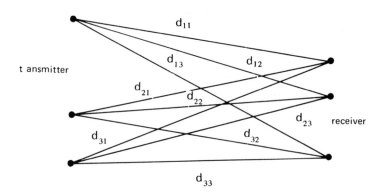

Figure 10.21   Typical System of Multiple Transmitters and Receivers

such that all receivers receive the signals from all of the transmitters. If the distances d_{ij}, i, j = 1,2,3 are properly chosen,
then it is possible to separate the received signals into the individual contributions due to the 3 transmitters. The concept
involved here is simply that of solution of 3 linear equations
in 3 unknowns. The output of each receiver is a weighted sum
of the signals received from the 3 transmitters. The weighting
can be determined from a knowledge of the distances, d_{ij}, and
the radiation patterns. In principle, the solution of the equations
can be done continuously using analog processing or discretely,
using digital processing. In a completely coherent system the
processing can be done after demodulation, but even so, a large
amount of computational ability is needed. Such a multichannel,
single frequency band system is only practical if the high channel
capacity justifies the complexity of the processing system. In addition to the disadvantage of an elaborate processing system, such
a communication system also suffers from noise resulting from
tropospheric irregularities which may have the effect of differen-
tially varying the effective d_{ij} as a function of time, thereby introducing a form of noise into the output signals; obviously, the
system would function better in space.

Note that these ideas could be extended to any number of transmitters and receivers, and the arrays of transmitters and receivers
could be two-dimensional. Assuming this to be the case and assuming very closely spaced transmitters and receivers, then, in

the limit, as the spacing decreases without limit, the communication system becomes analogous to an optical system projecting a picture.

The conclusion that may be drawn from this discussion is that communication system capacity may be increased in a guided wave system by increasing the number of waveguiding structures. In this manner, large increases in channel capacity are possible with no increase in the bandwidth utilized. In principle, the same is true of an unguided wave system. In practice, however, the complexity of the processing circuitry needed at the receiving end to separate the signals into those originating at the various transmitters is so great as to make such systems impractical until the cost of processing circuitry is much further reduced. The alternative is the use of a different frequency band for each of the side-by-side radio links. This may not be possible because of the unavailability of spectrum space. (At the present time this is a problem at microwave frequencies but not at optical frequencies.)

10.6 BROADCAST TYPES OF COMMUNICATION SYSTEMS

Broadcast types of communication systems as exemplified by the common radio and television broadcasting station fulfill a very different role in society than point-to-point communication systems. Their role is the dissemination of information in one direction, from one location to many. The objective of the flow of information may be education, entertainment, public service, or other. The nature of the information disseminated and the use of the so-called public airways for most broadcasting systems dictates that broadcasting be subject to close scrutiny by governmental regulatory agencies. Most point-to-point communication systems are closely regulated too, but only with regard to types of locations and costs of services rather than with respect to the nature of information handled. (Broadcasting stations are also regulated with regard to frequency bands, power output and modulation characteristics.) Trends in expansion of broadcasting systems are thus strongly influenced by government regulatory agencies, but the technical characteristics of such systems are also of importance. These characteristics will now be discussed.

10.6.1 Unguided Wave Systems

Most broadcasting systems are of the unguided wave type for
the reason that transmission of signals through the atmosphere
provides the desired communication at a very low cost. Costs
also influence other aspects of system design. A total broadcasting system of a given type consists of a relatively small number
of transmitters and a large number of receivers. Minimization of
total system cost dictates that a signal type which minimizes
receiver cost should be used. A clear cut example of this is
the use of double sideband amplitude modulation in commercial AM radio broadcasting. This type of signal permits the use
of very low cost receivers. This may be contrasted with point-
to-point radio communication systems in which single sideband
modulation is often used. Such a signal requires a more expensive receiver but it permits communication using only half as
much bandwidth as required for double sideband modulation.
In a point-to-point system only two receivers (one at each end)
are needed, and the added cost is sometimes felt to be justified
in view of the reduced bandwidth requirement.

Technical considerations relating to propagation of signals through
the troposphere have been of importance in the assignment of frequency bands for various types of broadcasting systems. Historically, commercial radio broadcasting was initiated at frequencies
for which equipment could be made at that time; it has continued
to use essentially the same range of frequencies ever since. It has
proven to be a satisfactory frequency band for such service because
it permits communication over substantial distances with moderate
power and can often achieve satisfactory communication along
propagation paths on which obstructions such as hills or man-made
structures would cause blockage of a higher frequency signal.

The desire for broadcasting systems with higher quality signals
dictated a higher bandwidth signal which, in turn, meant that
higher frequencies would have to be used to insure that there
would be enough bandwidth to make frequency assignments
to all of the stations that would be needed. The higher frequencies
meant that the signals could only be received over line-of-sight
paths, but this was the price to be paid for higher quality signals. After a detailed technical study, frequency assignments

were made for this new type of broadcasting system, the system of FM radio stations, in the frequency band from 88 to 108 megahertz. Extensive technical studies have also been made in connection with assignment of frequencies for television broadcasting.

Broadcast type (unguided wave) communication systems are being expanded in some parts of the world, primarily the developing countries, but expansion is minimal in the more highly developed countries. Technological developments in equipment for broadcasting do not seem to portend any significant changes in the near future.

10.6.2 Guided Wave Systems

Another type of broadcasting system, one that utilizes guided wave communication links, is in a state of rapid expansion at the present time. That is the cable television system, commonly called CTV or CATV. These systems receive television broadcasts directly from local television broadcasting stations and retransmit the signals by way of a guided wave system (usually coaxial cables) to system customers. Such systems were originally used in areas in which very tall antenna towers were needed to receive high quality signals; one such tower could thus serve the needs of an entire community. This usage gave the systems the name Community Antenna Television, a name which has been essentially replaced by the name "Cable Television." These systems brought television to many isolated communities where it would otherwise be impractical. More recently, CATV systems have widely expanded in large metropolitan areas where signals are easily available without the cable system. Cable systems in these locations are able to supply better quality signals than the potential customers could receive without cable and, in addition, can offer signals that are not broadcast locally. These additional signals may be television broadcasts or other video signals from distant cities transmitted to the cable system by use of a terrestrial or satellite, point-to-point, microwave communication links owned or leased by the company, or they may be local sporting events or recent movies which are not available from television broadcasting stations. Systems sometimes charge extra for some of these otherwise unavailable programs. It is apparent that

cable systems can offer customers substantially greater variety
than is available from only the local TV broadcasting stations. For
this reason the industry has been expanding rapidly in a number
of highly developed countries and the rate of expansion shows no
sign of abating in the foreseeable future.

Cable systems have cables extending from a central location outward in the direction of the customers, a branch of the cable connecting to the TV receiver of each. The video signals are transmitted
over the cable using the same type of modulation and the same set
of frequency bands (channels) as used by television broadcasting
stations. This permits the use of the same receivers on cable television systems as are used for over-the-air television signals. A television signal received by a cable system and then distributed to customers over the cable is often switched to a different channel for
cable distribution. This means a cable customer will receive a signal on a different channel than a non-customer will receive the same
signal over-the-air. The reason for the switching of channels is the
prevention of interference in the television receiver of a cable system
customer between a signal received from the cable and the signal present in the air. If television receivers were better (and more expen-
sively) made, the switching would be unnecessary.

The type of coaxial cable used in cable television systems is capable of transmitting signals with high quality in the frequency range
54 to 274 MHz, which is the range used for VHF television broadcasting. Such cable can be used, with suitably spaced repeaters, up
to distances of a few miles. Beyond that distance the distortion
caused by the cable dictates the use of more expensive equalizers
and repeaters, so systems are usually limited in size to avoid these
problems. More distant customers are served by another system
located such that the cable runs are sufficiently short to allow use
of less complex equipment.

UHF television stations broadcast signals in the 470 to 900 MHz
range, which is too high for transmission over existing cable television systems. (Such frequencies would require such close spacing
of repeaters and such short overall cable runs that they could not be
economically justifiable.) However, a UHF signal can be shifted down
in frequency and transmitted over one of the VHF channels not in

use for a VHF signal. A typical cable system can handle approximately 20 video signals. The cost of adapting it to handle more is such that it is probably more economical to install a second cable to handle the additional channels. A standard television receiver is capable of receiving only 12 VHF channels, so if a cable is carrying more than 12 video signals, a converter is required to shift frequencies of some of the signals to a channel, either VHF or UHF, for which the receiver is equipped.

Some of the newer cable systems have the capability for transmission of data at very low rates from the subscriber to the cable broadcasting center (called head-end of the system.) This limited two-way capability endows the system with the potential for a variety of new services such as remote library access, video "newspaper" and many other possible uses which have been discussed at great lengths by many writers in the past several years. In addition, the limited two-way capability will permit cable system operators to make special programs available to customers for an added fee; that is, for a charge in addition to the regular monthly fee for basic cable service. The two-way capability permits the system operator to determine which customers are receiving the special program and to determine which customers to charge for it. This would be a substantial improvement over the special programming that is now made available for an extra fee over many systems, but for which customers must purchase all such special programming for a month at a time. The service offering special programs over cable systems for an added fee is commonly called "pay cable", and it has been enthusiastically received by cable system customers. It is the most rapidly expanding service offered by cable systems. Its success has provided strong incentive for development of sources of programs suitable for pay cable and for systems for distribution of such programs to cable systems across the country. Such programs are now distributed by land based as well as satellite microwave systems.

Trends in the Design of Guided Wave Broadcast Systems. The large bandwidth of signals transmitted over cable systems necessitates the use of high quality amplifiers and other equipment. Technological developments which have the potential of permitting reduction in equipment costs or expansion of capacity are of great interest in the cable television industry. Optical fibers are a strong possibility as a replacement for the coaxial cable in cable systems. Their advantages lie in their great bandwidth and in their potentially low cost. However, they share one of the same disadvantages as coaxial cables, that being their relatively high dispersion which causes signal distortion. Distortion caused by dispersion can of course, be corrected with precision equipment.

Another development with significant potential is digital cable transmission systems. The use of digital signals increases bandwidth, but at the same time it greatly simplifies the repeaters and the problem of correcting distortion caused by dispersion. A disadvantage is the necessity of a converter to change the digital signal from the cable into an analog signal suitable for a standard television receiver. The advantages of digital systems are such that they are being evaluated experimentally at the present time. It is likely that after further development, both digital systems and fiber transmission lines, probably used together, will be used in cable television systems.

10.7 CONCLUSIONS

This chapter has been concerned with the technological aspects of the evolution of telecommunication systems. A foundation for the study was laid first by a review of the various forms assumed by communication signals and of the methods (modulation and multiplexing) for conversion among them. Next a review of information theory was undertaken. It provided a means for the measurement of the information in a signal and for the information handling capacity of a communication channel as well as providing guidelines for recoding a signal into a form adapted to a channel.

For purposes of studying trends in telecommuncation system design, systems were categorized in three different ways: point-to-point vs. broadcast systems, guided wave (cable or waveguide) vs. unguided wave (radio) systems, and finally analog vs. digital systems.

Point-to-point switched telecommunication systems (telephone and related types of systems) are being expanded, worldwide, as rapidly as available investment capital will permit. Demand seems destined to exceed supply indefinitely. Such communication systems can make use of guided wave communication links (cable or waveguide) or unguided wave links (radio). Although in principle it is possible to operate many closely spaced unguided wave links in the same frequency band, it is currently impractical which therefore demands the use of different frequency bands for different systems. As a result, the crowding of the frequency spectrum provides a strong argument for the use of guided wave systems, many of which can operate on the same frequency band, for expansion of telecommunication systems already having substantial capacity. Guided wave systems in contrast to microwave and other unguided wave systems, suffer from substantial dispersion-related signal distortion. Such distortion can be completely corrected. The correction can be accomplished by less complex and less costly equipment in a digital system than in an analog system. This fact, in addition to the fact that digital signal handling equipment of all kinds has recently become very attractive economically compared to equipment for analog systems, strongly suggests that digital guided wave systems will play a prominent role in the future expansion of point-to-point telecommunication systems. A significant added incentive for the use of digital systems is the versatility of such systems with regard to the variety of signals that can be handled. Analog systems, of course, are similarly versatile but with significantly less ease of adaptability.

Broadcast types of communication systems are predominantly of the unguided wave type (radio, television, etc.) Expansion of this type of telecommunication is occuring at a rapid rate in some parts of the world but at a slow rate in most developed countries. There are no apparent new technological developments of sufficient importance to significantly influence the current trends in the expansion of such systems. The other type of broadcast system, the guided wave type as exemplified by cable television systems, is presently in a state of rapid expansion in highly developed countries. Such systems seem destined to play a greatly expanded role in future society. Their importance lies in their ability to offer a much greater variety of programs than are available from over-the-air broadcasts and from their potential for providing new telecommunication services such as remote library access and public service broadcasts of such activities as school board meetings and city council meetings. The current development of optical fiber transmission links and of digital communication systems seems likely to significantly influence the design of cable television systems in the near future.

REFERENCES

[1] Beckmann, Petr: *Elementary Queuing Theory and Telephone Traffic,* Lee's abc of the Telephone, Training Manuals, Geneva, Ill., 1977.

[2] Beckmann, Petr: *Probability in Communication Engineering,* Harcourt, Brace & World, 1967.

[3] Carlson, Bruce A.: *Communication Systems,* McGraw-Hill, Inc. 1968.

[4] Karp, Sherman and Robert M. Gagliardi: *Optical Communications,* New York, John Wiley & Sons, 1976.

[5] Martin, James: *Future Developments in Telecommunications,* Englewood Cliffs, N.J., Prentice-Hall, Inc. 1971.

[6] Martin, James: *Telecommunications and the Computer,* Englewood Cliffs, N.J., Prentice-Hall, Inc. 1976.

[7] Martin, James: *Introduction to Teleprocessing,* Englewood Cliffs, N.J., Prentice-Hall, Inc. 1972.

[8] Martin, James: *Security Accuracy and Privacy in Computer Systems,* Englewood Cliffs, N.J. Prentice-Hall, Inc. 1973.

[9] Panter, Philip F.: *Modulation, Noise, and Spectral Analysis,* McGraw-Hill, Inc. 1965.

[10] Peterson, W.W.: *Error-Correcting Codes,* John Wiley & Sons., Inc. 1961.

[11] Pierce, J.R.: *Symbols, Signals and Noise,* Harper & Row Publishers, 1961.

[12] Pool, Ithiel de Sola: *Talking Back: Citizen Feedback and Cable Technology,* The MIT Press, Cambridge, Mass. 1973.

Chapter 11

THE TRANSMISSION OF ELECTROMAGNETIC SIGNALS

Warren L. Flock
Department of Electrical Engineering
University of Colorado
Boulder, Colorado

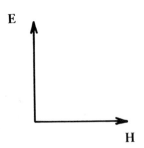

*Figure 11.1 The **E** and **H** vectors of a plane linearly polarized electromagnetic wave.*

11.1 ELECTROMAGNETIC WAVES AND THE ELECTROMAGNETIC SPECTRUM

11.1.1 Electromagnetic Fields and Waves

The practice of telecommunications is dependent almost exclusively on the use of electromagnetic waves. In some applications, the waves travel or propagate through the earth's atmosphere, as from one microwave repeater station to another or from an earth station to a satellite. In other situations the waves are confined to the interior of, or otherwise guided by, cables or waveguides. We discuss first some basic characteristics of electromagnetic waves and then their applications to telecommunications.

The simplest type of electromagnetic wave is a plane wave of infinite extent propagating through a vacuum. A plane wave travels everywhere in the same direction instead of spreading out over a range of directions, as do the ripples generated when a stone is thrown into a pond. Actual waves are not of infinite extent and are often spherical or of some form other than planar near their source. At large distances from sources, however, waves tend to be good approximations to plane waves, just as a portion of the earth's surface may appear flat, or planar, even though it is actually a portion of a spherical surface.

Plane electromagnetic waves are characterized by an electric field intensity, **E**, having units of volts/meter (V/m) and by a magnetic field intensity, **H**, having units of amperes/meter (A/m) and directed perpendicular to **E** as in Figure 11.1 (**E** and **H** are vector quantities that have directions as well as magnitudes. Another example of a vector quantity is force. Temperature has a magnitude but not a direction and is a scalar quantity.) An electromagnetic wave with **E** in the vertical direction is called a vertically polarized wave; if **E** is in the horizontal direction the wave is called a horizontally polarized wave.

The quantity, electric field intensity, can be defined by Equation (11.1) which states that a charged particle having an electric charge of q coulombs experiences a force **f** that is equal to q**E** and is therefore in the same direction as **E**

$$f = qE \qquad\qquad N \qquad\qquad\qquad\qquad (11.1)$$

If a force described by Equation (11.1) exists in a volume of space, an electric field of magnitude E exists in that volume. The unit of force in the SI-MKS system is the newton (N), which is the force required to accelerate a mass of 1 kilogram (kg) at a rate of 1 meter per second^2 (1 m/s^2). An electric field imparts motion to charged particles, and charged particles in motion constitute an electric current. In good conductors such as metals, one or a small number of electrons of each atom is relatively free and can move easily from one atom to the next; when an electromagnetic wave impinges on a metallic structure such as an antenna, an electron current flows.

A charged particle moving with velocity **v** in a magnetic field experiences a force as indicated by Equation (11.2)

$$f = q(v \times B) \qquad\qquad\qquad\qquad (11.2)$$

where **B** is magnetic flux density in webers/m^2 and is related to the **H** of Figure 11.1 in isotropic media by $B = \mu H$. For the case of nonmagnetic materials $\mu = \mu_0$ which has the value of $4\pi \times 10^{-7}$ henry/m. The force **f** involves the vector product **v** x **B**, and is perpendicular to both **v** and **B**. The force therefore imparts circular motion to the charged particle. The angular velocity of rotation ω_B of the charged particle in its circular orbit is given by $|\omega_B| = |qB/m|$, and the radius r of the circular motion is given by mv/Bq, where m is the mass of the charged particle. Equation (11.2) and the relation $B = \mu H$ define magnetic-field intensity in the same way that Equation (11.1) defines electric field intensity.

However, the force on a charged particle due to the electric field of an electromagnetic wave is greater than the force due to the magnetic field by the ratio c/v, where c the velocity of light is 3×10^8 m/s and v is particle velocity. The force due to the magnetic field of a wave itself is therefore usually of little importance, but the forces on charged particles due to the earth's magnetic field play an important role in ionospheric propagation.

In 1865 James Clerk Maxwell summarized all the then accumulated knowledge about electricity and magnetism and added exceedingly important insights of his own. The equations which he used for this purpose have become known as Maxwell's equations. They are usually considered to be four in number although some assert that two are contained in the other two and are unnecessary. We must forego a mathematical analysis, or even a complete statement of the equations or explanation of their notation, and instead refer the reader not yet familiar with electromagnetic fields to textbooks on the subject, such as those by Johnk[1]; Jordan and Balmain[2]; and Ramo, Whinnery, and Van Duzer[3]. Two of Maxwell's equations in the differential form applicable to "free space" are given below, however, as Equations (11.3) and (11.4).

$$\nabla \times \mathbf{E} = \frac{-\partial \mathbf{B}}{\partial t} \tag{11.3}$$

$$\nabla \times \mathbf{H} = \frac{\partial \mathbf{D}}{\partial t} \tag{11.4}$$

The quantities $\mathbf{E}$, $\mathbf{B}$, and $\mathbf{H}$, have already been defined. The symbol ∇ stands for the the vector operator del, which involves spacial derivatives. Partial derivatives with respect to time t appear on the right sides of the equations. The quantity $\mathbf{D}$, electric flux density, has units of coulombs/meter^2 (C/m^2) and is related to $\mathbf{E}$ for isotropic media by

$$\mathbf{D} = \epsilon \mathbf{E} \tag{11.5}$$

where*

$$\epsilon = \epsilon_0 K \tag{11.6}$$

---

*K is also denoted by ϵ_r, the relative permittivity.

with ϵ_0 = 8.854 x 10^{-12} farad/meter and K being the relative dielectric constant. K is a nondimensional quantity having values of 1 for a vacuum, close to 1 in the lower atmosphere, 2 to 10 for common dielectric materials, near 81 for water, and even higher values for certain materials such as titanium dioxide.

For present purposes, suffice it to say that the equations show that an electric field is generated by a magnetic field that varies with time and a magnetic field is generated by an electric field that varies with time. Thus time-varying electric and magnetic fields are directly related, and it is not possible to have one without the other. A combination of time-varying electric and magnetic fields, furthermore, results in an electromagnetic wave; manipulation of Equations (11.3) and (11.4) allows deriving the wave equations for and determining the properties of electromagnetic waves. Some basic characteristics of such waves will now be mentioned. One characteristic is that, for a wave in homogeneous, isotropic, lossless media, traveling in the +z direction, as indicated by + superscripts

$$\frac{E^+}{H^+} = \eta \qquad\qquad \Omega \qquad (11.7)$$

where η is the characteristic impedance of the medium and has units of ohms (Ω). η is determined by the electric and magnetic properties of the medium. For ordinary dielectric materials for which $\mu = \mu_0$

$$\eta = \sqrt{\frac{\mu_0}{\epsilon_0 K}} \qquad\qquad \Omega \qquad (11.8)$$

When K = 1, as is approximately true in the troposphere, η = 377 Ω. The velocity of propagation of an electromagnetic wave under the same conditions is given by

$$v = \frac{1}{\sqrt{\mu_0 \epsilon_0 K}} \qquad\qquad m/s \qquad (11.9)$$

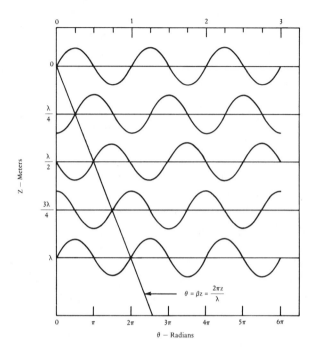

*Figure 11.2 Phase lag θ as a function of distance z, where
$\theta = \beta z = 2\pi/\lambda$. The individual sinusoidal curves, for
5 different values of z, show the instantaneous val-
ues of electric field intensity E as a function of time
for a wave having a frequency of 1 MHz.*

When $K = 1$, $v = 1/\sqrt{\mu_0 \epsilon_0}$ = c, where c has the value of 3×10^8 m/s
and is commonly referred to as the speed of light. A characteristic
of a wave of any kind is the relation between wavelength, frequency,
and velocity, namely

$$\lambda f = v \qquad\qquad\qquad\qquad (11.10)$$

where λ is wavelength in m, f is frequency in Hz (cycles per second),
and v is velocity in m/s. Finally for a wave in a lossless medium propagating in the +z direction

$$E = E_0 \, e^{-j\beta z} \qquad\qquad\qquad\qquad (11.11)$$

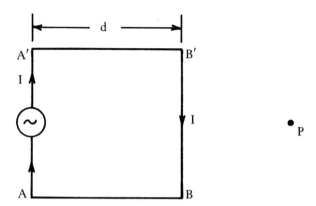

Figure 11.3 A simple low frequency circuit. Waves are radiated from the two arms AA' and BB' to the point P which is to be taken as being at a rather large distance from the circuit.

where e is the base of natural logarithms, β, the phase constant, is equal to $2\pi/\lambda$, and j is the square root of minus 1. This expression is based upon the assumption of a sinusoidally varying electric field intensity, and the factor $e^{-j\beta z}$ shows that the phase of the sinusoidal variations lags farther and farther with increasing distance z as shown in Figure 11.2.

Radiation. Maxwell's equations for **E** and **H** in free space predict the possibility of electromagnetic waves, but how are such waves originally initiated or launched? A partial answer, applying to the case of an electric circuit, lies in the facts that

1. When an electric current flows, a magnetic field is generated in the surrounding region, and if the current varies with time the magnetic field likewise varies with time.

2. A varying magnetic field generates a varying electric field.

3. The combination of varying electric and magnetic fields constitutes, and propagates as, an electromagnetic wave.

In the case of a low-frequency circuit as shown in Figure 11.3, however, the currents in the two arms AA' and BB' are in opposite directions in space. Thus if one considers that a wave is radiated from the arm AA' to P and that another wave is radiated from BB' to P, the fields of the two waves will tend to be of opposite polarity and to cancel. The exact degree of cancellation will be determined by the factor $2\pi d/\lambda$ which is the phase lag of the wave from AA' in propagating the distance d to BB'. At low frequencies for which $d \ll \lambda$, this phase lag is negligible, and the waves from AA' and BB' cancel out almost completely. If the frequency is raised and λ is correspondingly reduced, however, a condition may be reached for which $d = \lambda/2$, and the contributions from AA' and BB' add in phase. Under such a condition the radiation process in the direction of P is efficient. When the frequency is raised sufficiently, however, the situation is complicated by the fact that the current will no longer flow in the same direction with respect to the source throughout the circuit. We will not pursue further the subject of radiation from a simple circuit, but will discuss certain practical antenna structures in Section 11.2.3.

11.1.2 The Electromagnetic Spectrum as a Natural Resource

Electromagnetic waves exist over a tremendously wide range of frequencies which extend from nearly zero as a lower limit to as high as 10^{23} or more. Included in this range are radio waves and light. The term "spectrum" refers to a class or group of similar entities arranged in the order in which they possess a certain characteristic. One can refer to a group of people whose political opinions range from extremely conservative to extremely radical as representing a wide political spectrum. The term electromagnetic spectrum is used here to refer to electromagnetic radiation or waves of all possible frequencies arranged or displayed as a function of frequency. The availability of electromagnetic radiation is beneficial to man in various ways, and this radiation, or the electromagnetic spectrum, can be considered to be a major natural resource. The earth receives its energy from the sun in the form of electromagnetic radiation, and electromagnetic radiation is essential to all life on the earth. The electromagnetic spectrum is also used for communications and various other purposes, either in free space or in transmission lines or waveguides.

The electromagnetic spectrum and the environment are closely related in various ways, in addition to the fact that the spectrum is essential to life on the earth. Mankind lives in an environment permeated by electromagnetic radiation of both natural and manmade origin, and it is important to understand as well as possible just what effects the radiation has on people and biological matter. The designer and user of telecommunication and other electronic equipment, on the other hand, must consider the electromagnetic environment in which the equipment must operate. He must, for example, consider both the natural and man-made radiation in the frequency range of the equipment. Also he must take care to minimize interference caused by his equipment. The earth's atmosphere is part of our environment and affects our ability to use the electromagnetic spectrum. Electromagnetic waves that are incident upon the earth's atmosphere from the outside, as from the sun, tend to be sufficently attenuated or reflected that they have negligible intensity at the earth's surface, except in two wavelength bands that are referred to as the optical and radio windows (about 0.4 to 0.8 μm and 1 cm to 10 m, respectively). Wavelengths falling within the windows are affected also in some degree by the atmosphere, but usually not as drastically as those without. Other ways in which the environment and the electromagnetic spectrum are related involve the poles or towers and wires that may be employed to utilize the electromagnetic spectrum for telecommunications. The early years of telecommunications were characterized by very large numbers of unsightly telephone and telegraph wires, but these were reduced in number by development of underground cables and carrier systems. Further developments in communications involving more extensive utilization of broadband telecommunications services — cable television, closed circuit television, video teleconferencing, digital data transmission, etc. — may also have effects on the environment. Travel for some business and educational purposes might be reduced if more reliance were placed on the use of versatile broadband telecommunication facilities.

Some of the characteristics of the radio spectrum as a natural
resource, as specified by the Joint Technical Advisory Committee (JTAC) of the Institute of Electrical and Electronic Engineers
(IEEE) and the Electronic Industries Association [4,5] are:

1. The radio spectrum is utilized but not consumed. Present
   usage does not interfere with future usage or cause any
   degree of deterioration of the resource. A portion of the
   spectrum that is assigned or otherwise occupied for one
   use at a given place and time, however, may not be available for other uses in the same area and at the same time.

2. The resource has dimensions of space, time, and frequency
   and all three dimensions are interrelated.

3. The spectrum is an international resource.

4. The resource is wasted when assigned to tasks that can be
   done more easily in other ways or when it is not correctly
   applied to a task.

5. The spectrum is subject to pollution by man-made radio
   noise.

It has been argued in certain quarters that the radio resource should
be placed on the open market to be bought and sold as any other
commodity. A counter argument is that telecommunications has
the potential for providing social benefits which are not proportional to market value.

The above listing of characteristics mentions space and time as well
as spectrum itself. Hinchman[6] has pursued this view further and
has pointed out that efficient use of the spectrum involves time,
wave polarization, radiated power, antenna directivities, and terminal locations, as well as frequency. This reasoning led him to propose the term "electrospace" for the radio resource, electrospace
being a quantity that can be represented by an eight-dimensional
matrix involving frequency, time, polarization, power, direction
of propagation, and three spatial dimensions. Another variation is
the treatment presented in the JTAC report, Radio Spectrum Utilization in Space.[7] In this case, frequency, orbit and number of
earthward beams per satellite are regarded as the orthogonal axes
pertinent to satellite communications. It is suggested here that efficient use of the spectrum requires, among other factors, suitable

atmospheric characteristics (or a suitable medium or waveguide structure of some kind). Thus the concept of electrospace can be enlarged to encompass the subject of atmospheric effects on electromagnetic waves.

Utilization of the spectrum and electrospace requires governmental regulation, licensing, and international cooperation so that users can operate without interfering with other parties and without being interfered with. Obviously chaos would result without regulation and licensing procedures. The regulation of telecommunications is treated in Chapters 1 through 4 of this volume.

From the practical viewpoint, utilization of the electromagnetic spectrum involves communication systems, and these must have adequate signal-to- noise ratios for successful operations. Signals are generally man-made, but there are practical limits to the signal intensities which can be provided. Consequently, noise may be the factor which limits the use of the electromagnetic spectrum in a particular application. Noise is generated both in the receivers of communication systems and externally to the receivers. The external noise may be of natural origin or man-made. In this respect, what is noise to one party may be the desired signal to another party and vice versa. Man-made signals that are not in accordance with regulations and good practice or are otherwise unwanted may be considered to be a form of pollution.

When the layman thinks of the environment he perhaps thinks of air pollution or water pollution but very likely not of the electromagnetic spectrum and its pollution. The electromagnetic spectrum is an essential natural resource, nevertheless, and the electromagnetic environment is exceedingly important to the engineer who has the job of designing communication systems. It is largely the public which uses the communication facilities developed by engineers, however, and the electromagnetic spectrum and its utilization are thus important to the public as well as to engineers. Whether or not society thinks of telecommunications as involving environmental or pollution problems, it does appear to appreciate the importance and impact of telecommunications.

The world's population is increasing rapidly and its use of communication services is increasing more rapidly than population, as has been the case for energy and other resources. Arthur D. Little in

May, 1977, predicted the telecommunications equipment market to double (corresponding to an 8 per cent growth rate) in the decade ending in 1985. The communications industry appears to be in a better position to meet demands than in some other cases, but communication facilities use spectrum, energy, materials and land. The demands placed upon natural resources and the environment by the expansion of communications facilities and possible beneficial environmental effects resulting from such expansion need careful analysis.

11.2 TRANSMISSION LINES, WAVEGUIDES, AND ANTENNAS

11.2.1 Transmission Lines

Electromagnetic waves utilized for telecommunications travel from one location to another in transmission lines or waveguides or through the atmosphere. Even when most of the path from one location to another is through the atmosphere, transmission lines or waveguides almost invariably form the connecting links between the transmitter and the transmitting antenna and the receiver and the receiving antenna. The terms, transmission line and waveguide, are sometimes used in such a way as to be indistinguishable, but more commonly and in this chapter a transmission line is taken to be a two-or-more-conductor line such as a coaxial line, parallel-wire line, or microstrip line. The waveguides of interest in this chapter, however, are hollow metallic guides or dielectric guides.

Coaxial lines are an extremely important form of transmission line that has the advantage that the electric and magnetic fields are confined entirely to the space between the inner and outer concentric conductors, assuming the use of sufficiently high-quality construction and connectors. Coaxial lines can operate from dc (direct current or zero frequency) up to frequencies in the GHz range, but attenuation increases with frequency and ultimately becomes a limiting factor.

The type of wave normally utilized in coaxial lines is designated as a TEM wave. This notation indicates that both the electric field (E) and the magnetic field (M) are confined to the transverse plane (T) in the lossless case. The transverse plane is perpendicular to the length or axis of the line. The configurations

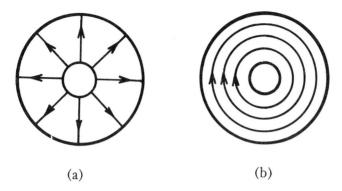

(a)   (b)

*Figure 11.4 (a) Lines representing electric field intensity **E** in a
coaxial line of inner radius a and outer radius b.
(b) Lines representing magnetic field intensity **H** in
the same coaxial line as in (a).*

of the electric and magnetic fields are illustrated in Figure 11.4.
Lines representing the direction of the electric field are radial, extending from one conductor to the other, as in Figure 11.4a, but
alternating in polarity with time and distance. Lines representing
magnetic-field intensity have the form of closed circles as in Figure 11.4b. The **E** and **H** fields are everywhere perpendicular to
each other, and their magnitudes vary inversely with radial distance
r from the center of the line.

In coaxial lines, however, reference is commonly made to the
voltage V between the inner and outer conductors, and the current I flowing in the conductors, rather than to E and H. The
relations between the magnitudes of these quantities are

$$E = \frac{V}{r \, \ln(b/a)} \; , \quad V = Er \, \ln(b/a) \qquad (11.12a\&b)$$

where the value of E is that at the distance r from the center of
the line, a is the radius of the inner conductor, b is the radius of
the outer conductor and $a \leqslant r \leqslant b$. Also

$$H = \frac{I}{2\pi r} \; , \quad I = H2\pi r \qquad (11.13a\&b)$$

The ratio of V to I for a positively traveling wave is

$$\frac{V^+}{I^+} = \frac{E^+}{H^+} \frac{\ln(b/a)}{2\pi} = \frac{\eta \ln(b/a)}{2\pi} = Z_0 \qquad (11.14)$$

η is the characteristic impedance of the medium between the inner and outer conductors, and Z_0 is the characteristic impedance of the line. An alternate approach to the analysis of two-conductor transmission lines that makes use of equivalent distributed circuits shows that Z_0 is also given by

$$Z_0 = \sqrt{\frac{\ell}{c}} \qquad (11.15)$$

where ℓ is inductance per unit length of line and c is capacitance per unit length of line. Coaxial lines that are available commercially have standard values of Z_0 such as 50 ohms, 75 ohms, etc.

The general solution of the wave equations which can be derived from Equations (11.3) and (11.4) describes waves that travel in both directions in a medium or along a transmission line. In most practical applications, a source is applied at one end of a line and travels in the assumed positive direction. Reflection at the end of the line or at discontinuities along the length, however, may cause a negatively traveling wave to exist as well as the original positively traveling wave. Reflection in transmission lines or of electromagnetic waves in general is similar qualitatively to the reflection of sound waves or water waves from obstacles in their path. But what determines the voltage of the negatively traveling wave in a transmission line? The answer can be found by writing equations for the voltage and current at the end of a transmission line, as shown in Figure 11.5.

The total voltage on the line, $V(z)$ [meaning the voltage V expressed as a function of the coordinate z], is the sum of the voltages of positively and negatively traveling waves as indicated by

$$V(z) = V_m^+ e^{-j\beta z} + V_m^- e^{j\beta z} \qquad (11.16)$$

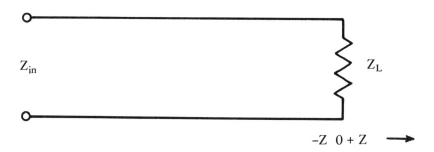

*Figure 11.5 Transmission terminated by a load impedance Z_L
     at z = 0.*

for the lossless case. The superscripts identify the peak voltages
of the positively and negatively traveling waves. A similar equation applies to current, I(z), but whereas $I_m^+ = V_m^+/Z_0$ it develops
that $I_m^- = -V_m^-/Z_0$ so that

$$I(z) = \frac{V_m^+}{Z_0} e^{-j\beta z} - \frac{V_m^-}{Z_0} e^{j\beta z} \tag{11.17}$$

If these equations are applied at z = 0 where $e^{\pm j\beta z} = 1$ and
$V(z)/I(z) = Z_L$, the load impedance, it is possible to solve for
the ratio, V_m^-/V_m^+. Identifying this ratio as a reflection coefficient, Γ_L,

$$\Gamma_L = \frac{Z_L - Z_0}{Z_L + Z_0} \tag{11.18}$$

This relation shows that if $Z_L = Z_0$, $\Gamma_L = 0$ and there is no reflected wave. Conversely if $Z_L \neq Z_0$, Γ_L has a finite value.
Γ_L will be real and either positive or negative if Z_L and Z_0 are
real. If instead of taking the ratio of the voltage of the negatively
traveling wave to the voltage of the positively traveling wave at
z = 0, we take the ratio for any arbitrary value of z we have

$$\frac{V_m^- e^{j\beta z}}{V_m^+ e^{-j\beta z}} = \Gamma(z) = \Gamma_L e^{2j\beta z} = \frac{Z(z) - Z_0}{Z(z) + Z_0} \tag{11.19}$$

and

$$V(z) = V_m^+ \, e^{-j\beta z} \left[1 + \Gamma(z) \right] \qquad (11.20)$$

$$I(z) = \frac{V_m^+}{Z_0} \, e^{-j\beta z} \left[1 - \Gamma(z) \right] \qquad (11.21)$$

The term $1 + \Gamma(z)$ can be represented by a phasor diagram as in Figure 11.6.

The horizontal phasor represents the unit phasor [the 1 of $1 + \Gamma(z)$], and the circle represents the locus of $1 + \Gamma(z)$ for a particular magnitude $|\Gamma(z)|$. The diagram also shows $\Gamma(z)$ and $1 + \Gamma(z)$ for one illustrative condition. Consideration of the diagram shows that the voltage $V(z)$ varies from a maximum value proportional to $1 + |\Gamma(z)|$ to a minimum value proportional to $1 - |\Gamma(z)|$. The ratio of maximum to minimum voltage along the line is called the standing-wave ratio (SWR) and is given by

$$SWR = \frac{1 + |\Gamma(z)|}{1 - |\Gamma(z)|} \qquad (11.22)$$

In most applications in telecommunications it is desirable to make $Z_L = Z_0$ in order to avoid having a reflected wave, and therefore a standing wave, on the line. As any obstacle, imperfection, or departure from a perfectly matched condition results in a small reflected wave, it is usually necessary to tolerate a SWR value greater than unity, such as 1.1. In certain applications, such as the use of a short-circuited stub (short auxiliary section of line) to produce an impedance match in the main line, a high SWR is deliberately produced.

Transmission line calculations for determining the value of Γ_L, SWR, and input impedance Z_{in} given the load impedance Z_L, or calculating Z_L from knowledge of the SWR and position of a voltage minimum, etc., are facilitated by the use of a graphical technique employing the well-known Smith chart. Reflection coefficient $\Gamma(z)$ plots directly on the chart in polar coordinates, and corresponding normalized

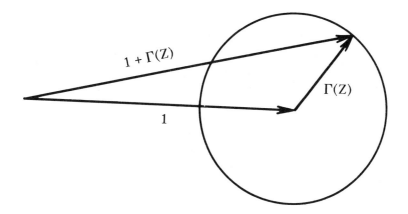

Figure 11.6 The circle shows the locus of 1 + Γ(z).

impedance values z(z) [where $z(z) = \dfrac{Z(z)}{Z_0} = r(z) \pm j\,x\,(z)$] can be determined from sets of super-imposed circles representing constant values of r and x. Coaxial lines can be used for long distance transmission, including transmission across an entire continent or ocean, when repeater stations are suitably placed. Coaxial lines, however, are subject to losses which increase with frequency. If the losses are not too great, the attenuation coefficient α which describes the decrease of voltage in accordance with $e^{-\alpha z}$ is given by

$$\alpha = \frac{r}{2}\sqrt{\frac{c}{\ell}} + \frac{g}{2}\sqrt{\frac{\ell}{c}} \qquad \text{Np/m} \qquad (11.23)$$

r is series resistance per unit length of line, and g is shunt conductance per unit length. r is given by

$$r = \frac{1}{2\pi}\left(\frac{1}{a} + \frac{1}{b}\right)\sqrt{\frac{\omega\mu}{2\sigma}} = \left(\frac{a}{2\delta} + \frac{b}{2\delta}\right) r_{dc} \qquad (11.24)$$

a and b are the inner and outer radii of the coaxial line, $\omega = 2\pi f$ where f is frequency, σ is conductivity of the inner and outer conductors of the line, and δ is skin depth, $1/\sqrt{\pi f \mu \sigma}$. The form of the expression involving δ is of interest because it indicates that the resistance is increased above the dc value by the factors shown. An

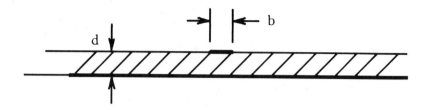

Figure 11.7 Microstrip line.

important point to notice is the increase of attenuation with frequency. This condition makes it necessary to convert to the use of
waveguides at higher frequencies in order to keep attenuation at a
reasonable level.

The familiar television lead-in line is an example of another type
of transmission line, the two-wire line. The characteristic impedance of this type of line, with air between the wires, is given by

$$Z_0 = 120 \, \ell n \left(\frac{2s}{d} \right) \qquad\qquad (11.25)$$

where d is the diameter of the wire and s is the center-to-center
spacing. Values of Z_0 for two-wire lines are commonly near 300
ohms. The electric and magnetic fields are not confined as in a
coaxial line, and radiation from two-wire lines restricts their use
to lower frequencies than can be utilized in the case of coaxial
lines.

Microstrip lines are coming into prominence for certain applications of microminiaturization. A microstrip line is constructed
as suggested in Figure 11.7, which represents a metallic base over-
laid with a thin dielectric layer of thickness d.

A transmission line is formed by a metal strip of width b on the
top surface of the dielectric layer. The characteristic impedance
of such a line is given approximately by

$$Z_0 = \sqrt{\frac{\mu_0}{\epsilon_0 K}} \frac{d}{b} \qquad\qquad (11.26)$$

where K is the relative dielectric constant of the dielectric.

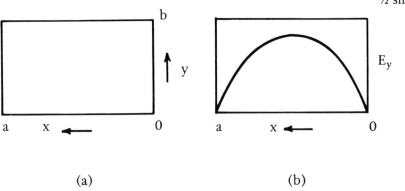

Figure 11.8 (a) Geometry of rectangular waveguide. (b) Plot of E_y versus x for TE_{10} mode.

11.2.2 Waveguides

Hollow metallic waveguides are practical for microwaves and millimeter waves, and dielectric fibers are coming into prominence for use at optical frequencies. Metallic waveguides may be rectangular, circular, or elliptical in cross section. Various configurations or modes of electric and magnetic fields can exist in waveguides. The modes are of the TE or TM variety, the designation TE indicating that the electric field has components in the transverse plane only and the designation TM indicating that the magnetic field has components in the transverse plane only. The direction along the length of the guide is normally taken as the z direction and the transverse plane is then the x-y plane. A TEM wave, the type found in coaxial lines, cannot propagate in a hollow metallic waveguide.

For discussing rectangular waveguide, consider Figure 11.8a which shows a guide with its wide dimension a in the x directions and with a smaller dimension b in the y direction. The widely used TE_{10} mode has a y component of electric field intensity E_y and x and z components of magnetic field intensity H_x and H_z as described by Equations (11.27) through (11.29).

$$E_y = E_{ym} \sin\left(\frac{\pi x}{a}\right) \tag{11.27}$$

$$H_x = \frac{-E_{ym}}{Z_{TE}} \sin\left(\frac{\pi x}{a}\right) \qquad (11.28)$$

$$H_z = \frac{jE_{ym}}{\eta}\left(\frac{\lambda_0}{2a}\right)\cos\left(\frac{\pi x}{a}\right) \qquad (11.29)$$

$Z_{TE} = \eta \Big/ \sqrt{1 - \left(\frac{f_c}{f}\right)^2}$ and is the characteristic impedance for
a TE mode. η is the characteristic impedance of the medium filling
the guide, and λ_0 is free-space wavelength. f_c is cutoff frequency,
and f is operating frequency. A plot of the magnitude of E_y as a
function of x is shown in Figure 11.8b, where the x and y axes are
drawn so that the z direction is into the paper. E_y must be zero at
x = 0 and x = a because it is tangent to the highly conducting walls
at these positions. The subscripts, 1 and 0 in this case, refer to the
cycles of variation of the electric and magnetic fields in the x and
y directions, respectively.

A valid picture of propagation in a waveguide is presented in
Figure 11.9 where the diagonal lines and arrows depict a wave
that is reflected back and forth from one side of the guide to
the other at an angle θ from the perpendicular to the walls. Under this condition the wavelength in the z direction λ_z is related
to the unlimited-medium wavelength λ_0 by

$$\lambda_z = \frac{\lambda_0}{\sin\theta} \qquad (11.30)$$

whereas

$$\lambda_x = \frac{\lambda_0}{\cos\theta} \qquad (11.31)$$

A wavelength is the distance between two surfaces of constant
phase that differ in phase by 2π radians or $360°$. Normally when
speaking of wavelength one refers to λ_0 but one can refer to
wavelength in any direction. In order to satisfy the requirement
that the electric field intensity E_y be zero at both x = 0 and x = a,
referring to the TE_{10} mode,

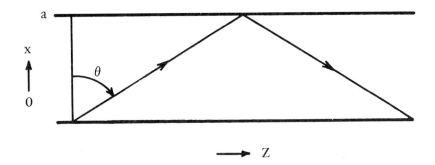

Figure 11.9 Representation of a wave that is reflected back and forth from one side of a guide to the other.

$$a = \frac{\lambda_x}{2} = \frac{\lambda_0}{2 \cos \theta} \tag{11.32}$$

and hence

$$\cos \theta = \frac{\lambda_0}{2a} \tag{11.33}$$

At a rather high frequency, or small wavelength λ_0, $\cos \theta$ will be rather small, corresponding to a rather large angle θ. If the wavelength is increased, however, a limiting value of λ_0 is reached corresponding to $\cos \theta = 1$ and $\theta = 0°$, which refers to a wave bouncing back and forth in the x direction but making no progress in the z direction. The particular wavelength satisfying this relation plays an important role in waveguide theory and is designated as λ_c, standing for cutoff wavelength. It can be seen that $\lambda_c = 2a$ and that for any angle,

$$\cos \theta = \frac{\lambda_0}{\lambda_c} \tag{11.34}$$

Referring now to Equation (11.30) for λ_z and making use of $\sin \theta = \sqrt{1 - \cos^2 \theta}$

$$\lambda_z = \frac{\lambda_0}{\sqrt{1 - \left(\frac{\lambda_0}{\lambda_c}\right)^2}} \tag{11.35}$$

and since $\lambda_z f = v_z$

$$v_z = \frac{v_0}{\sqrt{1 - \left(\frac{\lambda_0}{\lambda_c}\right)^2}} \qquad (11.36)$$

Alternatively as $\lambda_c f_c = \lambda_0 f$,

$$v_z = \frac{v_0}{\sqrt{1 - \left(\frac{f_c}{f}\right)^2}} \qquad (11.37)$$

Note that λ_z, also commonly designated as λ_g for wavelength in the guide, is greater than the unlimited-medium wavelength (which is the "free-space" wavelength if the guide is filled with air). Similarly the velocity v_z is greater than v_0 the unlimited medium velocity (c or 3×10^8 m/s if the guide is filled with air). v_z is a phase velocity, however, and not the velocity with which information or energy is transmitted. It is important to note that only frequencies greater than f_c, the cutoff frequency, corresponding to wavelengths less than λ_c, can propagate in a guide. A principal reason why the TE_{10} mode in rectangular guide is so widely used is that it has the lowest cutoff frequency of any mode. Thus, for a range of frequencies above its cutoff frequency, the TE_{10} mode is the only mode that can propagate, and the undesirable possibility of having more than one mode propagate at the same time is avoided. Equations (11.35) through (11.37), however, apply to any TE or TM mode.

Waveguides having elliptical and circular cross sections are utilized in addition to rectangular waveguide. Elliptical guide is difficult to analyze quantitatively but supports a mode that is similar to and compatible with the TE_{10} mode in rectangular guide. Commercially available elliptical guide has the advantage of flexibility with respect to rigid rectangular guide. Elliptical guide can be bent and twisted and for this reason does not have to be cut to precisely accurate lengths in order to join sections of guide together.

Associated with each mode in a metallic waveguide is a pattern of surface current which flows on the inner surfaces of the walls. The currents are integral features of the modes, and the modes

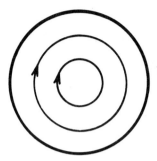

*Figure 11.10   The E_ϕ field of the TE_{01} mode in circular wave-
guide is represented by the inner circles and arrows.*

cannot exist without the currents. The waveguide walls are highly conducting but not perfectly conducting, and the flow of current results in the attenuation of waves propagating in the guide.
The attenuation of microwave frequencies is less than that for coaxial lines operating at the same frequency, but for most modes
it increases with frequency above a range of frequencies for which
attenuation is near minimum. A particular mode in circular metallic waveguide, however, has long been of interest because attenuation essentially decreases indefinitely with frequency for this mode,
the TE_{01} mode of circular guide. The subscripts for circular guide
refer to the cycles of variation of the fields in the circumferential
and radial directions. The TE_{01} mode has only a ϕ component of
electric field intensity (E_ϕ) as in Figure 11.10. The mode also has
radial and longitudinal components H_r and H_z of magnetic field intensity. The electric and magnetic fields are described by

$$E_\phi = j\eta \, \frac{f}{f_c} \, BJ_0 \, (k_c r) \tag{11.38}$$

$$H_r = - E_\phi / Z_{TE} \tag{11.39}$$

$$H_z = BJ_0(k_c r) \tag{11.40}$$

where $J_0(k_c r)$ is the Bessel function of zero order. $J_0'(k_c r)$ is the
derivative of $J_0(k_c r)$ and goes to zero at r = a, where a is the radius

of the guide. For a given power transmitted in the guide, or for a
given E_ϕ, Equation (11.38) shows that as the frequency f increases
the coefficient B decreases. As B decreases, H_z decreases, and the
surface current flowing in the wall in the ϕ direction also decreases,
as the surface current density is numerically equal in magnitude to
H_z. Thus the attenuation decreases with increasing frequency. The
problem with this mode is that it does not have the lowest cutoff
frequency of modes in circular waveguide. The modes having a
lower cutoff frequency all have an H_ϕ field, and associated with H_ϕ
there is a current flowing in the z direction along the length of the
guide. But this current is eliminated by constructing the inner wall
of a tightly wound helix of fine insulated wire[3]. The Bell System
has advertised that its circular waveguide, which is 2 inches in
diameter, can operate at frequencies to 100 GHZ and carry
230,000 conversations. Application of the guide has seemed close
at hand, but in recent years optical fibers have become highly
promising, and millimeter-wave systems utilizing the TE_{01} mode
in circular guide may thus be overshadowed by optical-fiber
systems.

Optical Fiber Communication. The invention of the laser spurred
interest in the use of light for the transmission of information, and
the production of optical fibers having losses under 20 dB/km in
1970 at the Corning Glass Works caused heightened interest in the
use of such fibers. By 1974 a loss as low as 2 dB/km had been a-
chieved in the laboratory and prospects for the application of
fibers appeared to be highly favorable[8]. Laboratory attenua-
tions of a fraction of a dB/km had been achieved by 1978, and
fibers were available commercially from several sources, including
Corning Glass Works, which sold fibers having an attenuation
$\leqslant 10$ dB/km.

Optical fibers for communications consist of a center glass core
having an index of refraction of about 1.5, surrounded by cladding (concentric and cylindrical) having a slightly lower index of
refraction. The need for cladding arises from the fact that electromagnetic fields are not confined entirely to the interior of a single
fiber but occur as evanscent fields outside the fiber.[3] Thus anything in contact with a single fiber would disturb its transmission
characteristics. When cladding is employed, however, the fields
are negligible at the outer surface of the cladding.

Although circularly symmetric TE and TM modes, like those in hollow metallic waveguides, can occur in fibers, the modes of most importance have three E and three H field components and are designated as $EH_{\ell m}$ and $HE_{\ell m}$ modes. All exhibit a cutoff frequency except the HE_{11}, which is called the dominant mode and is the mode used in single-mode operation. Analysis shows[8] that the number of modes that can be transmitted in a fiber, N, is given by

$$N \simeq v^2/2 \simeq (ka/\lambda)n_1^2 \Delta \qquad (11.41)$$

where $v = (2\pi a/\lambda) (n_1^2 - n_2^2)^{1/2}$, a is the radius of the core, λ is the wavelength, n_1 is the index of refraction of the core, n_2 is the index of refraction of the cladding, and Δ is defined by $n_2 = n_1 (1 - \Delta)$. In order to achieve single-mode operation a must be small, a value of $5\mu m$ being suitable. For multimode operation, a is typically near $60 \ \mu m$. For both single and multimode operation the outer diameters of fibers are usually in the 75 to 100 μm range. If the values of n_1 and n_2 are constant as a function of radius in the multimode case, significant dispersion can occur in fibers, with the result that when a single pulse is transmitted, for example, a smeared out version is received because the different modes have different velocities. If n_1 is made to vary appropriately with radius, however, the multimode dispersion is considerably reduced. The term, graded index, is applied to the variation of n_1 with radius. Waveguide dispersion (variation of velocity within a mode) and material dispersion (variation of index of refraction of glass with frequency) also contribute to total dispersion in both single-mode and multimode fibers. Whether it is practical to reduce multimode dispersion sufficiently to achieve total dispersion as low for multimode fibers as for single-mode fibers is not clear. The relative merits of single-mode and graded-index multimode operation were a subject of discussion in 1978.

Light-emitting diodes and injection lasers are the most prominent optical sources used with fibers; the former is an incoherent source and the latter is a coherent source. Photodiodes are used as detectors. Multimode fibers must be used when incoherent sources are employed, and lasers must be used for single-mode operation. Single-mode fibers pose more severe requirements upon connectors than do multimode fibers. Light-emitting diodes and injection lasers may be modulated internally, but external modulation using electro-optic or acousto-optic techniques allows higher bandwidths and bit rates, up to GHz and Gb/s values.

11.2.3 Antennas

Antennas are of may types [9,10] and only some of the varieties used most commonly in telecommunications will be mentioned. At microwave frequencies, paraboloidal-reflector and horn antennas are widely utilized, and at lower frequencies dipole and other thin-wire antennas are commonly employed. Directivity, gain, and antenna pattern are the basic antenna parameters that are considered here. Directivity is a measure of the degree to which an antenna concentrates radiation in a given direction instead of radiating it uniformly in all directions. Thus, by definition, directivity D is given by

$$D = \frac{4\pi}{\Omega_A} \tag{11.42}$$

where Ω_A is the solid angle of the antenna and 4π is the solid angle in steradians or radians^2 that surrounds a point in space. The solid angle Ω subtended by an area A at a distance r from a point of observation is $A_\perp/r^2$ where $A_\perp$ is the projection of the area of a plane perpendicular to the line of sight. Ω_A is given approximately by

$$\Omega_A = \frac{4}{3} \theta_{HP} \phi_{HP} \tag{11.43}$$

where θ_{HP} and ϕ_{HP} are half-power antenna beamwidths in two orthogonal directions, corresponding to the θ and ϕ coordinates of a spherical coordinate system. Gain G is related to directivity D by $G = \kappa_0 D$ where κ_0 is ohmic efficiency. In the theoretical case of a lossless antenna, gain and directivity have the same value but gain is less than directivity for an actual antenna.

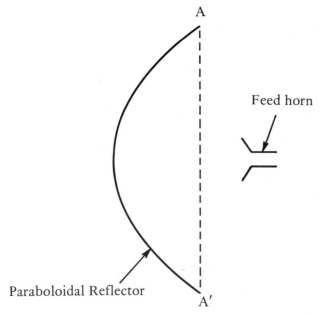

A

Feed horn

Paraboloidal Reflector

A′

Figure 11.11   Paraboloidal-reflector antenna.

In the case of paraboloidal-reflector antennas, an antenna feed is placed at the focus of a paraboloid with the result that a surface of constant phase is generated over the aperture which lies in the plane AA′ of Figure 11.11. For this type of antenna having an aperture area A

$$D = \frac{4\pi A'_{eff}}{\lambda^2} \tag{11.44}$$

and

$$G = \frac{4\pi A_{eff}}{\lambda^2} \tag{11.45}$$

where A'_{eff} is the effective aperture area for directivity, A_{eff} is the effective area for gain, and $A > A'_{eff} > A_{eff}$. The antenna feed blocks part of the aperture, and that is one of the reasons that $A > A'_{eff}$. Also the illumination of the reflector by the antenna feed usually is tapered intentionally so that the illumination is less intense near the edge of the reflector. Tapering minimizes antenna side lobes and minimizes the danger of spillover

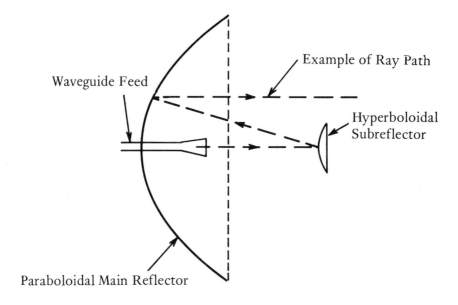

Waveguide Feed

Example of Ray Path

Hyperboloidal
Subreflector

Paraboloidal Main Reflector

Figure 11.12   Cassegrain antenna.

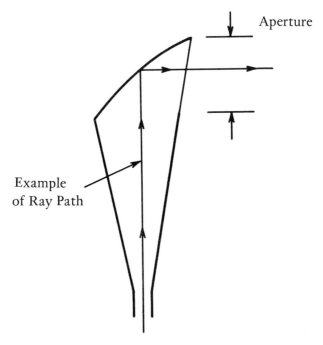

Aperture

Example
of Ray Path

Figure 11.13   Horn antenna.

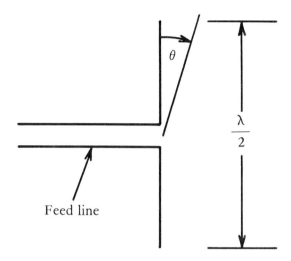

Figure 11.14   Half-wavelength electrical dipole antenna.

of radiation but results also in a decrease in effective area. (Spill-over refers to radiation from the antenna feed that misses the reflector.) A_{eff} is less than A'_{eff} because of ohmic losses. Values of A_{eff} generally vary between about 0.5 A and 0.7 A. A value of 0.54 A is commonly used for conventional antennas.

Other forms of microwave antennas are the Cassegrain and horn antennas, as illustrated in Figures 11.12 and 11.13. The Cassegrain antenna uses a paraboloidal main reflector and a hyperboloidal subreflector. The waveguide feed approaches the main relector from the rear, with the resultant advantage that a preamplifier can be placed there. Much the same advantage applies to the horn antenna.

A basic antenna type for use at lower frequencies than microwave frequencies is the electric dipole antenna, commonly $\lambda/2$ in length and fed at the center as in Figure 11.14. The electric field intensity of a $\lambda/2$ dipole at a distance r in the far field is in the θ direction and given by

$$|E_\theta| = \frac{60 I_m}{r} \frac{\cos\left(\frac{\pi}{2}\cos\theta\right)}{\sin\theta} \qquad (11.46)$$

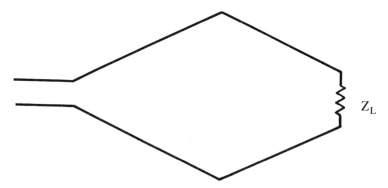

Figure 11.15   Rhombic antenna.

where I_m is the maximum (input) current of the dipole. $|E_\theta|$ is
seen to have its maximum value when $\theta = 90°$ or in the broad-
side direction. The dipole is an example of an antenna constructed
from relatively thin wires or rods. It has a broad beamwidth in
the θ direction and is omnidirectional (constant signal intensity)
in the ϕ direction (referring to a spherical coordinate system
with θ measured from the axis of the dipole). The directivity
of a single $\lambda/2$ dipole is 1.64, but dipole antennas can be arranged in linear or two-dimensional arrays to provide narrower
bandwidths.

For use of HF frequencies for long-distance commercial communications, rhombic and log-periodic antennas have been commonly used. The rhombic antenna can be regarded as an array of
four horizontal long-wire antennas arranged as in Figure 11.15.
The presence of the terminating impedance Z_L causes the antenna to have maximum radiation in the forward direction (to the
right in Figure 11.15) at an angle above the horizontal, and which is
a function of the height-to-wavelength ratio of the antenna. Large
rhombic antenna farms on the Atlantic and Pacific coasts of the
U.S. formerly handled a large fraction of overseas traffic before
the advent of satellites.

The log periodic antenna has the advantage of being able to operate over an especially wide frequency range. There are a variety
of forms for log-periodic antennas, one form being shown in Figure 11.16. Such an antenna is fed at the apex (at the left between

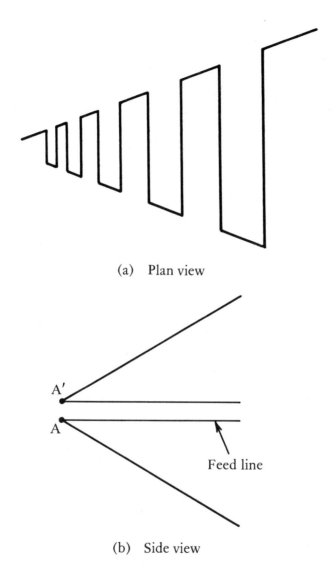

(a)   Plan view

(b)   Side view

Figure 11.16   *Log-periodic antenna. (a) Plan view of one ele-*
ment. (b) Side view showing relative position
of the two elements.

A and A' in Figure 11.16), and a wave propagates from the feed
point to the right until a resonant λ/2 condition is encountered.
Reflection then takes place and radiation from the antenna is to
the left.[2]

11.3 TROPOSPHERIC PROPAGATION

11.3.1 System Considerations

Electromagnetic waves that are utilized for terrestrial and satellite communications, and that propagate in or through the troposphere at frequencies that are too high to experience reflection in the ionosphere, follow paths that can be classified into line-of-sight, diffraction, and troposcatter categories. Line-of-sight paths are used extensively for terrestrial and earth-satellite communications. In some cases, such as a path over a mountain peak or other obstruction, communication is maintained with electromagnetic fields that reach the opposite side of the obstruction by a process of diffraction (scattering or reradiation) from near the top or edges of the obstruction. Troposcatter paths depend upon scattering from inhomogeneities in the atmosphere itself.

To a first approximation, radio waves travel in straight lines and one might think that little consideration need be given to the effect of the troposphere on communication systems. The atmosphere in a nonuniform or inhomogenious medium, however, and its properties vary with altitude, location and time with the result that radio waves tend to travel in curved paths, the exact path or paths changing with time. Also, higher-frequency radio waves experience attenuation and scattering due to precipitation and gases. In the case of paths in the clear atmosphere, the parameter that affects propagation is the index of refraction of the air. Variations in index of refraction are small in magnitude, but nevertheless affect propagation significantly. Raindrops, however, constitute inhomogeneities that scatter electromagnetic radiation and that have an index of refraction that differs from that of the surrounding air by approximately a factor of 9.

For satisfactory operation of a telecommunication system it is necessary to have a sufficient receiving system carrier-signal power to noise power ratio, sometimes designated by C/X, with C representing signal power and X representing noise power. The ratio is a function of system parameters, path geometry, and atmospheric effects. We first consider briefly the calculation of C in terms of system characteristics and path length alone for terrestrial and satellite systems. Attention will then be directed to the effects of the atmosphere and path geometry.

The carrier power C (or W_R) at the receiver input terminals for terrestrial line-of-sight systems is given by

$$C = \frac{W_T\, G_T\, A_R\, L_a}{4\pi d^2} \tag{11.47}$$

where W_T is transmitter power, G_T is transmitting-antenna gain, A_R is the effective area for gain of the receiving antenna, d is path length, and L_a accounts for losses such as the attenuation in the waveguides running to the transmitting and receiving antennas. Using Equation (11.45) to relate antenna gain and effective area, Equation (11.47) can be converted to the form

$$\frac{W_T}{C} = \left(\frac{4\pi d}{\lambda}\right)^2 \left(\frac{\lambda^2}{4\pi A_T}\right) \left(\frac{\lambda^2}{4\pi A_R}\right) = \frac{L_{FS}\, L_a}{G_T\, G_R} \tag{11.48}$$

Expressing the equation in dB values it becomes

$$L_{dB} = 10 \log_{10}\left(\frac{W_T}{C}\right) = \left(L_{FS}\right)_{dB} - \left(G_T\right)_{dB} - \left(G_R\right)_{dB} + \left(L_a\right)_{dB} \tag{11.49}$$

Also

$$C_{dBm} = \left(W_T\right)_{dBm} - \left(L_{FS}\right)_{dB} + \left(G_T\right)_{dB} + \left(G_R\right)_{dB} - \left(L_a\right)_{dB} \tag{11.50}$$

where dBm refers to power level in dB relative to a milliwatt. We shall see in Section 11.3.3, however, that an allowance must be made for fading of the received signal below this calculated value. A minimum value for C/X of 10 dB, after fading has been taken into account, is usually needed. The calculation for a satellite-earth path is similar to that for a terrestrial path, but it is common practice to utilize the quantity $C/T_{sys\ dBW}$ with the result that

$$\left(C/T_{sys}\right)_{dBW} = \left(EIRP\right)_{dBW} - \left(L_{FS}\right)_{dB} + \left(G_R/T_{sys}\right)_{dB} - \left(L_a\right)_{dB} \tag{11.51}$$

Here the transmitted power W_T and the transmitting antenna gain G_T have been combined into the quantity EIRP (for Effective Isotropic Radiated Power), expressed in this case as a power level in dB relative to a watt. The G_R/T_{sys} ratio is taken as a figure of merit for an earth receiving station. T_{sys} is the system noise temperature of the receiving system, and C and T_{sys} refer to values at the antenna terminals.

Noise, X, in a terrestrial line-of-sight system is referred to the receiver input terminals and calculated by use of

$$X = kT_0 BF \qquad (11.52)$$

where k is the Boltzmann constant (1.38×10^{-23} J/K), T_0 is the standard temperature in kelvins (290 K), B is bandwidth in Hz, and F is receiver noise figure. In the case of earth receiving stations, noise is referred to the antenna terminals and calculated by

$$X = kT_{sys} B \qquad (11.53)$$

Only a brief account of system considerations is given here, as this chapter is devoted mainly to how electromagnetic waves travel from one location to another. Further discussion of systems is found in Chapter 10, and systems and atmospheric effects are treated more fully by, for example, Flock [11], Freeman [12], and Panter [13].

11.3.2 Index of Refraction of the Troposphere

The index of refraction or refractivity n of a particular type of electromagnetic wave in a given medium is by definition the ratio of the velocity c, 3×10^8 m/s, to the phase velocity v_p. Thus

$$n = c/v_p \qquad (11.54)$$

In a lossless medium, n is related to K, the relative dielectric constant, by

$$n = \sqrt{K} \qquad (11.55)$$

Thus any effects upon propagation due to index of refraction can
be interpreted in terms of relative dielectric constant and *vice versa*.
(The symbols ϵ_r and ϵ' are sometimes used for relative dielectric
constant.) The index of refraction of the troposphere is only slightly greater than unity, and for a measure of refractivity it has become
standard practice to use N units such that the refractivity in N units
is given in terms of n by

$$N = (n - 1) \times 10^6 \qquad\qquad\qquad (11.56)$$

If n = 1.000300, for example, the refractivity in N units is 300.

The index of refraction of the troposphere is a function of pressure, temperature, and water vapor content and is given by

$$N = \frac{77.6}{T}\left(p + \frac{4810e}{T}\right) \qquad\qquad (11.57)$$

where p is atmospheric pressure in millibars (mb), T is temperature
in kelvins, and e is partial pressure of water vapor in mb. Note the
major effect of e in determining N. Sea level pressure p is 1013 mb
or 1.013×10^5 N/m^2. The water vapor pressure e can be determined
by taking the product of e_s, the saturation water vapor pressure, and
the relative humidity.

Propagation in the troposphere is influenced by the vertical index
of refraction profile. The profile can be determined at a particular
location and time by radiosonde data or by use of an airborne microwave refractometer. The latter utilizes a resonant microwave
cavity, having holes to admit the ambient air. The resonant frequency of the cavity is determined in part by the index of refraction of
the air filling the cavity. The refractivity tends to decrease with altitude, and in the absence of specific detailed information it is helpful to have available a model describing the typical decrease of refractivity with height. The CRPL exponential radio refractivity atmosphere [14] having the form

$$N = N_s e^{-h/H} \qquad\qquad\qquad (11.58)$$

has been found to correspond closely to average atmospheric conditions. N_s is surface refractivity, H is a constant, and h is height above the surface. The average value of N_s for the U.S. is said to be 313, and a value of H of 7 km is appropriate for the U.S. [14]. Average surface refractivities vary, however, from as low as about 240 for dry, mountainous or high-altitude areas to about 400 for humid tropical areas.

11.3.3 Ray Paths in the Troposphere

Tropospheric propagation can be analyzed by considering the paths taken by rays launched with initial angles β from the horizontal. When the index of refraction varies with altitude the ray paths tend to be curved rather than straight. By definition, curvature C is equal to $1/\rho$ where ρ is the radius of curvature. That is

$$C = \frac{1}{\rho} \tag{11.59}$$

(Curvature has no relation to signal power which was also designated by the symbol C.) If curvature is constant for a distance d along a ray path, the total change in direction in this distance is Cd. It can be shown [11,14] that when the index of refraction varies with altitude at a constant rate dn/dh the curvature of a ray is given by

$$C = -\frac{dn}{dh} \cos \beta \tag{11.60}$$

where h is height. For a nearly horizontal path, $\cos \beta \simeq 1$ and $C \simeq - dn/dh$.

For a path over the surface of the earth, the difference between the curvature of the earth's surface and the ray curvature is important. Neglecting surface topography, the curvature of the earth's surface is $1/r_0$, where $r_0 = 6370$ km, the radius of the earth. The difference in curvatures is

$$\frac{1}{r_0} - C = \frac{1}{r_0} + \frac{dn}{dh} \tag{11.61}$$

The same relative curvature can be maintained if, instead of using the actual earth radius and actual ray curvature, one uses an effective earth radius kr_0 and a ray of zero curvature as illustrated by

$$\frac{1}{r_0} + \frac{dn}{dh} = \frac{1}{kr_0} + 0 \qquad (11.62)$$

A typical value of dn/dh is – 40 N units/km, corresponding to a k value of 4/3. If it is assumed that this k value is applicable, path profiles can be drawn using 4/3 times the true earth radius. The advantage of this procedure is that ray paths can then be drawn as straight lines, corresponding to the effective value of 0 for dn/dh on the right side of Equation (11.62). This procedure has been widely used.

A limitation of this approach, however, is that dn/dh and k may take on a range of values such as those of Table 11.1.

Table 11.1 dN/dh and k values

$\dfrac{dN}{dh}$, N/km	k
220	5/12
157	1/2
78	2/3
0	1
– 40	4/3
– 52	3/2
–100	2.75
–157	∞
–200	–3.65

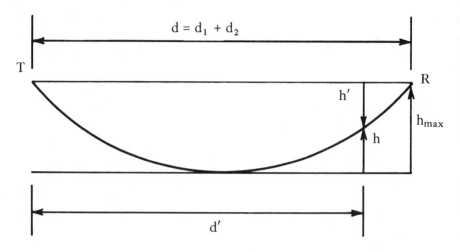

Figure 11.17   Flat-earth plot.

It is impractical to draw separate charts for each k value, but another procedure that also retains the correct relative curvature between the earth and the ray paths is useful. This approach involves making the earth flat and drawing a different ray·path for each k value of interest. The basis for flat-earth plots is as follows. As shown in the Appendix, a ray that is launched horizontally in a uniform atmosphere and propagates for a distance ℓ reaches a height above the spherical earth's surface given by $h = \ell^2/2r_0$, where r_0 is the radius of the earth. When refractivity varies with height corresponding to a particular k value, the relation becomes $h = \ell^2/2kr_0$. Or if ℓ is expressed in km and h in meters, we have

$$h = \frac{\ell^2}{12.75\ k} \tag{11.63}$$

Consider now the flat-earth plot of Figure 11.17. The curve is constructed by setting $h' = h_{max} - h$ where h_{max} and h are calculated by use of Equation (11.63). For calculating h_{max}, let ℓ in Equation (11.63) be $\dfrac{d_1 + d_2}{2} = d/2$, where d is the total path length and d_1 and d_2 are the distances from the two ends to any point along the path. For calculating h, let ℓ be $d_2 - \dfrac{d_1 + d_2}{2}$ or $d_1 - \dfrac{d_1 + d_2}{2}$.

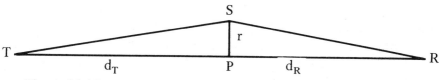

Figure 11.18   Geometry for calculation of Fresnel zone radii.

This procedure amounts to letting ℓ be the distance from the center of the path to the point for which the calculation is made. If the above procedure is followed, algebraic manipulation provides the expression

$$h' = \frac{d_1 d_2}{12.75\ k} \qquad\qquad (11.64)$$

As discussed more fully in a well-known reference by GTE-Lenkurt, [15] curves can be constructed for any k value of interest, and the path profile can then be drawn to the same scale to determine path clearance (the spacing between the ray path, or electromagnetic-wave path, and any topographic feature or other obstacle along the path). The GTE-Lenkurt reference includes a set of such curves for general use, but it is easy to construct such curves as needed by utilizing Equation (11.64).

Clearance and Reflections.   Obviously, it is necessary that a line-of-sight microwave path not be blocked by obstacles. But if a ray path just barely misses an obstacle, such as a mountain top or building, for the expected values of k, will the system operate satisfactorily? The answer to this question lies in an application of Huygens' principle, which says that every elementary area of an electromagnetic wavefront acts as a source of spherical waves. Thus the antenna of a telecommunications system tends to receive radiation from the full extent of the wavefront and not merely the radiation that follows the most direct path. This subject can be dealt with in terms of Fresnel zones.

Consider Figure 11.18 which shows two ray paths between a transmitter and a receiver. TPR = $d_T + d_R$ is the direct path and TSR = $r_T + r_R$, idealized as being composed of the diagonals of two right triangles, is greater than TPR by half a wavelength. That is

$$r_T + r_R = d_T + d_R + \lambda/2 \qquad\qquad (11.65)$$

The distance F_1 actually extends in all directions in the plane perpendicular to TPR and is defined as the first Fresnel-zone radius. Taking into account that $F_1 \ll d_T$ and $F_1 \ll d_R$, the following expression for F_1 can be obtained.

$$F_1 = 17.3 \sqrt{\frac{d_T\, d_R}{f_{Ghz}\, d}} \qquad (11.66)$$

with distances in km and F_1 in m.

The significance of F_1 is that all of the radiation passing within a distance F_1 of the direct path and reaching the receiver contributes constructively to the received signal power. Consider also a second path for which $r_T + r_R = d_T + d_R + \lambda$, with F_2 being the corresponding distance from the direct path. The radiation passing between F_1 and F_2 will interfere destructively with the radiation passing between the direct path F_1. The region within F_1 is known as the first Fresnel zone, that between F_1 and F_2 is known as the second Fresnel zone, etc. Radiation from adjacent zones interferes destructively.

There is no precise value for the clearance that is needed on a line-of-sight path but the GTE-Lenkurt reference [15] quotes a value of $0.6\, F_1 + 3m$ at $k = 1.0$ for light-route or medium reliability systems and at least $0.3\, F_1$ at $k = 2/3$ and $1.0\, F_1$ at $k = 4/3$ for heavy-route or highest reliability systems.

In addition to consideration of path clearance, possible reflections of rays from the earth's surface need to be taken into account. Reflected rays may add constructively or interfere destructively with the direct ray. Assuming perfect reflection and a reversal of phase of 180° upon reflection as is appropriate for a perfectly smooth, perfectly conducting earth, assuming equal power for direct and reflected paths, and neglecting the earth's curvature, it develops that the total electric field intensity, E, at a receiving location is given by

$$E = 2E_0 \sin\left(\frac{2\pi h_T\, h_R}{\lambda d}\right) \qquad (11.67)$$

where E_0 is the field intensity due to one path alone, h_T is the height of the transmitting antenna above the point of reflection, h_R is the corresponding height for the receiving antenna, d is path length, and λ is wavelength. Insofar as this equation is applicable, it indicates that E varies from 0 to $2E_0$, depending upon the values of h_T and h_R. The equation also indicates that if h_T is fixed, h_R can be chosen to maximize E. The limitations of the equation, however, should be kept in mind. The surface may not be perfectly smooth or highly conducting, in which case the amplitude of the reflected wave will be less than that of the direct wave. Also there may or may not be any reflection, depending on the k value. Finally when the curvature of the earth is taken into account, both the location of the point of reflection [15] and the phase of the reflected wave will vary with the k value.

Because of the possibility of destructive interference and variation of the phase and amplitude of the reflected wave, reflections should be avoided or minimized if possible by arranging for obstructions in the way of potentially reflecting paths or arranging for reflection to take place from rough, poorly reflecting surfaces rather than from smooth, highly reflecting ones.

Fading. In Section 11.3.1 a procedure for calculating the signal level on a terrestrial path was outlined. It is necessary to allow for fading of the received signal, however, the usual allowance being in the range of 35 to 45 dB. Fading may be caused by reflections from the earth's surface, as discussed in the previous paragraphs, or it may arise from strictly atmospheric effects. A common type of fading involves multipath propagation. Energy reaches the receiving antenna by two or more atmospheric paths, and the signals on the two paths alternately add constructively and interfere destructively. This type of fading tends to take place on nearly horizontal paths through or near temperature inversion layers, especially when the air below the inversion has a high moisture content as along the southern California coast. In that location fading tends to be intense when temperature inversions are present, but the signal is extremely steady when cyclonic storms eliminate the temperature inversion. Fading of this type can be minimized by avoiding horizontal paths and can be combatted if necessary by the use of space or frequency diversity. Space diversity involves using two receiving antennas, one spaced vertically above the other. Frequency diversity involves operation on two different frequencies, usually with the same receiving antenna.

A severe form of fading that may not be amenable to the remedy of space or frequency diversity has been referred to as blackout fading [16].This type of fading involves very large negative gradients of dN/dh which cause the beam to bend downwards sharply so that radiation misses the receiving antenna nearly completely or falls outside the main beam of the receiving antenna. A rather shallow surface layer of extremely high water vapor content, which decreases rapidly with height, is the usual cause of blackout fading. Possible means of avoiding blackout fading include avoiding the immediate area where it tends to occur, utilizing shorter paths than usual, and increasing antenna heights.

Paths between earth stations and satellites pass through the lower atmosphere at rather large angles from the horizontal and are not subject to fading to the same extent as terrestrial paths. A fading allowance of only about 4 dB has been considered to be sufficient. It has developed unexpectedly, however, that even the 4 and 6 GHz signals used for satellite operations are subject to scintillation of ionospheric origin in equatorial and auroral latitudes[17].

11.3.4 Attenuation Due to Precipitation and Atmospheric Gases

Although prominent radar backscatter echoes are received from precipitation at lower frequencies, including the S (10 cm) and L (23 cm) bands, attenuation due to precipitation and atmospheric gases does not begin to become serious until frequencies of 10 GHz (3 cm) or higher are reached. A water vapor absorption line causing attenuation up to about 2 dB/km is centered on 22.235 GHz and was responsible for the fact that the performance of K band (1.25 cm) radars developed during World War II did not live up to expectations. A strong oxygen absorption line centered on 60 GHz (0.5 cm) causes attenuation over 10 dB/km at sea level, and a second oxygen peak near 120 GHz causes attenuation near 2 dB/km. Attenuation due to water vapor reaches about this same value near 150 GHz.

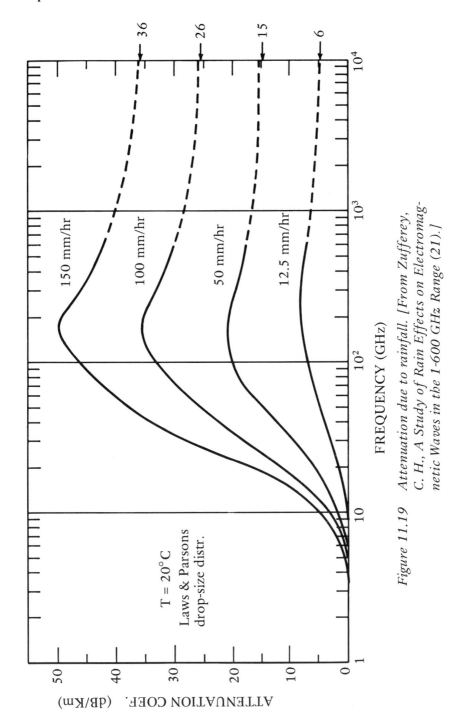

Figure 11.19 Attenuation due to rainfall. [From Zufferey, C. H., A Study of Rain Effects on Electromagnetic Waves in the 1-600 GHz Range (21).]

Attenuation caused by precipitation increases up to frequencies
near 150 GHz and is a factor that limits the application of millimeter waves. Much interest is presently directed, however, to the
use of higher frequencies for satellite communications. Paths between the earth and satellites have the advantage that they pass
steeply through regions of rainfall, rather than extending through
rainfall for a long, nearly horizontal distance. It has been asserted
also by Feldman[18] and others that because of normal delays in
placing phone calls, especially when rapidity of service is sacrificed
to obtain a lower cost, the occasional outages on millimeter paths
should not be allowed to preclude the use of millimeter waves.

The calculation of attenuation caused by given rates of rainfall is
complicated but tractable [19, 20]. Figure 11.19 shows the results
of calculations by Zufferey[21]. A problem in calculating attenuation is that of estimating the distribution of drop sizes, corresponding to a given rainfall rate. The calculations of Figure 11.19 are
based upon the use of the Laws and Parsons drop-size distribution
[22]. Another problem is the lack of sufficiently extensive statistical data concerning rainfall [23].

11.3.5 Troposcatter

Microwave telecommunication systems are an outgrowth of research on microwave radar systems during World War II. The first
commercial microwave service was put into operation in 1945. In
the early 1950's it was discovered that weak but reliable signals
were propagated beyond the horizon at microwave and lower
frequencies. Controversy existed for some time concerning
whether these troposcatter signals were reflected from layers or
scattered from atmospheric turbulence. The question has not
been resolved completely; the present trend is to emphasize
scattering from turbulence but to recognize that the turbulence
may be limited in spatial extent.

Signal intensity can be calculated as for a line-of-sight link except
that an additional loss is involved. This loss can be expressed as a
function of the "angular distance" θ which in the case of a smooth
spherical earth is simply d/r where d is the length of the path, and
$r = kr_0$, with r_0 the radius of the earth and k the k factor of
Table 11.1. The value of θ is modified as needed to take account

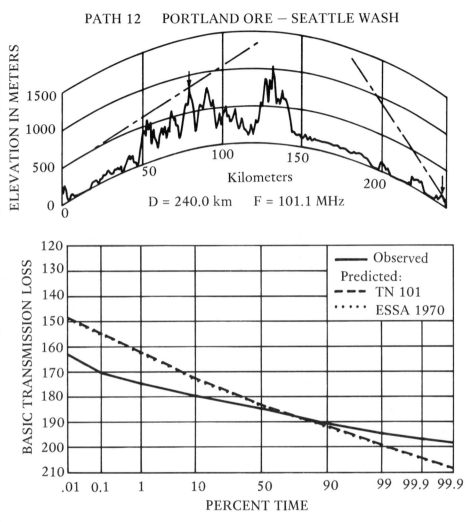

PATH 12   PORTLAND ORE – SEATTLE WASH

D = 240.0 km   F = 101.1 MHz

*Figure 11.20   Troposcatter path performance [From Longley,
A. G., R. K. Reasoner and V. L. Fuller, Measured
and predicted long-term distributions of tropospheric transmission loss (24).]*

of surface topography. Several methods for estimating the loss
factor have been utilized and are described by Panter [13] and
Freeman [12]. One approach developed by ITT is to use an
empirical curve which presents troposcatter loss at 900 MHz as a
function of angular distance [13].

Data concerning propagation on radio paths, whether line-of-sight or troposcatter, are best presented in statistical form as in Figure 11.20 [24]. The solid curve of this figure shows the observed percentage of time that the transmission loss is at a certain level or less. The basic transmission loss is, for example, 193 dB or less for 99% of the time, about 173 dB or less for 1% of the time, etc.

11.4 IONOSPHERIC PROPAGATION

11.4.1 Introduction

Before the advent of satellites, HF transmissions, propagated via ionospheric reflection, supplied a large fraction of long-distance communications. Satellites now handle an increasingly large proportion of long-distance communications, but HF and lower frequency systems that utilize or are affected by the ionosphere still fulfill important needs. Submarine cable previously shared transoceanic service in some cases with HF systems and now plays the same role with respect to satellites.

The earth's ionosphere [25], extending from perhaps 50 km as a lower limit to 1000 km as an upper limit, is a weakly ionized gas or plasma. This plasma is formed largely by ultraviolet and X-ray emissions from the sun. Different portions of the solar spectrum interact with different constituents of the atmosphere to form several different layers which have been designated as the D, E, and F layers, the latter often separated in the daytime into F_1 and F_2 layers. Ionization of the previously neutral molecules and atoms of the ionosphere results in free electrons and positive ions. We will be concerned primarily with the free electrons of the ionosphere as they interact the most readily with the electromagnetic waves that are used for telecommunications.

The lowest layer, the D layer, extends from about 50 to 90 km. It has a maximum electron density of approximately $10^3/cm^3$ between about 75 and 80 km in the daytime. This density is the lowest of the ionospheric layers, but a high electron collision frequency causes high attenuation of AM broadcast signals propagating in the D region in the daytime. The D layer essentially disappears at night. The E layer extends from about 90 km to 140 km with the peak electron concentration of $1 - 1.5 \times 10^5/cm^3$ occurring be-

tween about 100 and 110 km. The peak ionospheric electron concentration of up to 2 x $10^6/cm^3$ occurs in the F_2 region in the 200
to 400 km height range. The daytime F_1 region occurs at lower alti-
tudes with lower densities. Electron densities of the F_2 layer remain rather high, perhaps up to 4 x $10^5/cm^3$, at night. Long-range
HF propagation commonly involves reflection from the F_2 layer,
but electromagnetic waves are also reflected from the E and F_1
layers as well. The D region forms part of the earth-ionosphere
waveguide that influences the propagation of VLF (3 to 30 kHz)
waves.

11.4.2 Characteristic Waves

Propagation in the ionosphere is influenced by the earth's magnetic field, as well as by the free electrons of the ionosphere. Wave
propagation parallel to the magnetic field is different from propagation perpendicular to the magnetic field. In considering propagation in the ionosphere, the concept of characteristic waves is important, and to discuss characteristic waves it is necessary to discuss
wave polarization. It was mentioned above in Section 11.1.1 that if
the electric field intensity E is vertical the wave is vertically polarized, whereas if E is horizontal the wave is horizontally polarized.
In both cases, and whenever the E vector is in a fixed direction, the
wave polarization is linear. In this section we will not be much concerned with whether the polarization is vertical or horizontal but
whether it is linear or otherwise.

What kinds of polarization occur other than linear? Waves can have
circular or elliptical polarizations. What is meant by the term circular polarization? This term refers to a wave having an electric field
intensity vector that has a fixed length but rotates with angular velocity ω where $\omega = 2\pi f$ and f is the frequency. There are two possible directions of rotation, the right circular direction and the left
circular direction. Right circular rotation is in the direction of the
fingers of the right hand if they are pictured as encircling the thumb
when the thumb points in the direction of wave propagation. Left
circular rotation is in the opposite direction. Corresponding to these
two directions, we speak of electromagnetic waves that have right
circular (rc) and left circular (ℓc) polarizations.

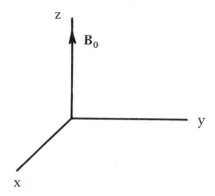

Figure 11.21   Coordinate system for considering propagation in plasma.

We now define what the term characteristic wave means. A characteristic wave is one that retains its same original polarization as it propagates. rc and ℓc waves have this characteristic when propagating parallel to the magnetic field in a plasma. An rc wave remains an rc wave as it propagates and likewise for an ℓc wave. What happens if a linearly polarized wave is launched parallel to the magnetic field in a plasma? It remains linear, assuming attenuation can be neglected, but the orientation of the E vector changes (rotates) as the wave propagates, and this change in orientation constitutes a change in polarization. That is, the concept of polarization involves not only the distinction between linear and circular but the direction of the E vector in the case of linear polarization. The characteristic waves for propagation parallel to the magnetic field in a plasma are rc and ℓc waves. A linearly polarized wave is not a characteristic wave and experiences rotation, known as Faraday rotation, as it propagates. The characteristic waves for propagation perpendicular to the magnetic field are two in number also but are linearly polarized. They retain the same linear polarization as they propagate.

To consider ionospheric propagation, further, we utilize Figure 11.21, which shows a rectangular coordinate system with the earth's magnetic field B_0 directed along the z axis. Propagation will be taken to be in the x-z plane. The simplest type of wave is a linearly polarized wave propagating in the x direction, perpendicular to the magnetic field, and having its electric field intensity vector in the z direction. The velocity v imparted to the free electrons of the plasma by the electric field is thus in the z direction; consequently the electrons are unaffected by B_0 as the force on electrons due to B_0 is proportional to $v \times B_0$ (Equation 11.2) the vector product of which is zero when v and B_0 are in the same direction. Thus the analysis proceeds as if there were no magnetic field. The force f on the electrons is due to the electric field alone and is equated to mass times acceleration, i.e. $f = qE = ma$; but $a = dv/dt$ and as E varies sinusoidally $dv/dt = j\omega v$ where v is now understood to be a complex quantity. Thus $mj\omega v = qE$ and

$$v = \frac{qE}{mj\omega} = \frac{-jqE}{m\omega} \tag{11.68}$$

The electrons in motion with velocity v constitute an electric current of density J having units of amperes/m², where $J = Nqv$, with N being the electron density. Therefore we obtain

$$J = \frac{-jNq^2 E}{m\omega} \tag{11.69}$$

This current density is 180° out of phase with the vacuum "displacement current density" $j\omega\epsilon_0 E$, whereas polarization current in a dielectric is in phase with vacuum displacement current. The various current densities appear in the right-hand side of Maxwell's $\nabla \times H$ equation which can be written as

$$\nabla \times H = J_t \tag{11.70}$$

where J_t represents the total current density (the sum of any and all current densities). The relative dielectric constant K is obtained by setting the right-hand side equal to $j\omega\epsilon_0 K E$. The resulting expression for K is

$$K_{ord} = 1 - \frac{\omega_p^2}{\omega^2} \tag{11.71}$$

where $\omega_p^2 = Nq^2/m\epsilon_0$ and the subscript "ord" indicates the "ordinary" wave, which is one of the linearly polarized characteristic waves that propagate perpendicular to the magnetic field in a plasma. The same value of K applies if there is no magnetic field or if the frequency ω is so high that ω_B (Section 11.1.1) is negligible by comparison. The second linearly polarized characteristic wave for propagation perpendicular to the magnetic field has a y component of electric field intensity and thus is affected by the magnetic field. The relative dielectric constant for this wave is given by

$$K_{ex} = \frac{K_\varrho K_r}{K_\perp} \tag{11.72}$$

where the subscript "ex" stands for "extraordinary", the name given to this wave. Expressions for K_ϱ and K_r are given in Equations (11.73) and (11.74) and $K_\perp = (K_\varrho + K_r)/2$.

It was stated previously that the characteristic waves for propagation parallel to the magnetic field are left and right circularly polarized waves. The K values for these waves will be designated as K_ϱ and K_r. They are given by

$$K_\varrho = 1 - \frac{\omega_p^2}{\omega(\omega + \omega_B)} \tag{11.73}$$

and

$$K_r = 1 - \frac{\omega_p^2}{\omega(\omega - \omega_B)} \tag{11.74}$$

Expressions for the relative dielectric constants for the two characteristic waves that propagate perpendicular to the magnetic field and for the two characteristic waves that propagate parallel to the magnetic field have now been presented. Indices of refraction n can be obtained from the dielectric constant values by use of $n = \sqrt{K}$ (Equation 11.55). Phase velocities can be obtained by using $v_p = c/n$ (Equation 11.54), and wavelength can be obtained by noting that $\lambda = \lambda_0/n$ where λ_0 is free space value. (As $\lambda f = v_p$ and f does not

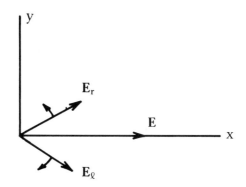

Figure 11.22   *Vector diagram showing instantaneous positions of*
E, E_ϱ, and E_r. The linearly polarized electric field
intensity E consists of ℓc and rc components, E_ϱ
and E_r.

depend on the propagation medium, λ must vary with n in the same
way as v_p.) Also as β, the phase constant, equals $2\pi/\lambda$, $\beta = \beta_0$ n
(Equation 11.11). Thus knowledge of the K values leads to the
needed information about the characteristic waves.

As the ℓc and rc waves propagating parallel to the magnetic field
have different K values, they also have different β values. It is the
difference in the β values that is responsible for Faraday rotation, as
a linearly polarized wave consists of and can be decomposed into
ℓc and rc components. This concept is suggested in Figure 11.22
which shows instantaneous positions of ℓc and rc waves at a particular instant of time. The directions of rotation are shown by the small
auxiliary arrows. As the ℓc and rc waves rotate their projections on
the x axis cancel, but their projections on the y axis add to give a
vector that varies sinusoidally in amplitude. In the usual vector diagram the E vector of a linearly polarized wave is shown as a vector of
fixed length, but the instantaneous amplitude does vary sinusoidally
with time.

If a linearly polarized wave propagates for a distance z parallel
to the magnetic field in a region with constant electron density
and magnetic field intensity, the rotation of E is given by

$$\phi = \frac{(\beta_\varrho - \beta_r)}{2}\, z \qquad (11.75)$$

A wave that is launched originally as a vertically polarized wave
can become horizontally polarized (or have any angle in between)
and *vice versa.* Thus, the occurence of Faraday rotation is of considerable practical importance. It affects both waves which are
used for communication by ionospheric reflection and waves which
have a sufficiently high frequency to pass through the ionosphere
without reflection. Faraday rotation is inversely proportional to frequency squared, however, for the higher frequencies, and frequencies in the microwave range that are utilized for satellite systems
experience no significant rotation.

In the discussion to this point we have referred to propagation as
being either parallel or perpendicular to the magnetic field. What
about propagation at an angle θ with respect to the field? Fortu-
nately it works out that if

$$4 \left(1 - \frac{\omega_p{}^2}{\omega^2}\right) \omega_B{}^2 \cos^2 \theta \gg \omega_B{}^4 \sin^4 \theta \qquad (11.76)$$

the quasi-longitudinal (QL) approximation applies and the characteristic waves are circularly polarized as for strictly parallel propagation. The expressions for K_ϱ and K_r can be utilized when the
QL approximation applies if ω_B in the expressions is replaced by
$\omega_B \cos \theta$. If the inequality is the reverse of that shown in Equation
(11.76) the quasi-transverse (QT) approximation applies and the
characteristic waves are linearly polarized.

11.4.3 Reflection of Electromagnetic Waves from the Ionosphere

In this section, we give emphasis to the ordinary wave, having a
relative dielectric constant given by Equation (11.71). One reason
for doing so is that near the reflection location, ω_p is close to ω
and the QT approximation applies, rather than the QL approximation that is specified by Equation (11.76). Thus near the reflection location, the ordinary wave is one of the two characteristic
waves. This wave has a K_{ord} value less than unity. Note that there
is nothing to prevent K_{ord} from decreasing to zero or becoming
negative. In this respect, ω_p in a plasma plays somewhat the same
role as f_c for a waveguide. Only waves having angular frequencies
ω greater than ω_p can propagate in a plasma as K becomes negative and the index of refraction becomes imaginary when ω is

less than ω_p. In a waveguide only waves having frequencies f which are greater than f_c can propagate. In both cases, when the operating frequency is above a critical value, wave propagation occurs with electric field intensity varying as

$$E = E_0 e^{-j\beta z} \tag{11.11}$$

When the frequency is less than the critical frequency, however, E varies as

$$E = E_0 e^{-\alpha z} \tag{11.77}$$

β in Equation (11.11) is a phase constant having units of radian/m. α in Equation (11.77) is an attenuation coefficient having units of Nepers/m. Equation (11.77) describes what is referred to as an evanescent "wave" or field. Attenuation in the evanescent mode does not involve the conversion of electromagnetic energy to thermal energy as in a resistive medium but takes account of the fact that reflection is occurring. The reflection process does not occur at a discrete surface as in the case of reflection from a perfectly conducting surface, but it involves some penetration into the evanescent region as indicated by Equation (11.77).

Consider an ordinary wave that propagates vertically upward from the surface of the earth into the ionosphere. As the wave travels upwards through the lower part of the ionosphere the electron density N and the plasma frequency ω_p increase with height. Depending on the frequency utilized, it is possible that the condition $\omega_p = \omega$ corresponding to $K_{ord} = 0$, may be realized. If so, the wave will be reflected at the level where $\omega_p = \omega$. (As stated in the previous paragraph reflection really does not all take place at a precise level, but that is of little consequence for present considerations; we will refer to reflection as taking place when $\omega_p = \omega$.) If the condition $\omega_p = \omega$ is never reached as a wave propagates upwards (if $\omega > \omega_p$ along the entire length of the ionospheric path), the wave will not be reflected but will continue on beyond the ionosphere.

A sounding system known as an ionosonde, actually a special-purpose radar system, is used to obtain information about electron density profiles in the ionosphere on the basis of the height at

which a wave of a given frequency is reflected. The typical ionosonde sweeps over about a 0.5 to 25 MHz frequency range in 10
to 15 s, transmitting pulses at a repetition rate of about 100 pps
during the time interval of the sweep. The echo pulse received
from the ionosphere forms a spot on a cathode-ray tube at a distance above a reference level proportional to the delay time of the
echo, and the delay time is porportional to the virtual height h′
of the ionospheric reflection layer. h′ differs from true height h
because the velocity of propagation in the ionosphere is different
from c, but procedures are available for recovering true height
from virtual height. (The Environmental Data Service of the National Geophysical and Solar-Terrestrial Data Center in Boulder,
CO will perform this function as a service.) Another technique
that can be used to obtain information about electric density in
the ionosphere, including the region above the peak of the F_2
layer for which the ionosonde provides no information, is the
incoherent scatter technique[26]. In this case a frequency too
high for reflection of the type we have discussed is used, but a
weak backscatter signal proportional to electron density is received from any height for which the system has sufficient sensitivity.

In carrying out communication from one location to another,
one does not utilize waves that are vertically incident upon the
ionosphere but waves that are obliquely incident. In this case a
wave of higher frequency can be reflected from the ionosphere
than for vertical incidence. In particular, reflection at a frequency of f can occur where the plasma frequency is f_p when f and
f_p are related by

$$f = f_p \sec \phi_0 \tag{11.78}$$

where ϕ_0 is the original launch angle of the wave measured from
the vertical. If f_p corresponds to the highest electron density of
the ionosphere, f is the maximum usable frequency.

The prediction of the performance of HF systems and the choice
of operating frequencies has long been of interest to the Institute
of Telecommunication Science (ITS) and to the National Bureau
of Standards (NBS) which had responsibility for such matters be-

fore the formation of ITS [27,28]. Oblique sounders, similar to ionosondes but operating with transmitter at one location and receiver at a distant location, are highly useful tools for obtaining both long-term and real-time data on propagation conditions. Oblique sounder terminals can be located at essentially the same positions as the terminals of a particular HF path and used to select suitable operating frequencies for the path.

Communication by ionospheric reflection at HF frequencies was formerly a principal means of long distance communication, and it is still an important and useful method. Communication by satellite, however, has taken over a large share of long-distance communication, and ionospheric propagation is not so widely used as it once was. Disadvantages of communication using HF frequencies and ionospheric reflection include: the variability of the ionosphere and possible disruptions of service due to ionospheric storms and disturbances, the low bandwidth available, and the susceptibility to mutual interference between various users. Advantages are low cost and the ability to communicate between a number of rather widely scattered locations. An example of an application that favors HF techniques is that of obtaining information from a number of widely scattered remote sensors such as low-cost data buoys which can be dropped at sea. The location of the buoys can be determined by HF radar and data can be transmitted from the buoys at HF frequencies. HF radar is also of interest for other over-the-horizon applications such as air-traffic control over the oceans and remote monitoring of sea-state.

Propagation by ionospheric reflection sometimes involves more than one hop so that the signals are alternately reflected by the ionosphere and the ground. Attenuation is experienced at each reflection. Actually the reflection process in the ionosphere is really a refraction process, and attenuation is encountered along the length of the path in the ionosphere. Data concerning the reflection and other loss factors, the recommended fade margins for HF systems, and noise in the HF band, together with sample calculations are given by Davies, Chapters 5 and 7 [26]. The propagation of ionospherically reflected waves is a large and interesting subject which is treated more fully by Budden [29], Davies [27], Kelso [30], and Ratcliffe [31].

REFERENCES

[1]   Johnk, C. T. A., *Engineering Electromagnetic Fields and Waves.* New York: Wiley, 1975.

[2]   Jordan, E. C. and K. G. Balmain, *Electromagnetic Waves and Radiating Systems,* 2nd ed. Englewood Cliffs, NJ: Prentice-Hall, 1968.

[3]   Ramo, S., J. R. Whinnery and T. Van Duzer, *Fields and Waves in Communication Electronics,* New York: Wiley, 1965.

[4]   Joint Technical Advisory Committee, *Radio Spectrum Utilization.* New York: IEEE, 1964.

[5]   Joint Technical Advisory Committee, *Spectrum Engineering — the Key to Progress.* New York: IEEE, 1968.

[6]   Hinchman, W. R., "Use and management of the electrospace; a new concept of the radio resource", IEEE, Int. Conference on Communications, *69C29-COM,* pp. 13-1 to 13-5. New York: IEEE, 1969.

[7]   Joint Technical Advisory Committee, *Radio Spectrum Utilization in Space.* New York: IEEE, 1970.

[8]   Miller, S. E., E. A. J. Marcatili and T. Li, "Research toward optical fiber transmission systems", *Proc. IEEE,* vol. 61, pp. 1703-1751, Dec., 1973.

[9]   Blake, L. V., *Antennas.* New York: Wiley, 1966.

[10] Kraus, J. D., *Antennas.* New York: McGraw-Hill, 1950.

[11] Flock, W. L., *Electromagnetics and the Environment: Remote Sensing and Telecommunications.* Englewood Cliffs, NJ: Prentice-Hall, scheduled for Jan. 1979.

[12] Freeman, R. L., *Telecommunication Transmission Handbook.* New York: Wiley, 1975.

[13] Panter, P. F., *Communication Systems Design.* New York: McGraw-Hill, 1972.

[14] Bean, B. R. and E. J. Dutton, *Radio Meteorology.* Washington, D.C.: Supt. of Documents, U.S. Government Printing Office, 1966.

[15]   GTE Lenkurt, *Engineering Considerations for Microwave
        Communication Systems.* San Carlos, CA: GTE Lenkurt,
        Inc., 1972.

[16]   Laine, R. V., "Blackout fading in line-of-sight microwave
        links", presented at PIEA-PESA-PEPA Conference, April
        22, 1975, Dallas, TX, pp. 1-15. San Carlos, CA: GTE
        Lenkurt, Inc., 1975.

[17]   Taur, R. R., "Ionospheric scintillation at 4 and 6 GHz",
        COMSAT Technical Review, vol. 3, pp. 145-163, 1973.

[18]   Feldman, N. and S. J. Dudzinsky, "A New Approach to
        Millimeter-Wave Communications," *R-1936-RC,* The Rand
        Corporation, Santa Monica, CA. April, 1977.

[19]   Kerker, M., *The Scattering of Light and Other Electromag-
        netic Radiation.* New York: Academic Press, 1969.

[20]   Kerr, D. E. (ed.), *Propagation of Short Radio Waves.* New
        York: McGraw-Hill, 1951.

[21]   Zufferey, C. H., "A Study of Rain Effects on Electro-
        magnetic Waves in the 1-600 GHz Range." M.S. Thesis,
        Department of Electrical Engineering, University of
        Colorado, Boulder, CO, 1972.

[22]   Laws, J. O. and D. A. Parsons, "The relation of drop size
        to intensity", *Transactions American Geophysical Union,*
        pp. 452-460, 1943.

[23]   Crane, R. K., "Prediction of the effects of rain on satel-
        lite communication systems", *Proc. IEEE,* vol. 65, pp.
        456-474, March 1977.

[24]   Longley, A. G., R. K. Reasoner and V. L. Fuller, "Mea-
        sured and predicted long-term distributions of tropospheric
        transmission loss," *OT/TRER 16,* ITS, Boulder, CO., Wash-
        ington, D.C.: Supt. of Documents, U.S. Government Print-
        ing Office, July 1971.

[25]   Rishbeth, H. and O. K. Garriott, *Introduction to Iono-
        spheric Physics.* New York: Academic Press, 1969.

[26]   Evans, J. V., "Theory and practice of ionospheric study
        by Thomson scatter radar", *Proc. IEEE,* vol. 57, pp. 496-
        530, April 1969.

[27]   Davies, K., *Ionospheric Radio Propagation*. Washington,
       D.C.: Supt. of Documents, U.S. Government Printing
       Office, 1965.

[28]   Haydon, G. W., M. Leftin and R. Rosich, "Predicting the
       performance of high frequency sky-wave telecommunica-
       tion systems," *OT Report 76-102*. Boulder, CO: Institute
       for Telecommunication Sciences, Sept., 1975.

[29]   Budden, K. G., *Radio Waves in the Ionosphere*. Cambridge:
       Cambridge University Press, 1961.

[30]   Kelso, J. M., *Radio Ray Propagation in the Ionosphere*.
       New York: McGraw-Hill, 1964.

[31]   Ratcliffe, J. A., *An Introduction to the Ionosphere and
       Magnetosphere*. Cambridge: Cambridge University Press,
       1972.

Chapter 12

THE COMPUTER AND TELECOMMUNICATIONS
SWITCHING OPERATIONS

Baylen Kaskey

Head, Billing Systems Department

Bell Laboratories

Columbus, Ohio

12.1 OPERATIONS IN THE TELECOMMUNICATION NETWORK

12.1.1 Scope

Over 300 million telephones throughout the world can connect into the United States telecommunications network. The network includes subnetworks that interconnect commercial television, computers, teletypewriters, and many other private or commercial data areas. The primary elements of this network are the transmission and switching facilities required to direct communications from one point to another. The network has evolved into a complex integrated design of unparalleled size involving many functional centers and organizations to manage the operations of the network. The functional centers and organizations are distributed throughout the system and may cover an area or district, a telephone operating company, or a particular structural element of the companies that operate the network. Activities in these organizations encompass general accounting, record keeping, work force management, construction and facilities planning, equipment maintenance and testing, traffic handling and management, evaluation of service quality, customer billing, installation of equipment, and operator services. Any one of these areas may be served by dozens of different computer-based systems. The growth of such system applications has been rapid, with the result that methods of operating the network have had to change rapidly to take advantage of the capabilities that computer-based systems provide.

One of the major changes has been the centralization of functional areas. Some of these centralized functional areas will be discussed in this chapter.

The chapter will provide an overview of the use of computers (primarily minicomputers) for engineering and network services with some details of computer applications given to demonstrate their use and impact on operations.

The applications selected for more detailed discussion are associated with switching operations in the Bell System. Switching operations involve operating and maintaining switching offices, billing customers for services, measuring traffic flow, and managing the flow of traffic in the network.

12.1.2   Operation of the Network

For every telephone or set of multiparty line telephones in the network, there is a line connecting that telephone or set of telephones to a switching office. These offices, called central offices, can interconnect any of the individual lines that connect directly to that office. If the call is to a telephone not served by that central office, it must be routed through connecting links, called trunks, to the terminating central office. Long distance connections are achieved automatically by a complex known as direct distance dialing. This complex depends on automated switching systems which select available routes using a switching and trunking plan with an associated transmission plan, and is based on a nationwide numbering system for the telephone address. The numbering system uses ten digits to provide unique identification for each telephone line. The first three digits, called the area code, generally designate a specific area of the country; the next three digits, a specific central office in that area and the last four, a specific telephone connected to that central office. The automatic switching system obtains the address information it requires by interpreting the signals generated by the dial of the caller's telephone. If the caller has a rotary dial, the system counts the pulses generated; if the caller has a TOUCH-TONE® telephone, the system interprets the pair of frequencies that the dialer transmits. In TOUCH-TONE dialing each digit has a unique pair of frequencies assigned. When a caller dials a telephone that is connected to a different central office, the call is not always routed directly to that office. Where there are large numbers of central offices in a local area, it is likely the call will be routed through a type of office known as a tandem. By connecting each office to a tandem rather than to every other office in the group, many fewer links are needed. The principle is illustrated in Figure 12.1 which shows a call from Office 764 connected through a tandem to a telephone in Office 683. By having a tandem act as the hub or home for 8 local offices, 8 rather than 28 links, interconnect all offices.

The principle of routing through tandems that can be considered the hub or home of other offices is extended into the toll switching network which is used for direct distance dialing. In extending the concept, tandems that are the hub for other offices are connected through offices that act as their hub or homing point. In this way the hierarchy of the network is set up. The network has five levels: local offices, toll centers, primary centers, sectional centers, and regional centers. The lowest level in the hierarchy is the local office which homes on the toll center, the toll center homes on the primary center, and so on to the highest level in the hierarchy, the regional center.

In the Bell Systems there are about 10,000 local offices, 800 toll centers, 230 primary centers, 70 sectional centers, and 10 regional centers. All offices which eventually home on a regional office constitute the offices in a region. Figure 12.2 shows the hierarchy of offices with some possible interconnections between the regions. The interconnections from one region to another are made through trunks that may connect the same level of office in the hierarchy or may connect different levels of offices. Therefore, there are many options available for setting up the interconnections of a call from one region to another. If one set of trunks is busy, an alternate set can be tried. The logic used by the automatic switching systems to attempt the alternate routes is based on a switching and trunking plan that employs groups of high usage trunks and final routes to end offices. To keep the network functioning smoothly, sufficient trunks must be in place between offices. This is accomplished by basing the engineering of the network on continuous measurements of the traffic flow during many hours of the day. The network must evolve to keep up with the changes in calling patterns that occur as business areas grow or contract and residential living patterns change. Even with a telecommunication

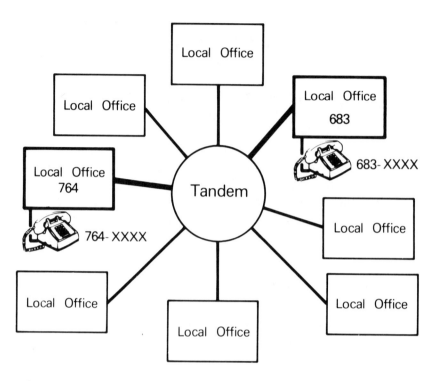

Figure 12.1.   Local Offices Served by a Tandem

network properly engineered and installed, there are events that occur that, if handled only by the logic built into the automatic switching systems, would cause blocking of portions of the network. Events such as hurricanes, earthquakes, and other natural occurrences may result in a volume of calls that can exceed the trunking capability of a localized area and require special action and real-time decisions to manage the flow of traffic in the network. Other events, such as Mother's Day, require special action which can be anticipated and planned for in advance.

Computer-based systems are used in the network to measure and analyze traffic flow and to maintain, manage, and control the routing of the traffic.

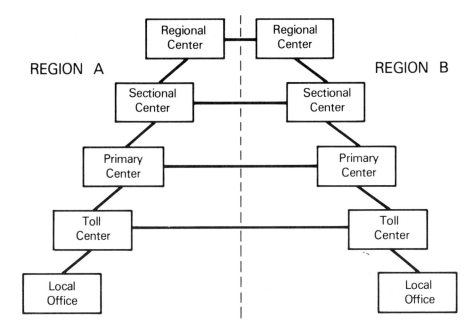

Figure 12.2.   Toll Network Hierarchy in Two Regions

The administration of traffic data collection is being centralized in Network Data Collection Centers. A major computer system supporting this traffic collection is called the Engineering and Administration Data Acquisition System (EADAS). This is a component of an overall Total Network Data System (TNDS), a multiple user system designed to perform all traffic data collection and processing. Overall monitoring and control of traffic in the network is the responsibility of Network Management Centers. Here, the primary computer support system is an EADAS Network Management System (EADAS/NM). A Network Operations Center (NOC) is responsible for controlling and maintaining service in the total network and for assuring the quality of network service.

12.1.3 Switching Systems

An automatic switching system recognizes that a telephone user is making a call when the handset is lifted from its cradle. The action causes contacts to close in a direct current circuit which operates a relay in the central office associated with the user's

line. Equipment in the switching system then connects other
equipment which will store or act on the information dialed
in by the user. Dial tone is heard by the user when the equipment is ready to receive the information. In one type of switching equipment, the dial pulses of the first digit set a selector
switch to a position corresponding to that digit, and connect
a second selector switch to the input signal. The second digit's
pulses set the second switch, and the process continues with
each digit dialed. Thus, each digit chooses one of a number
of outputs available from a selector switch, routing the call
through the interconnections within the switching system to
a unique output point corresponding in address to the digit
dialed. This system is one of the earliest and is known as step-
by-step. A typical step-by-step selector switch would give 100
terminals access to 10 trunks. Where large numbers of telephones are connected to a central office, it is useful to have
switches with greater access than is practicable with selector
switches. An early switch of this type is known as a panel
switch. It has 500 terminals with access to 90 trunks. The
switch contains rods with brushes that are moved by electric
motors over vertical banks of terminals to make connections.
Switching systems based on these switches are known as Panel
Systems. Panel Systems generally use a register to store the
dialed digit and then use other equipment to control the connection through the switches. Another concept in switching
is based on a coordinate system with horizontal and vertical
bars spaced uniformly across a switch; when bars at particular
horizontal and vertical coordinates are moved, they close a
set of contacts at their intersection. A switch of this type is
known as a crossbar switch. In 1938 the first system using this
type of switch was introduced in the Bell System and is known
as the No. 1 Crossbar System. This system uses markers to control the operation of switches at the input and output side of the
switching system. When registers record the dialing information
and markers and other equipment control the switches, the system
is generally said to be a common control system. The No. 5 Crossbar System was introduced about 10 years after No. 1 Crossbar. It
applies expanded common control features and uses a common set

of crossbar switches for both input and output from the system. This system was designed to meet the needs of direct distance dialing and provide automatic recording of call information required for billing.

These electromechanical switching systems are now usually augmented by computer-based systems to assist in their operations. For example, local crossbar offices can be equipped with computer-based systems for gathering billing information. These systems are called Local Automatic Message Accounting (LAMA).

Crossbar technology has also been applied to tandems. These tandems can act as a central point in charging arrangements with local offices. Crossbar tandems can be equipped with computer-based systems for automatically recording the centralized billing information. These systems are called Centralized Automatic Message Accounting (CAMA).

The No. 4A Crossbar System is the first system designed specifically for Direct Distance Dialing (DDD). It contains mechanical devices which have the capability of translating the area code and selecting the various alternate routes for long distance calls. Modern No. 4A Crossbar Systems have been modified to include electronic translation and routing of the area code information by using a stored program control processor with electronic peripheral circuits. These systems are called the No. 4A/ Electronic Translator System (No. 4A/ETS). A computer-based operations support system is now usually associated with the No. 4A/ETS. The system, known as the Peripheral Bus Computer (PBC), collects and operates on traffic data and equipment performance data. The system assists in maintenance by analyzing the data in cases where calls have not successfully reached their destination and by reporting on trouble patterns. Management and administration of the network are assisted by systems which consolidate and report trunk and equipment usage and performance data.

Most of the switching systems developed since 1960 have been designed with the central control being provided by a processor directed by a stored program. The switching elements generally

used in these systems employ reed switches which use small
strips of magnetic material enclosed in small glass bottles that
make contact when a magnetic field is applied. There are several
types of these electronic systems. The No. 1 Electronic Switching
System (No. 1 ESS) is designed for large metropolitan areas; No.
2 ESS is intended for smaller offices; and the No. 3 ESS better
fits the needs of rural communities. Newer and faster processors
are also in use in some of these systems. The No. 1A ESS can
handle more than twice the calls of a No. 1 ESS office; No. 2B
ESS can handle twice as many calls as No. 2 ESS.

The No. 4 ESS is a large toll switching system which has been
designed to handle digitally encoded transmissions in all electronic switches as well as analog voice transmissions.

Another type of system that is part of the network, but not a
local or toll switching office, is the Traffic Service Position System (TSPS). This system replaces the manual cord switchboard
that operators use, automating many of the tasks the operators
perform. The TSPS is called in whenever a customer needs operator assistance. The system is processor controlled and uses the
same type of system components that an ESS office would use.
The system is designed so that the groups of operators can be
located either at the TSPS location or in an office remote from the
TSPS. The remote capability means that operators don't have to
commute to the center of a city to handle the traffic of that city.
The system provides each operator with a desk-type automatic
console into which he or she can key information, dial international
and long distance calls, and obtain timing, billing, and other information automatically.

There are switching systems called Private Branch Exchanges
(PBX) that interconnect the telephones of a business customer
to each other and to trunks from a local central office. These
systems are usually located on the customer's premises and
may use any switching technology from step-by-step to ESS.
The PBX type of service may also be provided by common
control switching systems such as No. 5 crossbar and ESS,
eliminating the need for a customer to install a PBX on his
premises. When the local office provides the PBX capability,
it is called Centrex service.

Each of the switching systems is an individual design with unique characteristcs. Older systems such as Panel are being retired, but all of the other type systems mentioned here make up the bulk of the switching machines in the network.

There are differences in operation between each switching system type as well as major differences in the operations of the electronic switching systems compared to the electrome-chanical systems (Step-by-Step, Crossbar, etc.). As the network has evolved it has been necessary to make changes in early systems to keep them compatible with newer systems; however, all systems of one type may not have all of these changes implemented. In addition, there are many features and options available for each system type that are not universally applied. When computer-based systems are added to assist in the operations of the network, the design of the computer system must take these differences into account.

Some of the differences between switching systems affect the way telephone operating companies must assign responsibilities; this in turn affects to some extent the way computer-based systems for traffic, maintenance, and billing operations are assigned.

12.2   ORGANIZATIONAL ELEMENTS

12.2.1 Operating Company Responsibilities

The telecommunications network is operated by many independent telephone operating companies, 24 Bell System telephone operating companies, and the Long Lines Department of the American Telephone and Telegraph Company. Telephone operating companies are organized in many different ways to accomplish their responsibilities. Parts of the organization handle Personnel, Accounting, Public Relations, Legal Affairs, and the other normal business activities as well as the operation of their portion of the telecommunications network. This latter area of business can be considered to be made up of engineering, operation, and maintenance. The engineering includes forecasting needs for equipment and services, planning for the facilities and equipment to meet the needs, ordering the equipment, and analyzing the switching, traffic and transmission data from the

network as an input to the other steps in engineering. There
are many computer-based systems that support this engineering
function in the total network data system. The major system
for gathering and organizing traffic data is called EADAS which
will be described later.

The operation portion of network operations includes managing
and administering the flow of traffic, providing operator services,
and billing customers. The major responsibility for managing and
administering the flow of long distance traffic resides with the
Long Lines Department of the American Telephone and Telegraph Company (AT&T). The Long Lines Department provides
the long distance facilities and operates like a telephone operating
company. It has the same type of engineering, construction, operation, and maintenance responsibility as the other operating companies.

Operator services include directory assistance, operator help in
completing calls, preparing billing input, and consultation to
business customers. Computer-based systems are available to
assist in scheduling and managing the work force of operators.

An essential part of operating a business is collecting the revenue
for the service provided. The major computer-based systems used
for billing will be described later.

Maintenance operations involve the testing and maintenance of
facilities and equipment in the network. The availability of computer-based support systems has allowed these functions to be
centralized to a greater extent than would otherwise be possible.

Maintenance is now centralized in Switching Control Centers.
With the Switching Control Center (SCC) concept, the central
office personnel may report administratively to the management of the SCC.

12.2.2 Operating Company Organization

As a result of studies conducted by AT&T, many telephone
operating companies are moving towards an organization structure, shown in Figure 12.3, which has the following major divisions:

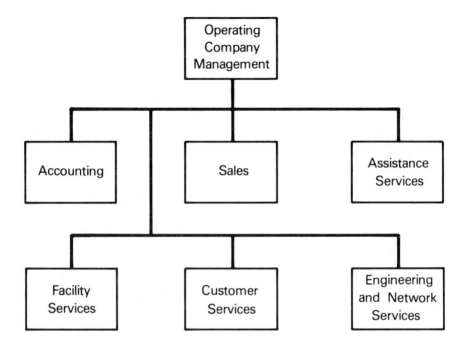

Figure 12.3. Operating Company Organization Structure

1. Accounting
2. Sales
3. Assistance Services
4. Facility Services
5. Customer Services
6. Engineering and Network Services

The functions of the Accounting and Sales Divisions include finance and marketing. Assistance Services covers the telephone directory and the operators; Facility Services covers outside plant including construction; Customer Services includes the telephone business office, data communications, and the testing, installation and repair of customer premises equipment; and Engineering and Network Services covers the major portion of operations of the network.

Operations in the Engineering and Network Services area can be classified into three groups:

1.   Administration: Activity to ensure that the network provides a high grade of service in the most efficient manner.

2.   Installation: Activities that respond to service demands by providing the resources for connecting equipment.

3.   Maintenance: Activities to ensure that the equipment provides the service and operates according to its design specification.

Bell System operating companies have been reorganizing their structure so that these functions can be accomplished in functional centers where related work functions can be performed, generally for a specific geographic area. The computer-based systems that assist these centers in accomplishing their designated functions are called operations support systems. These centers also use programs that are run in large central processors, usually in the batch mode. These involve large volumes of data and support nontime critical operations such as load balance and central office reports.

12.2.3 Central Office Operations

A Central Office (CO) includes switching systems, transmission and carrier equipment, trunk and line relay circuits, a distributing frame for making trunk and line cross-connections to the switching equipment, internal cabling, test equipment, power distribution equipment, batteries, and backup power facilities.

The telephone business involves a large volume of changes in assignment of telephones to customers; new telephone lines are always being installed as people move in and out of the area served by the CO. Craftspeople in the office accomplish some of these changes by moving and reassigning lines and cross-connections on connecting equipment known as distributing frames. In toll offices, trunks are added and changed in response to toll traffic needs using trunk distributing frames. Craftspersons in offices are called on to assist installers of outside plant equipment and customer lines. When a customer's telephone problem is being investigated, a repairperson will often need the assistance of a craftsperson in the Central Office for making and completing operational and transmission tests. The switching system in the office requires preventive maintenance activity on the part of the craftsperson as well as requiring periodic testing. Critical elements of the CO systems are arranged to sound alarms in the event of a malfunction, and craftspeople may be engaged in responding to these alarms, performing troubleshooting, corrective maintenance and repair. Administering the many individual items of work activity requires numerous reports and records, the majority of which lend themselves well to computerized administration.

12.2.4 Central Office Data Bases

Record keeping in central offices falls into many categories. Records of inventory, personnel, and training can be supported by special programs on general purpose computers. Similarly, records of the various drawings that pertain to the equipment can be maintained and updated using standard computerized file management techniques. In the area of telephone line circuit assignments, the large number of connections and the frequent changes make record keeping a difficult task. The design of most switching systems assumes a distribution of load over the entire system, and at times changes in cross-connections may be required to balance the load. A computer-based system called the Computer System for Mainframe Operations (COSMOS) is being successfully used to assist in this area. It has eased the problems of record keeping and congestion, and improved the load balance and user line equipment distribution across the switching equipment. It allows preferential assignment of line equipment and real-time access to the records for assignment updating. The system also provides off-line management reporting. COSMOS is a centralized system that can serve a number of central offices in a time-share mode.

Many operations support systems assist in the administration of central offices. A number of these operate with the centralized Switching Control Center which is one of the major centers in the Maintenance category.

Other computer-based systems have been found to be necessary to change the format of data used by the telephone operating company into the proper form required to update the record bases of the variety of operations support systems now being implemented. One such system is called the Record Base Coordinator System (RBCS).

12.3   DEVELOPMENT CONSIDERATIONS FOR OPERATIONS SUPPORT SYSTEMS

12.3.1   Extension and Change Capability

The telecommunications network has been evolving over many years using facilities and equipment of many different vintages and capabilities. As a result, a computer-based operations support system may not be able to perform its function for all of the facilities and equipment that may be involved. For example, the Step-by-Step, Crossbar and Electronic Switching Systems all require entirely different maintenance functions. Even the individual types of ESS have many differences among themselves.

When a maintenance operational support system is first introduced into the network, it may be designed specifically to assist in the maintenance of No. 1 ESS and not be applicable to other switching systems. Once the worth of the system for No. 1 ESS has been established through experience in its operation, it may be expanded to handle other systems. Thus, most computer-based operations support systems continue to expand their system coverage capability.

As experience is gained in the use of an operations support system, many new features are identified that would be desirable for these systems, and changes are identified to improve the operation of the system. At the same time that the operations support systems are expanding and improving, the switching systems to which they connect are improving and adding new features. The result is that in many cases the support system must be changed for compatibility.

To accomplish the required changes, portions of the software in the computer are changed, and each system developer provides to the system users in the field new generic programs which incorporate these changes. In many cases these generic

programs are released on an annual basis, but the frequency
of release depends on the need for the changes, the size of the
program, and the complexity of the operation. In all cases it
is desirable to minimize the development interval in order to
keep development costs low and make the features rapidly
available in the network.

12.3.2   Minimization Of Development Interval

Minimization of the development interval of large software
packages requires efficient ways of developing software; this
usually means writing programs in a higher level language than
the basic assembly language of the computer.

This use of a higher level language is desirable even though it
usually results in the computer real-time and memory use not
being optimized. In Bell Laboratories a higher level language
has been developed and is generally used in Bell System computer support systems. This language is called "C". For this
language Bell Laboratories has developed a powerful program
development support system. An interactive environment for
program development is also provided. The development system includes internal aids to assist in the documentation of
the system.

As more and more features are added to an operations support system, it often becomes necessary to change the minicomputer on which the system operates to a larger, more
capable unit. In some cases more cost-effective minicomputers become available. By choosing minicomputers that are
part of families of upward compatible computers, the use of
higher level languages helps to ease the job of transferring
the system to the new computer.

12.3.3   Reliability Requirements

The reliability requirements of an operations support system
depend on the function being performed. Support Systems
for circuit maintenance and for recording billing data demand
the highest reliability. Lower reliability is acceptable in routine
maintenance where there is no immediate urgency. In the Bell
System the automatic switching systems themselves are designed
to operate everyday for 24 hours a day with no more than 2

hours of lost operation in a period of 40 years. These switching systems achieve a minimum of lost operation by duplication of equipment, internal tests and audits that point out failing pieces of equipment, and an overall design philosophy that allows preventative and corrective maintenance to be made without turning off the switching system.

Some operations support systems use duplex computers to achieve high availability of the system. However, operation support systems are always used outside the main switching functions and, therefore, their failure will not cause lost telephone calls. As a result their required reliability is not as high as those of the actual network switching elements. In the case of some operational support systems that involve essential functions such as circuit maintenance or centralized recording of billing data, the reliability requirements may be set at less than one hour of lost service a year.

In the Bell System the performance of all operations support systems are monitored, and data are fed into a Reliability Reporting System. The system is based on a computer program designed to accept manually inputted data from reports generated each time the operations support system experiences trouble which requires some maintenance action. Output reports from this system go to the telephone operating companies so they can compare performance among their various support systems and among different locations and organizations. Reports are also sent to the system developers so that they can assess the performance of the systems they have designed.

12.3.4 Acceptance By Telephone Operating Company Personnel

The operations support system developer must not only take into consideration the functions the system is to accomplish, but he must also take into account the capabilities and training of the telephone company personnel who will operate the equipment and the administration of that personnel and the system. The design of the computer-based system and its software is usually referred to as Computer Subsystem (CSS) development, and the design of the operation of the system taking into account personnel and administration is usually referred to as Personnel Subsystem (PSS) design.

Part of the Personnel Subsystem design process involves using operating personnel from telephone operating companies. Such personnel may operate early feasibility designs of portions of the system, or they may be observed in their daily work of accomplishing the functions the system provides. In many cases, before a system is released for general application a trial installation will be made with both the performance of the Computer Subsystem and the Personnel Subsystem observed and evaluated before the system is released to the field. Some systems have made provisions for automatic feedback to the developer on the acceptance of the system by operating personnel.

12.4   SWITCHING CONTROL CENTER

12.4.1 SCC Functions

The Switching Control Center (SCC) is one of the major Maintenance centers in the telecommunications network and, therefore, has been selected for further discussion here. The Switching Control Center maintains and administers local, tandem and toll switching offices and the Traffic Service Position System. The center is responsible for the operation and maintenance of switching, transmission, power and billing equipment in the offices and is responsible for the administration of the personnel of the offices. To accomplish its tasks, the Switching Control Center remotely controls equipment in the Switching office, alters its configuration, and performs corrective tasks that will keep an office operational even during some trouble conditions in the office. This requires the capability for alarm surveillance, analysis, testing, trouble isolation, and control to be performed remotely for all offices administered by the Center.

Because of the high reliability of automatic switching equipment, few system failures occur. When they do occur, they usually require highly skilled personnel to correct the situation. Rather than locate such personnel at all offices, a small group of skilled personnel can be centrally administered and dispatched when needed to solve individual switching office problems.

In some cases a switching office may be completely unattended except for personnel dispatched to that office from the Switching Control Center; in other cases, particularly in large offices, a work force is assigned to an office on a permanent basis even though the work force reports administratively to the Switching Control Center. In a number of cases an office is manned and controlled locally during the day and controlled from the Switching Control Center at night and on weekends.

Because electromechanical (Step-by-Step, Crossbar) systems require entirely different operations and skills for their maintenance and administration than are required by Electronic Switching Systems, two types of Switching Control Centers are used, one for each of these types.

12.4.2 SCC Operations

The Switching Control Center usually contains several rooms where administrative functions are performed and one large room where switching offices are monitored and controlled. The administration functions include preventive maintenance, installation activities, work force administration, interfaces with other centers, change information for data bases, and administration of the minicomputers that assist the Switching Control Center to perform its operations. The center also provides assistance to other centers that may be involved in network management or services.

There are over a dozen operations support systems that assist the SCC in accomplishing its operation, but the primary systems that make the centralization possible are those that allow remote monitoring and control of the switching offices. In the SCC for Electronic Switching Systems, the primary system is the No. 2 Switching Control Center System (No. 2 SCCS); in the SCC for electromechanical systems, it is an Automatic Trouble Analysis (ATA) system and a Telecommunication Alarm Surveillance and Control (TASC) system.

12.4.3 Switching Control Center For Electronic Switching Systems

The No. 2 Switching Control Center System (No. 2 SCCS) allows the implementation of the centralized maintenance of ESS offices. Figure 12.4 shows one of each major type of switching system that may connect to a No. 2 SCCS and lists two other operations support systems that assist the SCC. The number of switching offices that can be controlled by a single No. 2 SCCS varies depending on the type and size of the switching offices, but usually a dozen or more are connected to one system. The system sounds alarm tones and displays the alarm status of all connected offices by lighting words on a wall display called a Critical Indicator Panel and by displaying words on large cathode ray tubes called Alarm Monitors. These are mounted forward and above head height so that they are readily visible to personnel assigned to work stations used for the control or analysis of incoming information.

A cathode ray tube terminal on the desk at each work station can be requested by the craftsperson to display information from any of the switching offices connected to the system. The terminal is connected to the central minicomputer, which in turn is connected by data links into the processors of the connecting offices. The connection in the switching office is made in the same way local teletypewriters in an office are connected to the switching system central control processor. The terminal performs the functions of a maintenance teletypewriter in an office providing the capability to arrange system hardware intercommunications, initiate tests, and receive data and reports. The No. 2 SCCS, by having its own computer capability, is able to store data, analyze data, as well as display data at the work station terminal and on the Alarm Monitor.

The system operator has the option of setting the threshold for the alarms that will appear on the Alarm Monitor. Some abnormal situations in a switching office are self-correcting and no alarm needs to be sounded. However, if the situation occurs several times in a short period, self-correction may

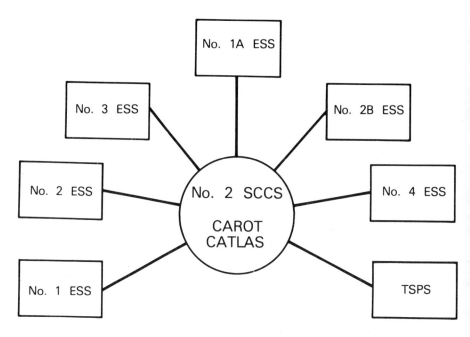

Figure 12.4.   *Switching Control Center for Electronic Switching Systems.*

not be properly taking place. For each of these possible situations, the system operator presets the number of occurrences that he wants to occur in a given period before an alarm situation is indicated.

The computer system of the No. 2 SCCS performs functions in real-time, allows interactive timesharing, and accomplishes batch processing. Equipment associated with the computer to assist in these functions includes a tape unit, a disk unit, communications multiplexer, and a line printer. Messages from all the connected offices are analyzed in real-time and alarm messages are displayed on the Alarm Monitor. The upper portion of the Alarm Monitor shows the alarm conditions for all offices and the lower portion shows the last series of messages that relate to these alarm conditions.

A craftsperson can use a work station terminal to browse through all messages entering the computer from a particular office of interest. When the terminal screen is filled with messages, as a new one enters at the bottom of the screen, the top message moves off the top. The work station craftsperson can reverse the flow and review earlier messages. This person can also request the computer to select certain types of inputs and sort or analyze the information, display results, or print out reports on the line printer.

A craftsperson working in a switching office often needs to take out manuals and look up information to understand the meaning of some of the coded information available on the output teletypewriter. In No. 2 SCCS this information can be included in the computer memory and the work station terminal can provide the expanded message. Pattern analysis of situations can be made to provide guidance for the actions required.

Electronic Switching Systems include a control capability located in an equipment area called the Master Control Center (MCC). This center is of a different design for each type of switching system. Direct control, manual override, trunk and line test, and reconfiguration of the switching system is accomplished by using the MCC. The No. 2 SCCS achieves the MCC capabilities not through its minicomputer, but rather by the addition of telemetry equipment in the switching office. The MCC is modified so that relays controlled through the telemetry equipment operate the controls available at the MCC, and control information can be forwarded to the Switching Control Center. At the SCC end, the telemetry inputs information to sound alarm tones and display alarm status on the Critical Indicator Panel (CIP). Telemetry polling of each connected office updates the CIP about once a second.

At each work station the telemetry information for any office can be connected by a central office selector unit to a special console designed to actuate the monitor and control functions of the MCC. Because each type of switching system has its own unique design of MCC, unique designs of consoles for No. 1 ESS and the Traffic Service Position System (TSPS) are available. Designs for other switching system types could be made available; however, a more universal console is available for this function. This console, called Control Console No. 1A, can function with any type of electronic switching system or the TSPS.

The Control Console No. 1A can be considered an operations support system in itself since its design is based on a microcomputer controlled cathode ray tube. In operation the face of the cathode ray tube displays simulated pictures of the lamps and keys that are needed to accomplish the monitor and control functions of the particular MCC. The Control Console No. 1A consists of a small table on wheels with a cathode ray tube terminal and keyboard on top, and in its base it contains the microcomputer, a floppy disk driver, and interface equipment. By using reverse video, small lighted areas with a designation printed inside appear on the face of the cathode ray tube and have the appearance of lamps and keys. Keys are operated using the keyboard.

In use, the Control Console No. 1A automatically determines the type of the connected office and displays the appropriate simulated control panel. This console can store information that allows messages to be expanded and guide the craftsperson.

Part of the activity normally accomplished with the Master Control Center is trunk testing. In the No. 2 SCCS the work station craftsperson uses the work station terminal to request the switching system to connect trunks for test to test terminations in the switching office. A console is provided for the operator to dial-up the trunks and run the required tests and checks.

Switching Control Centers have been providing centralized maintenance and administration for Electronic Switching Systems in the Bell System since 1974, and now the majority of these systems are covered by SCC operations.

12.4.4 Switching Control Center For Electromechanical Systems

The Automatic Trouble Analysis (ATA) System and the Telecommunications Alarm Surveillance and Control (TASC) System combine to allow centralized maintenance of electromechanical systems. Figure 12.5 shows one of each type switching system that can connect to the SCC and the operations support systems that can assist in its operation. In an electromechanical office, preventive maintenance is important in keeping the switching system operational, and therefore another system, Central Office Maintenance Management System-Preventative Maintenance (COMMS-PM), can be considered one of the primary systems involved. COMMS-PM is a computer system that maintains a record of equipment that requires preventative maintenance and a record of required equipment tests. It provides the maintenance work orders, keeps track of completion of work, and provides management reports. The user inputs the description of the equipment in the switching office so that the computer knows which tests and maintenance tasks to assign for a particular office.

In Crossbar systems when for some reason the progress of a call is stopped due to a trouble, even though there are automatic sequences in the switching system that make subsequent attempts to complete a call, the trouble is registered automatically by equipment called a Trouble Recorder or Indicator that provides a record with information indicating the circumstances of the call, the particular units of equipment involved in the call attempt, and where the call was stopped in the sequence of events. In an office, several times a day these cards may be sorted and analyzed by the central office staff to see where individual units of equipment are malfunctioning. For the Switching Control Center this operation is automated and performed remotely by the Automatic Trouble Analysis (ATA) system.

This is accomplished by placing in the crossbar switching office a microcomputer unit called a Maintenance Data Transmitter (MDT) that scans the office leads used to provide the Trouble Recorder or Indicator with information. It formats this into messages that it transmits by data link to a central computer to be stored and analyzed. The computer replaces the manual

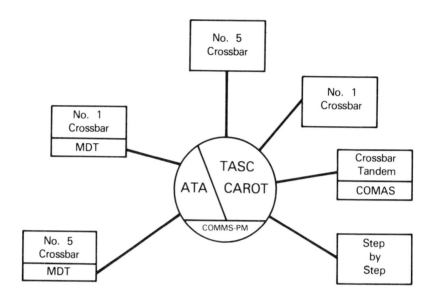

*Figure 12.5.   Switching Control Center for Electromechanical
            Systems*

sorting and analysis function, but due to its greater memory
and analysis capability, it can sort into many categories and
can use previous trends in analyzing problems. The interconnection information for each office is stored in the central
computer and, based on what is called an analysis tree for each
office, the system can pinpoint areas needing attention.

ATA is an exception reporting system; that is, it reports troubles
of a certain type occurring within selected equipment units only
when they have occurred a number of times determined by preset thresholds within the system. When a consistent pattern of
trouble is identified, as indicated by a threshold being exceeded,
an ATA report will be issued for corrective action by switching
maintenance personnel. In addition to real-time reports, ATA
provides summary and measurement reports to indicate the quantity and type of troubles occurring in the switching offices. Maintenance personnel may also interact with the system to retrieve

from memory information on previously issued reports and add relevant information on troubles that have not resulted in an automatic report.

The TASC system uses a central minicomputer with telemetry equipment to poll alarm and status information from electromechanical switching systems. This system also monitors alarms in transmission, building, and power systems. The system actuates alarm tones in the Switching Control Center, prints out teletypewriter messages, provides video displays in summary form on switching system status, and allows the SCC to remotely remove faulty equipment form service. There is a communications link between the TASC and ATA computers which permits craftspersons to interact with either or both systems from common work station terminals in the SCC.

12.4.5 Other Maintenance Support Systems

Switching Control Centers use input and analysis from other systems for maintenance purposes. In trunk maintenance work the center uses inputs from Centralized Automatic Reporting On Trunks (CAROT). This system performs routine and demand tests of trunks and groups of trunks. It provides exception reports, analysis results, and historical data summaries.

For maintenance of crossbar tandem switching systems, there is a Computerized Maintenance and Administration Support (COMAS) system. This system analyzes ineffective attempts by the switching system to switch calls and makes tests to check the ability of the switching system to route calls.

A system that assists in centralized maintenance of ESS is the Centralized Automatic Trouble Locating and Analysis System (CATLAS). This system uses pass/fail result patterns of automatic trouble diagnosis to make recommendations for circuit pack replacement. The data base contains the output possibilities of diagnostic programs in addition to the probable circuit pack failures that would cause that output. The system uses the results of repairpersons' experience in fixing earlier troubles in making recommendations for correcting new problems.

The Engineering and Administrative Data Acquisition System (EADAS) is the principal support system for centers that deal with traffic data in the network, but also supports the Switching Control Center by reporting maintenance related abnormalities that it detects. As the design of other operations support systems evolve, they also can be expected to report maintenance related problems to the SCC.

12.5   TELEPHONE BILLING OPERATIONS

12.5.1 Automatic Message Accounting

With the advent of automatic direct distance dialing, it became necessary to provide a means of automatically recording telephone call billing information. The first system designed for this purpose was introduced with the No. 5 Crossbar System. It is known simply as Automatic Message Accounting (AMA). That first system used electromechanical equipment to record the call information on punched paper tape. This system is now being replaced by operations support systems that record the information on magnetic tape.

Switching offices that perform an AMA function locally are called Local AMA offices or LAMA offices. Tandem and toll offices that perform an AMA function as the hub of several local offices are called Centralized AMA offices or CAMA offices. Computerized operations support systems that now assist in these functions are LAMA-C, CAMA-C, and No. 1 AMARC. The No. 1 AMARC (No. 1 AMA Recording Center) System performs the AMA function remotely for many offices without requiring that the offices home on a switching system located with the AMARC.

12.5.2 AMA Control Center

Extending the centralization concept results in an AMA Control Center (AMACC) whose responsibility would be to ensure that accurate billing data are collected and delivered to revenue accounting offices in a timely fashion. The center would be responsible for supporting the systems involved in collecting billing records. Functions include record base maintenance, detection of billing problems, and interaction with other centers to correct problems in the billing network, the data processors, the switching network or the message billing equipment. The rapid deployment of AMARC makes this centralization concept feasible.

12.5.3 Billing Data Collection

In order to bill a customer for telephone service, it is necessary for the operating telephone company to know certain call details such as the calling number for each call attempt and whether or not the called party answered. If the customer is charged a flat rate, not depending on the number of calls made, it is only necessary to assign charges. Toll billing and some local billing is based on a measured rate, that is, call duration and call destination information are also required. Early switching systems (panel and step-by-step) were not equipped to automatically provide this information. Most offices of these types are now modified, and all major switching systems in the Bell System can now provide information for billing.

The first step in adding automatic billing capability to switching systems was to add equipment that provided Automatic Number Identification (ANI). This equipment would obtain the calling line number and forward it to centralized AMA tandem or toll switching offices as the call was switched through those offices. Crossbar tandem and No. 4A toll crossbar were equipped with perforated paper tape systems, and electronic switching systems generally contained their own magnetic tape recording capability for billing.

To allow remote collection and recording of data by the No. 1 AMARC system, other equipment has been added to electromechanical systems. A Call Data Accumulator (CDA) has been added to Step-By-Step Systems for this purpose. The CSA scans leads at the first selector switch in these offices to detect dial pulses, answer, and disconnect information. The CDA uses the ANI equipment to identify the calling party. The required information is stored in the CDA until it can be transmitted by data link to the No. 1 AMARC.

A microcomputer controlled Billing Data Transmitter (BDT) is added to Crossbar systems performing LAMA and CAMA functions to replace paper tape perforators. The BDT receives billing data by scanning the recorder leads. This information is then encoded and transmitted to the AMARC.

12.5.4 Computer-Based Local AMA (LAMA-C)

The computerized LAMA-C system performs the AMA function for No. 5 crossbar offices. The system collects billing data for telephones that connect directly to the office in which it is installed, and it delivers the billing information by data link to the No. 1 AMARC system where it is automatically recorded on magnetic tape. For reliability, two interconnected minicomputer systems which function independently are used. Both systems track the progress of all calls, but only one is in an active state at a time. The active system assembles and forwards the call information to AMARC while the other remains in a standby state. If a trouble condition occurs in the active system, there is an automatic switchover to the standby system. Each system includes scanners, a disk memory unit, and control and interface equipment.

The data required for billing purposes is obtained by scanning both the marker equipment for the initial call start information and the associated trunk circuits for answer and disconnect information. The information is stored in memory and held until the entire call information can be assembled and formatted into a single entry. The information is then held on disk storage until it is transmitted to the AMARC.

LAMA-C makes extensive checks on the data before it is transmitted, and timing is established by an internal clock that is checked and reset, if required, by a time signal from AMARC.

Operator control of the system is through a control panel and a teletypewriter terminal. The operator can perform start, stop, and normal operations from the terminal. The system can be manually switched from one state to another via a control panel. Maintenance-related data is printed out by the system on a maintenance teletypewriter. In addition, teletypewriter terminals can be used to request tests and print output reports. The system automatically gives an output error message when problems are detected during call processing. These messages are designed to localize the trouble to a particular part of the system.

Without computer systems, the AMA function required a wired-in correspondence between a directory number and equipment numbers to which the customer's line was connected. The computerized system eliminates this need by providing translation tables in its data base. Also the computer allows the customer's billing treatment (flat rate, message rate) to be included in the data base and automatically added to the billing records.

12.5.5 Computer-Based Centralized AMA (CAMA-C)

The computerized CAMA-C system performs the AMA function for Crossbar tandem and No. 4A toll crossbar offices. The system collects billing data for any telephone whose local office does not collect billing data, but which homes or switches through the tandem or toll office in which the CAMA-C equipment is located. The CAMA-C system contains its own equipment for magnetic recording of billing data. These tapes are forwarded directly to the telephone company accounting center.

In CAMA-C, a two-processor system is used, one is the primary system and the other the backup system. In operation both systems are operating and collecting billing data, but the primary system is in the active state and the backup in standby. Billing data is recorded on tape by the active system only. The backup system becomes active when a trouble is detected in the primary system, and in this case switchover to the backup system is automatic. When the primary system is repaired, it is made active again by a manual reset. Both the primary and backup systems include a disk unit, a tape unit, input-output interface equipment, an operator's console, and scanners.

CAMA-C replaces the perforated paper tape system and obtains its billing data by scanning the recorder circuits and trunks that were associated with the paper tape system. All parts of the call data are placed in memory as they become available and then are assembled into single entry type data. At that time the entire call information is put on disk until it is scheduled to be written onto the magnetic tape.

To maintain a high level of billing integrity, the system makes extensive checks on its data and uses a combination of a hardware and a software clock for timing.

Operator control of the system is through the teletypewriter terminal which provides for all normal functions. A separate maintenance teletypewriter provides messages that relate to AMA circuit failures and allows for tests to be conducted.

CAMA-C is capable of detecting irregular signaling in the network. It can detect signals originating from unauthorized points and collect network data following the detection of the irregularities. It associates these data with billing details recorded on tapes for use in analyzing the irregularities. The data collected also can be printed out on remote teletypewriter locations.

12.5.6 Automatic Message Accounting Recording Center

The centralized No. 1 Automatic Message Accounting Recording Center (No. 1 AMARC) system collects, assembles and records on magnetic tape call billing data from switching offices. Figure 12.6 shows one of each type switching office that may connect to the No. 1 AMARC. The switching offices include step-by step, No. 1, No. 4A, and No. 5 crossbar LAMA offices and the No. 3 Electronic Switching System. The information is normally transmitted to AMARC over dedicated data links, but in case of trouble on the link, the AMARC can re-establish a connection by automatically dialing a backup connection over the network.

As in the case of CAMA-C and LAMA-C, two interconnected minicomputer systems which function independently for increased reliability purposes are used in AMARC. No single hardware trouble in these systems will cause loss of data processing capabilities.

Both systems process all calls, but recording is performed by only one, the active processor; the other is in the standby state. The output of both processors are continuously compared. If a

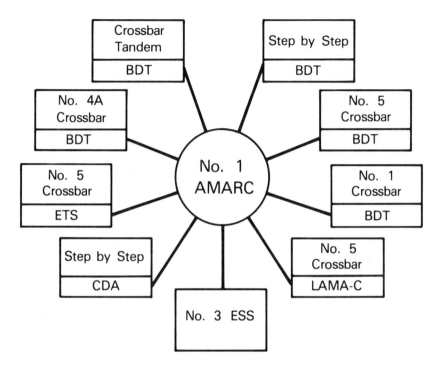

Figure 12.6.   AMA Recording Center System

trouble is detected in the active processor, an automatic switch changes the state of the second processor from standby to active, removes the active from service, and recording takes place on a tape unit associated with the newly active processor.

AMARC includes input data handling equipment, data processing equipment, and maintenance equipment. In addition to the processors and tape units, the system includes alarms and status displays and input/output terminals.

The data links to the switching offices are operated in an asynchronous mode; the AMARC system sends a polling command message requesting each block of data. The polled office responds by sending a block of call billing data. When that block is received and checked, the AMARC repeats the sequence. If no data is ready to be sent, the switching office sends a message that indicates no data is available. Each message sent is formatted with information that allows checking the validity of the information. For example, before the data is transmitted the complement of each bit is generated and included in the data block. When the AMARC receives the data, it computes the complement and checks for discrepancies. If an error is detected, AMARC requests the data to be retransmitted.

The data transmitted from offices equipped with the Call Data Accumulator or the Billing Data Transmitter is in multiple entry form; that is, the calling number, answer, and disconnect information for a call may not be in the same block of data. In these cases, AMARC assembles the data into single entry format.

The No. 3 ESS system has been designed to use the AMARCs capability to assemble multiple entry data. In offices equipped with LAMA-C or No. 5 crossbar offices equipped with an Electronic Translator System, the data is already in single entry format.

Operator control of the system is through an input/output teletypewriter terminal which provides for all normal functions. Information on alarms and status of call processing is printed out automatically on teletypewriter terminals. The system can monitor test calls generated in connecting offices to detect troubles in switching or billing and sound alarms in those offices. It can detect troubles on the data links as well as the processor complexes themselves. Tests can be requested and results printed out for maintenance purposes. Diagnostic programs in the system can isolate problems to one of a few circuit packs or modules.

The telephone companies have tended to group several No. 1 AMARC systems into one central location where billing data for hundreds of offices can be recorded on tapes.

12.6   NETWORK TRAFFIC DATA COLLECTION

12.6.1 Network Traffic

In the local network area, switching systems may switch calls directly to a final local office or the call may be switched through a tandem office. This arrangement is a two-level hierarchy. The toll network uses a five-level hierarchy with local offices, toll centers, primary centers, sectional centers, and regional centers. Local offices normally home on the next level or toll center, toll centers normally home on primary centers. The trunks between one office and another act as a group servicing the total call traffic between these offices. Depending on traffic patterns, a local office may have groups of trunks that go directly to more than one toll center and within a region, local offices may have groups of trunks going to offices at other levels in the hierarchy. The same is true with respect to interconnections between offices at higher levels in the hierarchy. Between regions, trunk groups may go from toll center to toll center, primary center to primary center, etc.; but in addition, trunk groups may interconnect offices at different levels in the two regions. The trunk group that connects a lower ranking office to the next higher center on which it homes is called a final trunk group. Trunk groups between regional centers are also final trunk groups. The trunk groups making other connections between offices are called high usage trunk groups.

If the group of trunks that are used first as a route between offices are all busy, other alternate high usage trunks may be tried. The last alternate route planned is the final trunk group. Trunk groups are sized to meet the needs of the traffic between the offices they connect. In the local network, a final group is usually sized to have no more than a one percent chance of being completely busy during the busy hour. When there are large numbers of trunks provided between two points, it may not be necessary to provide an alternate route, in this case the trunk group is called a full trunk group. Call traffic is variable; there are busy hours during each business day and often on Sunday evening. Peaks of traffic occur on holidays as well as with nonpredictable events such as disasters. Traffic patterns vary with the growth and decline of business and residential areas. Extensive measurements are continuously required to engineer the network and route the traffic through it.

12.6.2 Traffic Measurement

One measure of the traffic is the number of calls that are being carried. In early days of the telephone, operators counted the calls by moving a peg from one hole to another in a numbered row of holes. The measurement of the number of call attempts is still called peg count. Before computerized operations support systems were available, calls were recorded on electromechanical traffic registers which indicated the number of call events. These were either read manually or photographed.

A second measure of traffic is usage of trunks; here the item of interest is the amount of time the trunk is in use, not the number of times it is used. Periodically scanning the trunks to determine if they are busy or idle and recording the number of trunks that have been found busy, gives the indication of usage. Traffic Usage Recorders (TURs) perform this function and the data can be displayed on traffic registers.

A measure of network service can be obtained by examining delays in the switching equipment. A Dial Tone Speed (DTS) machine records the portion of test calls that have dial tone delayed three seconds or more after the handset is picked up. Other data useful in the traffic area is obtained from the AMA equipment which shows the volume of calls from one telephone office to another.

12.6.3 Traffic Data Utilization

The operational areas that make use of traffic data can be classified into four groups: (1) Traffic Engineering, (2) Network Administration, (3) Network Management, (4) Work Force Administration.

Traffic Engineering covers the area of determining where, when, what, and how much equipment and facilities will be required to handle the expected traffic. Projections need to be made as far in advance as five years to allow for planning, engineering, ordering, and installing the equipment and facilities. Determinations are made regarding whether new trunk groups are required, how many trunks need to be in the new groups, and what changes are required in existing groups.

Network Administration assures that sufficient and accurate traffic data are made available and controls the assignment of lines and trunks. Part of the administrative function is to keep the load balanced on the switching systems. Efficient utilization depends on an even distribution of load over the equipment.

Network Management is responsible for the efficient handling of abnormal traffic. This is effected by application of controls to reroute through less congested portions of the network or restriction of selected areas of traffic. Under severe overload conditions, a portion of the traffic with a low probability of completion is restricted to avoid loss of switching capability which would prevent other calls from being carried.

Work Force Administration uses data to project the near term work assignments of operators and to forecast hiring programs.

12.6.4 Centralization Of Functions

Because the applications of traffic data are broad, no single center is currently envisioned for all traffic operations, but some centers are expected to be implemented. A Network Data Collection Center (NDCC) will be responsible for administration of network data collection activities. A Network Service Center (NSC) will be responsible for assuring the quality of network services.

At present, EADAS, the major traffic gathering operations support system, provides for some centralization of the tasks performed.

Traffic data is assembled and formatted by a batch computer system called the Traffic Data Administration System (TDAS). This system administers the traffic flow and acts as a warehouse for data. It accepts input data in magnetic tape, paper tape or punched card form and provides data for administrative reports. Other batch programs use TDAS to provide engineering and administration reports. The complete set of traffic data collection and processing hardware and software is called the Total Network Data System (TNDS).

12.6.5 Engineering And Administration Data Acquisition System (EADAS)

The Engineering and Administrative Data Acquisition System (EADAS) is a traffic data gathering system that automatically collects data from electromechanical and electronic switching system. Figure 12.7 shows one of each type of switching offices

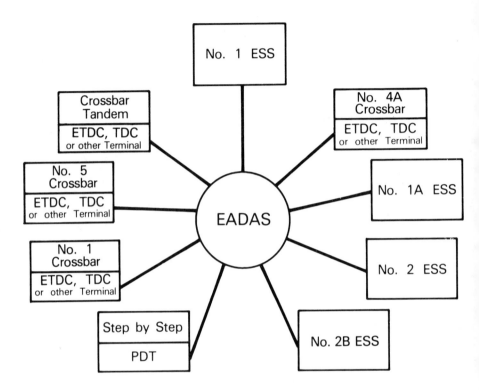

Figure 12.7. *Engineering and Administration Data Acquisition System*

that may connect to EADAS. The system uses a central computer complex which receives peg count and usage data over dedicated data facilities or over dialed data links. The system includes disk units, tape units, high speed printers, and input/out-
put teletypewriter terminals. Data from large offices are sent
over dedicated data links; in smaller offices data are sent when
polled over dial-up links. The system accumulates data totals
on disk for each input and performs real-time calculations on
this data.

Data is made available for network management every 5 minutes.
Analysis of traffic conditions and switching system performance
can be made regularly on 15 and 30-minute packets of data, and
hourly status reports can be scheduled. Magnetic tapes are gener-

ated for downstream data processing by the Traffic Data Administration System (TDAS). Calculations are performed on summarized data by EADAS to provide information on the quality of service being provided by the switching offices. Exception reports are generated that delete the normal data and emphasize abnormal data or trouble conditions. The EADAS operator can use input/output terminals to examine the data and reports at any time.

The system continually monitors its own hardware and software functions looking for trouble conditions. It can initiate alarm tones and lights and print out maintenance-type information on a teletypewriter.

EADAS collects data over dedicated links directly from Electronic Switching Systems. These systems collect, hold, and transmit data in blocks of 5 minutes for network management and in blocks of 30 minutes for other uses.

Electromechanical systems have had data collection devices added for use in EADAS. An EADAS Traffic Data Converter (ETDC) collects, encodes, and transmits peg count and usage data to EADAS over dedicated data links. One or more ETDCs can be located in a telephone switching office. They collect data that is normally placed on traffic registers in an office. The ETDC contains circuitry that permits either local or remote trouble isolation in itself. The output of Traffic Usage Recorders (TURs) is connected to an ETDC to transmit data to EADAS and to allow EADAS to control the TURs. In smaller electromechanical offices, Pollable Data Terminals (PDT) are used. These terminals control TUR scanning and accumulate peg count and usage data. During nonbusy periods EADAS uses dial-up links and requests that data be transmitted.

A Traffic Data Recording System (TDRS) Traffic Data Converter (TDC) may also be used in EADAS. This converter can be arranged to provide either peg count or usage data.

12.7 NETWORK MANAGEMENT

12.7.1 The Network Management Function

Due to unexpected traffic or major equipment failures, portions of the network may become overloaded. Network management involves changing the alternate routing algorithms, rerouting traf-

fic, or restricting traffic flow in portions of the network to keep
traffic flowing smoothly. The goal is to maximize the number of
call completions while providing equitable treatment for all customers.

Using switching congestion data and network performance and
traffic data, the network manager orders reroutes and other controls on traffic flow. With the availability of computerized operations support systems, the manager has more information available in real time to make decisions, and he or she can review sim-
ulations of possible reroute and control actions to evaluate and
forecast more accurately what will happen as a result of the action. The centralized system allows actions to be initiated at the
EADAS/NM which are automatically carried out at the switching
offices. In the past, manual action was required at individual
switching offices in response to telephone instructions.

12.7.2 Network Management Centers

The EADAS Network Management (EADAS/NM) system is the
major operations support system for network management. Figure 12.8 shows one of each type of interconnection with the
EADAS/NM. It is located in Network Management Centers (NMC)
that have wide jurisdiction over the various regions in the network. About 30 of these centers are expected to handle the national network along with one Network Operations Center which
performs overall network management administration.

The Network Operation Center is managed by the Long Lines
Department of AT&T, while both Long Lines and the telephone
operating companies own and manage the Network Management
Centers. A special operations support system provides the surveillance capability for the NOC.

The Network Management Centers have the responsibility for
network monitoring and control, EADAS/NM data collection
and record base maintenance, and preplanning control activities.

Preplanned controls are generated to cover special control
strategies for holidays such as Christmas and Mother's Day

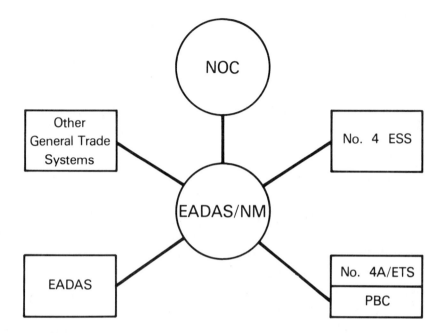

Figure 12.8.   *EADAS Network Management and the Network*
Operations Center

as well as other mass calling situations such as telethons. In
addition, preplanning must exist for major switching and
transmission facilities failures.

12.7.3 Reroute And Controls

The flow of traffic through the network is controlled by the
switching systems. Dialed digits from each originating call enter the translator portion of the switching system where they
are translated into primary and alternate routing information.
Controls can be instituted that change this routing information, make certain trunk groups unavailable, or restrict traffic
access to the network.

These controls may be instituted locally in the switching system
or remotely from a Network Management Center. The remote
control capability is built into the programming of Electronic

Switching Systems and crossbar systems with Electronic Translator Systems (ETS). In other electromechanical systems, it is accomplished locally or by the addition of telemetry equipment.

Rerouting can be implemented by a control that routes traffic onto an alternate trunk group that is not normally used. For example, because of the time difference between the east and west coasts, the busy hours do not generally coincide; therefore, during events that cause congestion in the network some calls going north and south near one coast may be routed through a switching center on the other coast.

Alternate routes normally used can be deleted by a control when it is desirable to reduce the load on other switching centers. Another way to eliminate a normal route is to use a control that makes the trunk group appear busy to the switching office attempting to use trunk in the group.

When calls have very low probability of completion, restrictive controls may be used. This keeps the network from being congested by ineffective attempts. Restricting traffic flow into the network is accomplished by code blocking. Here a control routes the call to a circuit that indicates that all trunks are busy or provides a recorded message recommending that the call be tried later.

Some controls can be programmed into Electronic Switching Systems or systems with Electronic Translator Systems to operate when needed. When the number of calls being processed exceeds a preset threshold, automatic signals can be sent to offices that are sending traffic so that controls may be automatically implemented there.

12.7.4 EADAS Network Management

The EADAS Network Management (EADAS/NM) system provides centralized real-time network monitoring and control assistance to the Network Management Center. It uses traffic data from all electromechanical offices and ESS offices received through EADAS and data directly from No. 4A crossbar offices equipped with the Peripheral Bus Computer. It exercises traffic controls through EADAS, directly through telemetry, or through an Electronic Translator System. The EADAS/NM System is processor controlled and uses disk units, tape units, cathode ray tube terminals, input/output terminals, and a wall display board. The wall display is a standard design, but it is arranged and labeled to provide overview status and can be customized to meet the needs of each particular Network Management Center.

The system will eventually be able to exchange data over dedicated data links with other EADAS/NM Systems through the data transfer point at the Network Operations Center (NOC).

When the system identifies network problems, it updates the wall display board to alert network managers to the problems. The cathode ray tube terminals allow network managers to interactively examine the data and perform further detailed analysis. Network management control actions can also be initiated through the cathode ray tube terminals. Data and analysis results can be printed out when requested.

The system makes checks on its hardware and software elements; when trouble is detected, audible and visual alarms are triggered. Maintenance type information can be printed out on teletypewriter terminals.

12.7.5 Network Operations Center

The Network Operations Center is located in Bedminster, New Jersey. Its function is the management and administration of the entire North American network including two regions in Canada and surveillance of the international telephone network as viewed from the North American switching systems. The center is supported by a computer-based operations support system known as the NOC system. The system receives input from all Network Management Centers by dedicated data links. It obtains status information on selected major toll switching offices and trunk groups in the network including those going overseas. It surveys the traffic flow in real time and analyzes reroute problems, when required automatically. It can make recommendations for solutions to problems. When reroutes are in progress, it monitors these and can recommend when to remove them. The system uses disk units, tape units, printers, cathode ray tube terminals, and a wall display board. In order to make the system workable, a common data base for all parts of the network was

established. Because of the large number of independent EADAS/
NM sites, a semi-automatic data base system was designed to keep
the NOC record base in step with the EADAS/NM record bases.
Within seconds of when a Network Management Center activates
or deactivates a control, a message is sent to the NOC. At times
the NOC may request an audit of all controls in effect. The audit
is made by EADAS/NM for all controls including any controls
that were made in offices manually, thus assuring that the NOC
has a complete set of information.

If the network manager is alerted to exceptional network conditions by the wallboard, a cathode ray tube terminal can then
be used to view the supporting data in expanded form and from
many aspects. Terminals display recommendations for reroutes
as they are necessary. Problems with no computer solutions will be
displayed. Problems that are not severe enough for the system to
analyze automatically can be analyzed when requested by the network manager through the terminal.

The major portion of the reroute analysis performed by the NOC
concerns final trunk groups, especially those between sectional
and regional centers. The routing algorithms identify other regional centers or specific sectional centers as alternate routing
paths or via offices.

The system makes checks on its hardware and software and,
when troubles are detected, audible and visual alarms are ac-
tivated. Messages are printed out on teletypewriter terminals
to assist in maintenance activity.

12.8　MINICOMPUTER CENTER

12.8.1　Minicomputer Center Functions

The large numbers of operations support systems now in use
in the network have prompted AT&T to recommend that all
Bell System telephone operating companies centralize the administration of the minicomputers they employ. Most companies have now centralized into one or more administrative units
the majority of their minicomputer complexes.

The first function assumed by these centers has been the administration of reliability reporting. All individual reports are

forwarded to the center as maintenance activities occur. At the end of each monthly reporting period, each minicomputer system manager provides a summary report on the number of maintenance activities as a check on the individual reports. The center is responsible for forwarding reports to be processed and distributing the output reports of the reliability reporting system. Functions that can be assumed by the Minicomputer Centers include ordering systems as they are required, stocking spare parts, installing hardware changes, coordinating maintenance, and administering the operation of the systems. The centers may also be involved in gathering and using data to forecast and plan new system acquisitions and upgrades of existing systems.

12.8.2 Record Bases Required

In order to accomplish their task, Minicomputer Centers have assembled complete inventory records of all equipment associated with the systems under their jurisdiction. The records can also include the status of new equipment in the process of being made operational.

A second category of records required are the manuals and user documents for each equipment element. These should be recorded along with the information indicating the proper issues to be used with each vintage of equipment. These records can be used to assure that proper documentation is being maintained at each system location.

The centers will maintain complete maintenance history logs for each installation and combine and sort these to identify problems that appear to be generic to each particular code of equipment unit. In addition, these records should include a complete list of changes made available by the minicomputer developer with descriptions of problems that these correct or the improvements they provide.

Finally, these centers will require personnel records that can be used to administer the operational and maintenance work forces.

12.9 SUMMARY AND TRENDS

12.9.1 Network Elements

The telecommunications switching network consists of many types of switching systems and transmission facilities for lines and trunks that connect telephones, customer equipment, and the switching centers. The switching systems are of two general types: (1) electromechanical, which use electrically driven motors, relays, and other components to make and control connections, and (2) electronic, which use processors under stored program control to operate magnetic or electronic switches. The network is operated jointly by many independent telephone companies and Bell System telephone companies and departments. The Bell System companies tend to operate in similar ways with similar equipment. For that reason, statements can be made concerning network operation which generally apply to these companies. The same statements may apply to the independent companies, but on a less uniform basis. All of the network elements discussed in this chapter are Bell System elements. These elements have been selected from a category identified as engineering and network services for the switching operations portion of the network.

12.9.2 Network Evolution

Before 1900 the first telephone exchanges were in operation, and soon after 1900 dial telephones were in use, operating with step-by-step switching systems. Step-by-step was followed by panel switching, crossbar, and since 1960, electronic switching systems of several types have become part of the network. In 1970 minicomputers began to spread throughout the operating companies; initially each was specifically programmed to provide assistance to some individual operation or function. Soon thereafter minicomputer systems became available that were more broadly applicable across operating companies and system types.

The network operates today using examples of nearly every type switching system ever put into service. The majority of the early systems have been augmented to keep up with the growth of telephone use and have been altered to operate in the modern network. With the variety of features and options available, very few switching systems are identical.

Early in the evolution of the network the central office had become the center of operations and administration of the network. In the past few years with the availability of many broadly applicable computer-based operations support systems, centralization began to move maintenance, control, and administration responsibility away from the central office.

12.9.3 Functional Centralization

By 1975 the concept of centralizing the maintenance function for electronic switching systems was being implemented in the network using an operations support system, No. 2 SCCS, to assist in this function. This concept has proven to be cost-effective and well able to maintain the extremely high reliability requirements of the network elements. It is expected that most Bell System switching systems will be maintained by Switching Control Centers.

The availability of other operations support systems that can assist in the centralization of functions such as billing, traffic measurement, and traffic control have increased the probability that all major elements of network operation will be centralized in functional areas.

While some centers are already in operation, such as the AMA Recording Center, these centers do not necessarily have the direct overall administrative responsibility for the function they serve. An AMA Control Center has been proposed to serve the overall administration needs of the billing function. Some telephone companies are combining the functions of a Switching Control Center with network administration responsibilities over a central office district. This includes monitoring of service and load in the network for that district. The resulting center is called a District Operations Center. It is expected that functional centralization of network operations will continue to evolve as long as the result continues to be greater operational efficiency and better service.

12.9.4 Operations Support Systems

There are a large number of different operations support systems assisting in the operation of the network. Some of these systems are one-of-a-kind and others may be applicable to only one telephone company. Many of the systems assist normal business functions such as personnel, accounting, and sales. Not all network operations have been covered here, nor are the system examples used a complete set of support systems for these operations; rather, they are the major support systems for the major operations involved.

Listed below by way of summary in Table 12.1 are the computer-based operations support systems covered in this chapter. The operations area is listed first, followed by the functional center involved, followed by the operations support system and an abbreviated description of its function.

1. MAINTENANCE

A. SWITCHING CONTROL CENTER FOR ESS
No. 2 SCCS - (No. 2 Switching Control Center System): Maintenance support for electronic switching

CATLAS - (Centralized Automatic Trouble Locating and Analysis System): Recommendations for circuit pack replacement

B. SWITCHING CONTROL CENTER FOR ELECTROME-CHANICAL SYSTEMS
ATA - (Automatic Trouble Analysis): Analysis of switching system troubles

TASC - (Telecommunications Alarm Surveillance and Control): Monitoring and control of office alarms

COMMS-PM - (Central Office Maintenance Management System/Preventive Maintenance): Administration of preventive maintenance

COMAS - (Computerized Maintenance and Administration Support): Analysis of crossbar tandem call routing

CAROT - (Centralized Automatic Reporting on Trunks): Trunk testing (also supports the SCC for ESS)

2. BILLING

    A.   AMA CONTROL CENTER

        No. 1 AMARC - (No. 1 Automatic Message Accounting Recording Center system): Remote collection of billing data

        CAMA-C - (Centralized Automatic Message Accounting-Type C): Collection of billing data of subtending offices

        LAMA-C - (Local Automatic Message Accounting - Type C): Collection of billing data in No. 5 Crossbar

3. TRAFFIC DATA

    A.   NETWORK DATA COLLECTION CENTER

        EADAS - (Engineering and Administration Data Acquisition System): Collection of traffic data

4. NETWORK CONTROL

    A.   NETWORK MANAGEMENT CENTER

        EADAS/NM - (EADAS/Network Management): Monitoring and control of network traffic

        PBC - (Peripheral Bus Computer): Administration of No. 4A/ETS and support of No. 4/ETS Traffic Control

    B.   NETWORK OPERATIONS CENTER

        NOC - (Network Operations Center system): national level traffic network surveillance and control

Table 12.1.   *Summary of Computer-Based Operations Support Systems*

12.9.5 Operations Support Systems Development

The need for rapid change to support the network as it evolves has been translated into the development of a software design support system that allows rapid introduction of new features and expanded capabilities in operations support systems. In the Bell System this involves the use of a higher level software language and a powerful set of development utilities and aids to the software development.

Where the network operation must be achieved without errors or must be continuous, the operations support system requires high reliability. To achieve this reliability, systems are using duplex or redundant computer complexes. At present circuit maintenance and billing systems are in this category, but as the telephone operating companies become more and more dependent on the continuous, reliable operation of the support systems, spare or duplex processors or complexes may be added to other types of support systems.

The increasing number of systems that are being used in network operations has given the telephone craft personnel opportunities to make cross comparisons of systems and judge where improvements can be made. System developers are recognizing that not only does the computer subsystem need to be designed, but the personnel subsystem needs to be designed rather than just letting it happen. Craft personnel are being used in the development stages of systems to help evaluate the personnel subsystem design. The result that can be expected is improved operations in the network.

12.9.6 Operational Efficiency

Many types of equipment and many functions make up the central office an its operations. The central office craftsperson in the past was often required to be a Jack or Jill-of-all-trades. As switching and transmission systems have become more complex, it has become more difficult to locate in each central office the level of expertise required to maintain the operation. A first attempt to solve this problem was the centralization of highly trained and skilled maintenance personnel. These personnel were assigned to a Technical Assistance Center (TAC) and could be called in when the central office craftsperson could not solve the problem. The Switching Control Center has proven to be a widely accepted concept where all levels of maintenance capability are administered centrally. The extension of this functional centralization concept which is assisted by computerized operations support systems can be expected to result in significant increases in overall network operating efficiency.

REFERENCES

1. Bell Telephone Laboratories, Incorporated: *Engineering and Operations in the Bell System*, 1977.

2. Macurdy, William B. and Alistair E. Ritchie: "The Network: Forging Nationwide Telephone Links," *Bell Laboratories Record*, Vol. 53, No. 1, p4, 1975.

3. Joel, Amos E. Jr.: "Switching," *Bell Laboratories Record*, Vol. 53, No. 1, p30, 1975.

4. Mack, John E. and William B. Smith: "Centralized Maintenance and Administration of Electronic Switching Systems," *Proceedings of the IEEE*, Vol. 65, No. 9, p1374, 1977.

5. Cuilwik, Anthony: "Minicomputers Give Operations People a Maxi-Assist," *Bell Laboratories Record*, Vol. 54, No. 1, p14, 1976.

6. Lyons, John R.: "Automation Improves Maintenance of Local Crossbar Offices," *Bell Laboratories Record*, Vol. 56, No. 4, p99, 1978.

7. Neville, S. M. and Robert D. Royer: "Controlling Large Electronic Switching Systems," *Bell Laboratories Record*, Vol. 54, No. 2, p30, 1976.

8. Armstrong, Roderick J., Robert Gottdenker and Robert L. Kornegay: "Servicing Trunks by Computer," *Bell Laboratories Record*, Vol. 54, No. 2, p39, 1976.

9. Watson, George F.: "Getting No. 4 ESS on Line on Time," *Bell Laboratories Record*, Vol. 54, No. 4, p82, 1976.

10. Bowyer, L. Ray: "Developing Accurate Equipment Records for TIRKS," *Bell Laboratories Record*, Vol. 54, No. 4, p97, 1976.

11. Byrne, Charles J. and Donald C. Pilkinton: "Toward Automated Local Billing," *Bell Laboratories Record*, Vol. 54, No. 4, p104, 1976.

12. *Bell Laboratories Record*, Vol. 54, No. 6, 1976.

13. Mandigo, Paul D.: "No. 2B ESS, New Features from a More Efficient Processor," *Bell Laboratories Record,* Vol. 54, No. 11, p304, 1976.

14. Almquist, Richard P., David L. Carney, and Robert A. Estvander: "1A ESS, Newest, Largest-Capacity Local Switch Cuts Over Early," *Bell Laboratories Record,* Vol. 55, No. 1, p15, 1977.

15. Arnold, Thomas F.: "TSPS Goes to the Country," *Bell Laboratories Record,* Vol. 55, No. 6, p146, 1977.

16. Nance, Robert C. and Baylen Kaskey: "Initial Implementation of Common Channel Interoffice Signaling" *Conference Record-International Switching Symposium,* Vol. 2, p413-2-1, 1976.

17. Jacobsen C. R. And R. L. Simms: "Toll Switching in the Bell System," *Conference Record-International Switching Symposium,* Vol. 1, p132-4-1, 1976.

18. Gerard, A., A.F. Rehert, and F.J. Webb: "Applying Automated Centralized Maintenance Techniques to Electromechanical Systems," *Conference Record-International Switching Symposium,* Vol. 2, p433-4-1, 1976.

19. Averill, R.M. Jr. and R.E. Machol: "A Centralized Network Management System for the Bell System Network," *Conference Record-International Switching Symposium,* Vol. 2, p433-3-1, 1976.

APPENDIX

BASIC MATHEMATICAL AND ENGINEERING CONCEPTS

L. Lewin
Department of Electrical Engineering
University of Colorado
Boulder, Colorado

A.0. INTRODUCTION TO THE APPENDIX

It has been our experience that many students, not excluding those with degrees in technical subjects, often lack a necessary fluency in relatively elementary operations in mathematics and physics. It is not that these individuals haven't been exposed to the material at some time or other (although occasionally this may have been so) but that oftentimes it has been little used, or has been largely forgotten, or perhaps was never properly understood in the first place. This imposes an undesirable handicap on the individual, and in order to remedy the shortcoming, and also to provide essential material for revision, a course entitled "Introduction to Communication Systems Theory" was instituted. The material of this course is summarized in the nine sections of the following appendix. It should be made clear that the purpose is to introduce the use of mathematics as a working tool, and that the key feature of the course is to gain a working familiarity with the subject. It is no part of the intention to provide rigorous mathematical proofs; and in fact it is probably the case that for many students a premature involvement in rigorous theorem proving may have been off-putting, and the cause of much aversion to mathematics. This is a pity, because mathematics can be used very powerfully as a tool in many ways, and a competency in this area can prevent both misunderstandings and also an awareness of possible misuse. (The suspicion that many people have for statistics is but one example of this.) Here is probably not the place to discuss the general teaching of mathematics, but the remarks by one student that "for the first time I can see what the calculus is all about — the explanations are not all bogged down in a mass of 'epsilonics' and other things." may indicate that, for engineering students, at least, a more pragmatic approach to teaching mathematics could be potentially rewarding. In line with this, the "proofs" in the following sections should be viewed more as heuristic demonstrations of plausibility, and as indications as to where results come from, and what their relevance to the subject matter may be. It is not my wish in any way to disparage the rigorous mathematical treatment; in advanced work the subtleties are often needed, and failure to observe the finer points is often a cause of error. But I feel that mathematical rigor, for many students, is something to be built up later on a foundation that has maybe been laid in a different way. To some degree this appendix may be helpful to those students whose earlier involvement in this area may have been incomplete, or whose fluency in the subject may be deficient.

A.1. BASIC PHYSICAL RELATIONSHIPS

A.1.1 Units and Dimensions

The three basic mechanical units are of length, mass and time. In the MKS system they are measured in meters (m), kilograms (kg) and seconds (s) respectively. Occasionally, particularly in magnetic units, one meets features derived from the earlier cgs (centimeter, gram, second) system, and this difference is largely responsible for powers of 10 that enter into some formulas.

Although these basic units are now defined in terms of atomic constants they are specified from semi-arbitrary physical standards.

Two other units needed are of temperature and electricity. The former is the degree Celsius (°C) and the latter the ampère (A).

These five basic units are all independent in character; none can be derived from the others. In contrast, many other units, e.g., velocity, can be expressed in terms of them. In particular, no independent magnetic unit is needed, since magnetic effects are expressible in terms of the others. Units so expressible are called *derived* units.

A.1.2 Multiples and Submultiples

The following multiples and subdivisions of units are the ones most commonly encountered. They are based on powers of a thousand, and are used in order to avoid small numbers, strings of zeroes or powers of ten in statements of results. When using a formula which is self-consistent in a certain set of units (practical units) the subdivision or multiple should ALWAYS be first converted into the relevent unit times a power of ten. Otherwise the formula, based on a self-consistent set, will yield incorrect results. (This stipulation does not apply to some practical engineering formulas where the dimensions may be *explicitly* required to be in some other unit, e.g. MHz, miles, etc.). Needless to say, non-metric units must first be converted to metric values before MKS formulas are used.

When using a formula, the units of the quantity being derived should be stated *after* the formula or calculation. It is a source of confusion to allow the units themselves to be caught up in the calculations, and is also unnecessary since the formula, being self-consistent in character, ensures that the units of the calculated quantity are automatically looked after.

Multiple			Submultiple		
k	kilo	10^3	m	milli	10^{-3}
M	mega	10^6	μ	micro	10^{-6}
G	giga	10^9	n	nano	10^{-9}
T	terra	10^{12}	p	pico	10^{-12}

Do not confuse M as designated here with M as used for mass (in a formula) or (as occasionally used) to denote miles. Similarly avoid confusing m for milli with m used to denote meter. (They occur together as mm, millimeter.) Submultiples are denoted by lower-case letters and multiples by capital letters (k for kilo is an historical exception).

Occasionally, other subdivisions are encountered, such as centi (c) = 10^{-2}, as in centimeter (cm); deci (d) = 10^{-1}, as in decibel (dB).

The unit of mass in the MKS system is the kilogram, and is treated *as if* it were a basic integral unit. (Historically the gram was the basic mass unit and the kilogram is therefore really composite.)

A.1.3 Derived Mechnical Units

The important derived mechanical units are

a) Frequency, denoted by f, measured in hertz, Hz. A hertz is one cycle per second, s^{-1}.
b) Velocity, measured in meters per second, m/s or ms^{-1}.
c) Acceleration, measured in meters per second per second, (m/s)/s or m/s^2 or ms^{-2}.
d) Force, measured in newtons, N. One newton is the force necessary to produce an acceleration of one meter per second per second in a mass of one kilogram; its units are $kg\, ms^{-2}$.
e) Work or Energy denoted by W, measured in joules, J. One joule of work is done if a force of one newton is applied, in the direction of the force, for a distance of one meter; its units are $kg\, m^2\, s^{-2}$.

The cgs unit of energy is the erg. Since there are 10^2 centimeters in a meter and 10^3 grams in a kilogram, one joule = 10^7 ergs.
f) Power, denoted by P, measured in watts, W. Power is the rate of doing work. One watt is the power when one joule of work is performed per second; its units are $kg\, m^2\, s^{-3}$.

A.1.4 Derived Electrical Units

The speed of light in a vacuum is a mechanical measure which provides a link with electromagnetic phenomena. It has the value

$$c = 2.99792458 \times 10^8 \text{ m/s} \tag{A.1}$$

which is independent of the frequency (color) of the light, and is also independent of the state of motion of the observer. The value in (A.1) is usually approximated by the abbreviated value $c = 3.10^8$ m/s, and is the source of the number 30 (= $c.10^{-7}$) that occurs frequently in electromagnetic formulas. The main electrical units are defined in relation to mechanical units as follows:

a) Electric current, denoted by I or i, measured in ampères, A. (This is the *basic* electrical unit). Two parallel wires carrying equal currents in the same direction feel a mutual force of attraction. If the wires are each one meter long and spaced one

meter apart, and if the current is adjusted such that the attractive force is made equal to 2×10^{-7} newtons*, then the strength
of the current in each wire is defined as being one ampère. Experimentally it is found that this corresponds to a flow of
6.23×10^{18} electrons per second.

b) Electric charge, denoted by Q, measured in coulombs, C. If a
   current of one A flows for one second, the total charge carried
   is defined as one coulomb. Its units are A s. Since a coulomb is
   equal to the charge on 6.23×10^{18} electrons, the charge on one
   electron, usually denoted by e, is equal to 1.60×10^{-19} C. It is
   conventionally taken as *negative*. Since a (positive) current is
   considered to be a flow of positive electric charge, this means
   that the direction of current flow is *opposite* to the actual
   direction of *electron* flow. Note that a positive current flow
   from, say, left to right, is equivalent to a negative current flow
   from right to left. Each is an *equally valid* representation. In any
   case, with alternating current, the current changes direction
   twice per period, so there is no absolute sense in which one can
   specify the direction of current flow.

c) Potential difference, denoted by V, (voltage, electromotive
   force) is measured in volts, V. It is the cause of electric current
   flow. If a potential difference causes current to flow, then work
   is being done at a certain rate. The volt is defined as that value
   of potential difference which results in work being done at the
   rate of 1 joule per second when the current strength is one amp.
   Hence we can write

   $$1 \text{ watt} = 1 \text{ volt} \times 1 \text{ amp}$$

   This equation can be re-written dimensionally in the form

   $$J/s = V \cdot C/s$$

   $$\text{or } J = [V/m)C] \, m$$

d) Electric fieldstrength, denoted by E, is measured in volts per
   meter, V/m. In the previous equation the left-hand side is
   work = force $\times$ distance (dimensionally), so it is apparent that
   (V/m)C is a force. The quantity V/m is called electric fieldstrength and is the force *per unit charge* caused by a voltage
   difference operating over a distance.

*The factor 2×10^{-7} comes from the need to provide continuity with prior definitions
of the amp. It equals $\mu_0/2\pi$, where $\mu_0 = 4\pi \times 10^{-7}$ is defined in section A.1.5 (c).

Coulomb's law says that the force in free space between two electric charges Q_1 and Q_2 is proportional to their product and inversely proportional to the square of the distance between them: Force $\propto Q_1 Q_2 / r^2 = (Q_2 / r^2) Q_1$. Clearly, then, the expression Q_2 / r^2 appears here as a force per unit charge, and can be thought of as due to the electric field produced by the charge Q_2 at distance r. The relation can be written with an equality sign provided a suitable constant is incorporated to take into account the units used. In the rationalized MKS system the equation is written

$$\text{Force} = \frac{Q_1 Q_2}{4\pi\epsilon_0 r^2} \tag{A.2}$$

where the force is in newtons, Q in coulombs and r in meters.

e) Permittivity. The quantity ϵ_0 in (A.2) is called the permittivity of free space, and from the equation it is clear that its units are C^2 / Nm^2, but this representation is not usually utilized. Rather, a new unit, the farad, F, is defined, and the units of ϵ_0 are F/m. The definition of F is given later under section f) on capacitance. The numerical value of ϵ_0 is

$$\epsilon_0 = \frac{1}{4\pi 9 \times 10^9} \quad F/m$$

$$= 8.854 \text{ pF/m} \tag{A.3}$$

In a more general medium, (A.2) is modified by the replacement of ϵ_0 by ϵ where ϵ is the permittivity of the medium. It is usually between about 2 to 100 times ϵ_0, and this multiplying factor is called the *relatively permittivity* and is denoted by ϵ_r. It is dimensionless, since it is defined by

$$\epsilon_r = \epsilon/\epsilon_0 \tag{A.4}$$

f) Capacitance, denoted by C, is measured in farads, F. If a body is charged, its potential increases. If the potential becomes one volt when the charge is one coulomb, the capacitance is defined to be one farad. Capacitance is the charge per unit potential, and in practice one volt would produce only a minute charge on most bodies. Thus the farad is a *very* large unit, and most capacitances are measured in microfarads or picofarads. Very approximately a sphere 1 cm in radius and removed more than a few cm from neighboring bodies, has a capacitance of about 1 pF.

g) Resistance, denoted by R, is measured in ohms, Ω. If a circuit requires a voltage of one volt to drive a current of one amp, it is said to have a resistance of one ohm. If the response of the circuit is linear (i.e. doubling the voltage doubles the current, etc.) we have Ohm's law:

$$V/I = R \text{ or } V = IR \text{ or } I = V/R \tag{A.5}$$

Since the power P = VI we also get the corresponding relationships

$$P = VI = I^2 R = V^2 /R \tag{A.6}$$

for the power dissipated in a resistor.

h) Resistivity, denoted by ρ, is measured in ohm meter, Ωm. A resistor made from a certain material has a resistance proportional to the length of the resistor and inversely proportional to the cross-section area. Hence we can write

$$R = \rho \cdot \text{length/cross section area} \tag{A.7}$$

The proportionality constant ρ is called the *resistivity* and is a property of the material, not the size and shape of the resistor. (It will also vary with the temperature, however.)

i) Conductivity, denoted by σ, is measured in mho/meter, $1/\Omega$m, where mho, the inverse of ohm is also called Siemens. It is defined as the inverse of the resistivity, so that $\sigma = 1/\rho$. For pure copper at room temperature its value is approximately $\sigma = 5.8 \times 10^7$.

j) Electric Displacement, denoted by D, is measured in coulombs per square meter. It is equal to ϵE, and may be thought of as the density of electric flow lines emanating from an electric charge.

A.1.5 Derived Magnetic Units

It might at first sight appear that a separate basic unit for magnetic quantities is needed, and in the earlier development of magnetism this was implicit. It was, moreover, based on the cgs system of units, and two of these units, the oersted and the gauss, are still with us. They are not part of the MKS system, and since the self-consistent equations require all units to be in the MKS system, measurements in the cgs system must be converted. A further difficulty of the earlier development of magnetism is that it presumed the existence of an isolated magnetic pole analogous to the electric charge, for which Coulomb's law (A.2) applies. In practice, magnetic poles always occur in equal and opposite pairs, though the properties of one end of a long bar magnet approximate those of the hypothetical isolated pole.

We commence, therefore, with an analogue of (A.2):

$$\text{Force} = M_1 M_2 / r^2 \tag{A.8}$$

where M is a measure of a (hypothetical) magnetic pole. The formula can also be written force $= (M_1 / r^2) \cdot M_2$ where (M_1 / r^2) can be interpreted as the force per unit pole and is the analogue of the electric field which is the electrically caused force per unit charge. The quantity (M/r^2) is therefore called the magnetic field and, if the units in (A.8) are based on the cgs system, the unit of magnetic field is called the oersted, Oe. As mentioned above, it is not part of the MKS system.

a) Magnetic field, denoted by H. In the MKS system the unit (analagous to volt/meter for electric field) is the amp per meter. It is the field at a distance of one meter from a long straight wire carrying a current of a certain strength. It would be neat if this current strength were one amp, but in fact it has to be 2π amps. The factor 2π comes from the rationalization process used in setting up the MKS system. Basically it comes from the circumference of a unit circle, which is 2π meters, and this factor 2π (and also 4π, from the surface area of the unit sphere) has to appear in one place or another in the system. The conversion factor between the two magnetic fields is

$$1 \text{ A/m} = (10^3/4\pi) \text{ Oe}$$

b) Magnetic induction, or magnetic flux density, denoted by B. It is analagous to the electric displacement D, and is a measure of the density of magnetic flux lines. In the cgs system it is called a gauss, and in the MKS system the unit is weber/(meter)^2, where the weber is the analogue of the coulomb, and is a measure of magnetic pole strength. Its magnitude is such that, placed a distance of 1 meter from a long wire carrying a current of 2π amps, its experiences a force of 1 newton. The conversion factor between gauss and weber/m^2 is

$$1 \text{ gauss} = 10^{-4} \text{ Wb/m}^2$$

c) Permeability, denoted by μ. The relation between magnetic induction and magnetic field can be written (at least for simple materials) $B = \mu H$ where the proportionality quantity μ is either a) independent of the field and equal to, or very nearly equal to, the value μ_0 in free space, or b) is a highly non-linear function of the field and is typically many times the value of μ_0. Most substances fall into the first category, while many substances containing iron or certain other metals, fall into the second. The material of a permanent magnet would be a typical example of this group.

According to the above definition, the units for μ would be (Wb/m^2)/(A/m) but this representation is not usually utilized. Rather a new unit, the henry, H, is defined, and the units of μ are H/m. The definition of H is given later under section d) on inductance. The numerical value of μ_0 is

$$\mu_0 = 4\pi10^{-7} \text{ H/m}$$
$$= 1.256 \ \mu\text{H/m} \tag{A.9}$$

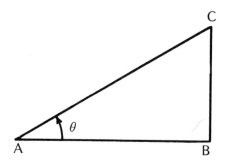

A.1   Triangle for Definition of the Sine: Acute Angle

d) Inductance, denoted by L, is measured in henries, H. If a magnetic flux density penetrates (at right angles) a circuit of area S, then the total flux is BS. If this changes with time its rate of change produces a voltage around the circuit. This law was discovered by Faraday. A change of 1 weber/sec induces a voltage change of 1 volt.

Now the change of magnetic flux could be due either to the movement of a magnet near the circuit or a change of current in the circuit, since the current itself produces a magnetic field. The inductance is a property of the circuit which relates the voltage induced by a change of current in it. If the current changes at 1A/s and induces a voltage of 1V, then the inductance of the circuit is said to be 1 henry.

A.2   TRIGONOMETRIC FUNCTIONS

A.2.1   The Sine

The sine (abbreviated to sin) of an angle in a right-angled triangle is defined as the ratio of the *opposite* side to the *hypoteneuse.* It is a function of the angle only, and does not depend on the absolute size of the triangle. Thus, in Figure A.1 ABC is a triangle, with right-angle at B. If the angle at A is denoted by θ, then

$$\sin \theta = BC/AC \tag{A.10}$$

As $\theta \to 0$, BC $\to 0$, so sin 0 = 0
As $\theta \to 90°$, BC $\to$ AC, so sin 90° = 1

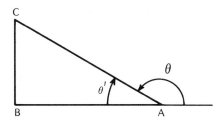

A.2   Triangle for Definition of the Sine: $90° <$ Angle $<180°$

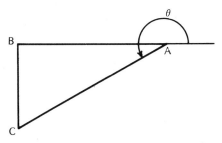

A.3   Triangle for Definition of the Sine: $180° <$ Angle $< 270°$

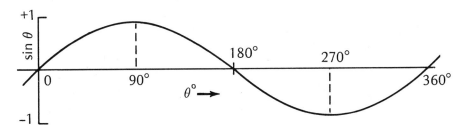

A.4   Graph of the Sine Function

This definition can be extended beyond 90° by thinking of AC as a rotating arm, pivoted about A, and CB as a "plumb line" dropped onto AB. Thus, for angles up to 180° we get Figure A.2 with $\sin \theta = BC/AC$ as before. Also, from the triangle, $BC/AC = \sin \theta'$ where $\theta + \theta' = 180°$.

Hence $\sin \theta = \sin \theta' = \sin (180° -\theta)$. Accordingly $\sin \theta$ is symmetrical about $\theta = 90°$. Beyond $\theta = 180°$ the rotor arm is below AB, so the line CB has to extend in the opposite direction as before. Hence $\sin \theta$ is negative for $180° < \theta < 360°$, as in Figure A.3. Otherwise the values repeat the 0 to 180° range. Thus $\sin (\theta) = -\sin (\theta - 180°)$. The graph of $\sin \theta$ repeats every 360°, and is as in Figure A.4. Notice that $\sin \theta$ always lies between the limits ± 1, and that $\sin (-\theta) = -\sin \theta$.

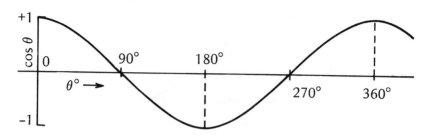

A.5   Graph of the Cosine Function

A.2.2 The Cosine

The cosine (abbreviated to cos) of an angle in a right-angled triangle is defined as the ratio of the *adjacent* side to the *hypoteneuse*. In Figure A.1

$$\cos \theta = \frac{AB}{AC} \qquad\qquad\qquad (A.11)$$

As $\theta \to 0$, AB $\to$ AC, so cos 0 = 1
As $\theta \to 90°$, AB $\to$ 0, so cos 90° = 0

The definition can be extended beyond 90° by noting that AB reverses sign the other side of the vertical, so that cos θ is negative from 90° to 270°. The graph of cos θ is as in Figure A.5. Note that this is like a graph of sin θ shifted 90° to the left. It repeats every 360°.

A.2.3 The Tangent

The tangent (abbreviated to tan) of an angle in a right-angled triangle is defined as the ratio of the *opposite* side to the *adjacent* side. In Figure A.1

$$\tan \theta = \frac{BC}{AB} \qquad\qquad\qquad (A.12)$$

As $\theta \to 0$, BC $\to$ 0, so tan 0 = 0
As $\theta \to 90°$, AB $\to$ 0, so tan 90° = BC/0 = ∞ (infinity)

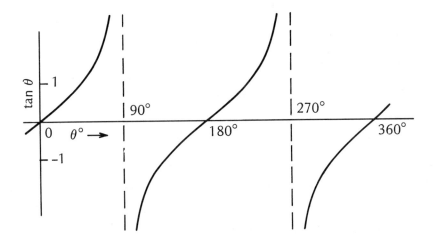

$$\text{A.6}\quad \text{Graph of the Tangent Function}$$

From 90° to 180°, BC is positive and AB is negative. Hence the tangent is negative. From 180° to 270° both BC and AB are nega tive; their ratio is therefore positive. From 270° to 360° BC is negative and AB is positive; the tangent is therefore negative. The graph of tan θ is as in Figure A.6. It repeats every 180°. The dotted line at 90° is known as an asymptote: tan θ is $+\infty$ on one side and $-\infty$ on the other.

A.2.4 Important Relations Between These Functions

a) $\tan \theta = \dfrac{BC}{AB} = \dfrac{BC}{AC} \cdot \dfrac{AC}{AB} = \dfrac{BC/AC}{AB/AC} = \dfrac{\sin \theta}{\cos \theta}$ (A.13)

b) In Figure A.7 cos θ = AB/AC and sin θ = AB/AC. But these ratios are the same. Hence sin θ' = cos θ. Now the three angles of a triangle add up to 180°. This gives $\theta + \theta' + 90° = 180°$, or $\theta' = 90° - \theta$. Hence
sin $(90° - \theta)$ = cos θ, or   (A.14)
cos $(90° - \theta')$ = sin θ' (A.15)

c) From Pythagoras's theorem we get

$$(AC)^2 = (AB)^2 + (BC)^2 \qquad (A.16)$$

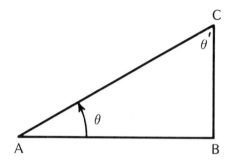

A.7   Triangle Illustrating Complementary Angles

Dividing by $(AC)^2$ throughout gives

$$\left(\frac{AC}{AC}\right)^2 = \left(\frac{AB}{AC}\right)^2 + \left(\frac{BC}{AC}\right)^2 \text{, or}$$

$$1 = \cos^2\theta + \sin^2\theta \tag{A.17}$$

d) If θ and ϕ are any two angles, the sin or cos of combinations of them can be expressed in terms of sines and cosines of the separate angles. There are an unlimited number of such relations, but the following are *particularly important:*

$$\sin(\theta \pm \phi) = \sin\theta \cos\phi \pm \cos\theta \sin\phi \tag{A.18}$$

$$\cos(\theta \pm \phi) = \cos\theta \cos\phi \mp \sin\theta \sin\phi \tag{A.19}$$

Note that the + sign for the cosine on the left of (A.19) goes with the − sign on the right, and *vice versa.*

Particular cases come from taking $\theta = \phi$, and using the + sign.

$$\sin(2\theta) = 2\sin\theta \cos\theta \tag{A.20}$$

$$\cos(2\theta) = \cos^2\theta - \sin^2\theta \tag{A.21}$$

Two variants of (A.21) come from using (A.7) in (A.21)

$$\cos(2\theta) = 2\cos^2\theta - 1 = 1 - 2\sin^2\theta \tag{A.22}$$

This last equation can be presented in a form which gives the *squares* of sines or cosines in terms of the cosine of the double angle,

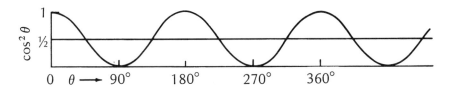

A.8   Graph Illustrating the Average Value of the Cosine Squared

$$\cos^2\theta = \frac{1+\cos(2\theta)}{2} \ , \quad \sin^2\theta = \frac{1-\cos(2\theta)}{2} \tag{A.23}$$

It is seen from Figure A.8 how $\cos^2\theta$ behaves like a half-amplitude $\cos(2\theta)$ curve, displaced in level by 1/2.

Since $\cos(2\theta)$ consists of equal sections that are alternatively positive and negative, it averages out to zero. Hence, from the figure, we see that the *average value* of $\cos^2\theta$ is exactly 1/2.

A.2.5   Special Values

Certain simple values recur frequently. The sines and cosines of 0 and 90° have already been given. If, in Figure A.9, AB = BC, then the angles at A and C are equal, from symmetry, and each will be 45° since the sum of all the angles has to be 180°. Let AB and BC

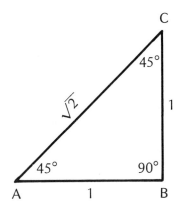

A.9   Triangle for Calculating Functions of 45°

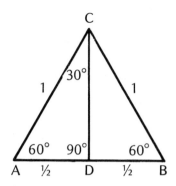

A.10   Triangle for Calculating Functions of 30° and 60°

each be of unit length. Then, from Pythagoras's theorem, $(AC)^2 = (AB)^2 + (BC)^2 = 1 + 1 = 2$. Hence $AC = \sqrt{2}$. From the triangle, and considering the angle at A,

$$\sin 45° = \frac{BC}{AC} = \frac{1}{\sqrt{2}} = 0.707 \ldots$$

$$\cos 45° = \frac{AB}{AC} = \frac{1}{\sqrt{2}} \qquad\qquad (A.24)$$

$$\tan 45° = \frac{BC}{AB} = 1$$

Consider now an *equilateral* triangle, of unit side as in Figure A.10. The angles are all equal, and hence are 60° each. If we construct a perpendicular bisector from C to AB, cutting AB at D, then $AD = 1/2$ and the angle at C in the triangle ACD will be 30°. Moreover, $(AC)^2 = (AD)^2 + (DC)^2$ gives $1 = 1/4 + (DC)^2$. Hence $(DC)^2 = 3/4$ and $DC = \sqrt{3}/2$. Hence

$$\sin 60° = \frac{DC}{AC} = \frac{\sqrt{3}}{2} = 0.866 \ldots = \cos 30°$$

$$\sin 30° = \frac{AD}{AC} = \frac{1}{2} = 0.5 \qquad = \sin 60°$$

$$\tan 30° = \frac{AD}{DC} = \frac{1}{2} / (\sqrt{3}/2) = 1/\sqrt{3} = 0.577\ldots \qquad \text{(A.25)}$$

$$\tan 60° = \frac{DC}{AC} = \frac{\sqrt{3}}{2} / \left(\frac{1}{2}\right) = \sqrt{3} = 1.732\ldots$$

A.2.6 Reciprocal Functions

These are the secant, cosecant and cotangent, abbreviated to sec, cosec and cot respectively. They are defined by

$$\sec\theta = \frac{1}{\cos\theta}, \quad \operatorname{cosec}\theta = \frac{1}{\sin\theta}, \quad \cot\theta = \frac{1}{\tan\theta} \qquad \text{(A.26)}$$

Their graphs are shown in Figure A.11. Clearly they have the same periodicity as their corresponding functions have.

A.2.7 Combinations of Sinusoidal Functions

Provided they have the same periodicity, sines and cosines can be combined to form a composite function which is also sinusoidal. Consider the equation

$$A \cos\theta + B \sin\theta = C \cos(\theta+\phi) \qquad \text{(A.27)}$$

Is it possible to choose a value for C and for ϕ such that this is true for *all* values of θ? It is not immediately obvious that this is so. Using (A.19) on the right-hand side we get

$$A \cos\theta + B \sin\theta = (C \cos\phi) \cos\theta - (C \sin\phi) \sin\theta.$$

For this to be true for *all* values of θ we *must* have

$$A = C \cos\phi, \text{ and, } B = -C \sin\phi \qquad \text{(A.28)}$$

By squaring and adding we get

$$A^2 + B^2 = C^2 \cos^2\phi + C^2 \sin^2\phi = C^2(\cos^2\phi + \sin^2\phi) = C^2 \text{ by (A.17).}$$

Hence

$$C = \sqrt{A^2 + B^2} \qquad \text{(A.29)}$$

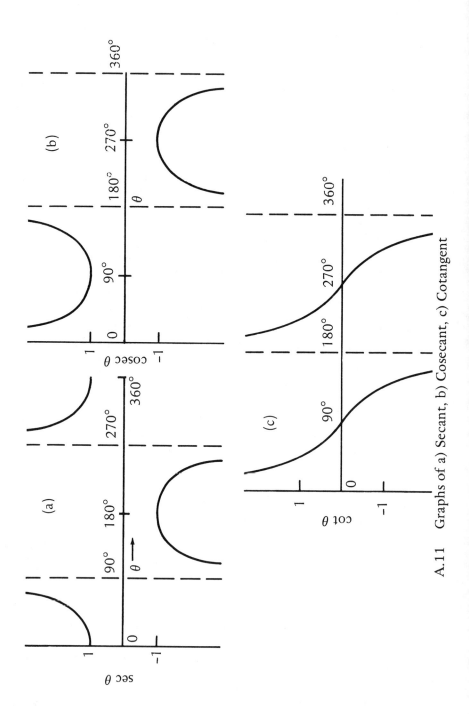

A.11   Graphs of a) Secant, b) Cosecant, c) Cotangent

It is usual to take the positive square root in this equation. By dividing the two relations in (A.28) we get

$$\frac{B}{A} = \frac{-C \sin \phi}{C \cos \phi} = -\tan \phi \tag{A.30}$$

This equation will give *two* values of ϕ separated by 180°. But only *one of these* will satisfy (A.28). *It is always necessary* to return to (A.28) *to check* which is the correct value.

More generally, let

$$A \cos \theta + B \cos (\theta + \psi) = C \cos (\theta + \phi) \tag{A.31}$$

Applying (A.19) to both the B and C terms we get
$$\cos \theta \, (A + B \cos \psi) - \sin \theta \, (B \sin \psi) = \cos \theta \, (C \cos \phi) - \sin \theta \, (C \sin \phi).$$

Hence

$$\begin{aligned} A + B \cos \psi &= C \cos \phi \\ B \sin \psi &= C \sin \phi \end{aligned} \tag{A.32}$$

Squaring and adding gives

$$C^2 = (A + B \cos \psi)^2 + (B \sin \psi)^2 = A^2 + 2AB \cos \psi + B^2 (\cos^2 \psi + \sin^2 \psi)$$

whence

$$C = \sqrt{A^2 + 2AB \cos \psi + B^2} \tag{A.33}$$

Equation (A.29) is the special case $\psi = -90°$, since $\cos (\theta - 90°) = \sin \theta$. Dividing the two relations of (A.32) gives, similarly to (A.30),

$$\tan \phi = \frac{B \sin \psi}{A + B \cos \psi} \tag{A.34}$$

As before, this equation gives *two* values of ϕ, separated by 180°, and they have to be checked with (A.32) to select the correct one.

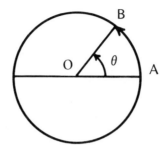

A.12   Circle for Definition of Radian Measure

A.2.8   Radians

Angular measure can be expressed in cycles, degrees or radians. One cycle = $360°$ = 2π radians. To see the relation between degrees and radians, consider the circle in Figure A.12, for which the ratio of the arc AB to the radius is AB/OA = θ where θ is *defined* as the angle in radians. For a complete cycle the arc AB becomes the circumference of the circle, = 2π times the radius. Hence 1 cycle = $360°$ = 2π, where π = 3.14159 . . .

$$\pi \text{ radians} = 180° \text{ or } 1 \text{ radian} = 57.3° \ldots \tag{A.35}$$

A.2.9   Angles in Excess of $90°$

Tabulated angles usually cover the range 0 to $90°$ only. From Figure A.13 we can get sines of angles greater than $90°$. Thus $\sin \theta = \sin (180° - \theta) = -\sin (180° + \theta) = -\sin (360° - \theta)$, etc. Similarly, if we imagine a new axis erected at $\theta = 90°$, the curve from there is a cosine curve. Thus $\sin (90° + \theta) = \cos \theta$. Sines or cosines shifted by $180°$ or $360°$ become sines and cosines, with a possible sign change that needs to be ascertained. Sines or cosines shifted by $90°$ or $270°$ become, respectively, cosines or sines, again with a possible sign change. Note, in particular,

$$\sin (-\theta) = -\sin \theta \qquad\qquad \cos (-\theta) = \cos \theta \tag{A.36}$$

$$\sin (90° - \theta) = \cos \theta \qquad\qquad \cos (90° - \theta) = \sin \theta \tag{A.37}$$

The quadrants where the respective functions $\sin \theta$, $\cos \theta$ and $\tan \theta$ are *positive* are shown in Figure A.14. Note that the angles are conventionally measured from the horizontal axis in a *counter-clockwise* sense.

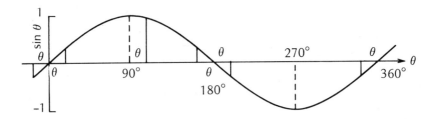

A.13   Graph for Sine Function of Angles in Excess of 90°

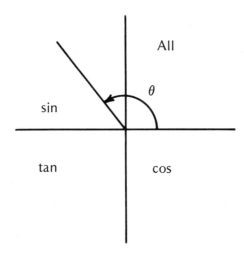

A.14   Quadrants where the Trigonometrical Functions are Positive.

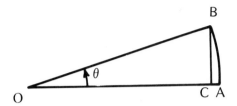

A.15   Approximate Equality of Arc to Chord for Small Angles

A.2.10   Approximations for Small Angles

It can be seen from Figure A.15 that the arc AB and the side BC of the triangle OBC are nearly equal when the angle is small. This leads to the approximation

$$\sin \theta \approx \theta, \quad \theta \text{ small and measured in radians} \tag{A.38}$$

From (A.17) it can then be shown that

$$\cos \theta \approx 1 - \theta^2/2 \tag{A.39}$$

and from (A.13)

$$\tan \theta \approx \theta \tag{A.40}$$

It should be emphasized that these three equations require that the angle be measured in radians. If it is in degrees, θ must be replaced by $(\pi\theta/180)$ on the right hand sides.

A.3   ELECTROMAGNETIC WAVE TRANSMISSION

A.3.1   Maxwell's Equations and Electromagnetic Waves

Under static conditions it is known that the magnetic field produced by a current flowing through a small area is proportional to that current. Maxwell surmised that, under time-varying conditions, the current should be supplemented by the time rate of change of the electric displacement through the area. This term is somewhat analogous to the term in Faraday's law which says that the voltage round a circuit is equal to the time rate of change of magnetic induction through the circuit. These two laws form the basis of

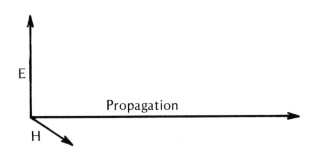

A.16   Direction of Propagation

Maxwell's equations which, when set up mathematically, lead to solutions which are interpreted physically as electromagnetic waves, a form of oscillatory energy in which the electric and magnetic fields are coupled together and propagate with a velocity given by $(\epsilon_o \mu_o)^{-\frac{1}{2}}$ in free space. This velocity turns out to be 3×10^8 m/s, the velocity of light, and leads to the conclusion that light is an example of an electromagnetic wave.

A.3.2   Plane Electromagnetic Waves

An idealized form of electromagnetic wave is the *plane wave*. A long distance away from a transmitter in free space the wave is a good approximation to a plane wave. It has the following important properties.

a) It propagates in a given direction with the velocity of light, denoted by c. In vacuum c has the value 2.9979×10^8 m/s. It is usually approximated by 3×10^8 m/s, and this is good enough for most purposes except for very accurate conversion of wavelength to frequency. It is very slightly slower in air, depending on, among other things, temperature, pressure and moisture content.

b) In a plane transverse to the direction of propagation the value of electric field E is everywhere the same, as is also the value of the magnetic field H.

c) E is perpendicular to H, and they are related such that a right-handed rotation from E to H points in the direction of propagation, as indicated in Figure A.16.

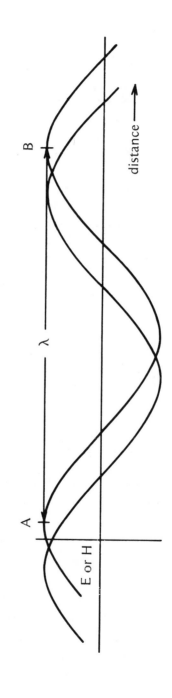

A.17   Propagation of a Radio Wave

d) The ratio E/H has the dimensions $(V/m)/(A/m) = V/A$ or ohms. In free space it has the value $(\mu_0/\epsilon_0)^{1/2} \approx 377$ ohms. Here, ϵ_0 is the permittivity of free space and has the approximate value $1/4\pi \cdot 9 \times 10^9$ F/m, and μ_0 is the permeability of free space and has the value $4\pi \times 10^{-7}$ H/m. Hence the square root of the ratio can also be expressed as 120π, an expression frequently encountered.

e) We shall not consider magnetic materials for which the permeability differs from μ_0 since they barely affect practical transmission. But for materials whose relative permittivity is ϵ_r we have $E/H = 377/\sqrt{\epsilon_r}$, and wave velocity $v = 3.10^8/\sqrt{\epsilon_r}$. For various soils ϵ_r varies from about 4 to 20. For sea-water it is about 80, though dropping at microwave frequencies.

A.3.3 Propagation

A typical plot of the instantaneous field in a plane wave is given in Figure A.17. The peak-to-peak separation AB is called the wavelength, and is denoted by λ. At a later instant the wave has moved to the right, as shown by the dotted curve. The time it takes to move one wavelength is the *period* of the wave, and is denoted by T. The frequency, f, is the number of cycles per second (hertz) so that 1/f is the time for one cycle, and therefore equals T. Hence

$$T = 1/f \qquad \text{or} \qquad fT = 1 \tag{A.41}$$

Since the wave moves a distance λ in a time T, the velocity is

$$\begin{aligned} v &= \lambda/T \\ &= \lambda f \qquad \text{from (A.41)} \end{aligned} \tag{A.42}$$

A.3.4 Sine Waves

The angular frequency ω is the number of radians per second. Since a complete cycle represents 2π radians we get

$$\omega = 2\pi f \quad \text{radians/sec} \tag{A.43}$$

At a fixed position the wave amplitude can be written

$$E = E_o \cos (\omega t + \phi) \tag{A.44}$$

where E_o is the maximum value of the field

  t is the time (seconds)

  ϕ is the initial phase of the wave.

The total angle is usually expressed in radians but for the convenience of use of tables it can be converted to degrees by using 2π radians = $360°$.

At a position distant D from the reference implied in (A.44) the wave is delayed by the time it takes to propagate the distance D, i.e. D/v. Hence the wave at that position has t replaced by t – D/v. The wave representation is therefore

$$E = E_o \cos [\omega (t–D/v) + \phi] \tag{A.45}$$

From (A.42) and (A.43) we get $\omega/v = 2\pi f/\lambda f = 2\pi/\lambda$. Hence (A.45) can also be written,

$$E = E_o \cos [\omega t + \phi - 2\pi D/\lambda] \tag{A.46}$$

The following formula combining cosines of different angles should be noted:

$$\cos A + \cos B = 2 \cos \left(\frac{A+B}{2}\right) \cos \left(\frac{A-B}{2}\right) \tag{A.47}$$

Thus if two equal waves arrive at a point having transversed distances D_1 and D_2, the composite wave is represented by

$$E_o \cos [\omega t + \phi - 2\pi D_1 /\lambda] + E_o \cos [\omega t + \phi - 2\pi D_2 /\lambda]$$
$$= 2E_o \cos [\omega t + \phi - 2\pi (D_1 +D_2)/2\lambda] \cos [\pi (D_1 -D_2)/\lambda] \tag{A.48}$$

The final factor, $\cos[\pi(D_1 - D_2)/\lambda]$, would be unity if the distance difference were an integral multiple of the wavelength, leading to a total signal of double amplitude. But if the difference were an odd multiple of the half-wavelength, the cosine would be zero. This result exemplifies the phenomenon of *wave interference*. Since the effect depends on λ, it is frequency-dependent, and the interference is known as frequency-selective fading.

A.3.5 Power Flow

The Poynting vector gives the power flow per unit area. In a plane wave, if the area S is perpendicular to the direction of propagation, the power flow is

$$P = EHS \quad \text{watts (W)} \tag{A.49}$$

In free space this gives the value

$$P = \frac{E^2 S}{377} \tag{A.50}$$

Hence the power is proportional to the square of the electric fieldstrength.

In these equations, E or H vary sinusoidally with time, so that E^2, say, varies between its maximum value and zero. The time average value is half the maximum, so that if E is given by, say, (A.45) then the *average* power flow is seen, as in Figure A.8, to be given by

$$P_{av} = \frac{1}{2}\frac{E_o{}^2}{377} \cdot S \tag{A.51}$$

A.4 INDICES, LOGARITHMS AND THE EXPONENTIAL FUNCTION

A.4.1 The Law of Addition of Indices

Let us consider the following sequence which both defines the index or exponent for positive integers, and extends the definition to zero and negative integral values.

$$1000 = 10^3$$
$$100 = 10^2$$
$$10 = 10^1$$
$$1 = 10^0$$
$$0.1 = 1/10 = 10^{-1}$$
$$0.01 = 1/100 = 10^{-2}$$
and so on.

In the relation $10 \times 10 \times 10 \ldots \ldots \times 10 = 10^n$ (in which n tens are multiplied together), 10 is called the *base* and n the *exponent*, or *index*. Each time we divide by ten we reduce the exponent by unity. The relations $10 = 10^1$ and $1 = 10^0$ can be thought of as self-consistent definitions of 10^1 and 10^0. Similarly the relation

$$1/10^n = 10^{-n} \tag{A.52}$$

forms a self-consistent continuation of the meaning of the index to negative integers.

From the relation $(10 \times 10 \times 10 \ldots \ldots \times 10)_{n \text{ terms}} \times (10 \times 10 \ldots \ldots \times 10)_{m \text{ terms}} = (10 \times 10 \times 10 \ldots \ldots \times 10)_{(m+n) \text{ terms}}$ we get the important equation

$$10^n \times 10^m = 10^{n+m} \tag{A.53}$$

This is the law of addition of indices, and it clearly holds for any base b

$$b^n \times b^m = b^{n+m} \tag{A.54}$$

A.4.2 Extension to Fractional Indices

Clearly (A.54) holds for any number of terms; e.g., $b^n \times b^m \times b^p = b^{n+m+p}$, etc. To extend the definition to fractional powers, suppose $y = b^{p/q}$ where p and q are integers. Then $y^q = (b^{p/q} \times b^{p/q} \ldots \times b^{p/q})_{q \text{ terms}} = b^{(p/q + p/q + \ldots + p/q)}|_{q \text{ terms}} = b^p$. Hence $y^q = b^p$ or $y =$ the qth root of b^p.

$$b^{p/q} = \sqrt[q]{b^p} \tag{A.55}$$

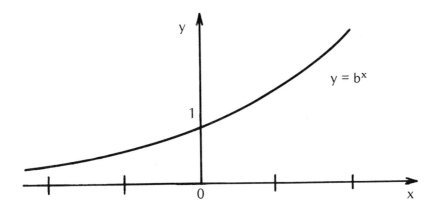

A.18   The Curve $y = b^x$

In particular $b^{\frac{1}{2}} = \sqrt{b}$, an important special case.

Note also

$$\left(b^m\right)^n = b^{\underbrace{(m+m+ \dots +m)}_{n \text{ terms}}} = b^{mn} \qquad \text{(A.56)}$$

A.4.3 The Exponential Function

Figure A.18 shows a plot of the function $y = b^x$ where the base, b, is any number greater than unity. (If b were less than unity the graph would have been the mirror image about the y-axis).

The curve is seen to go through the point (x=0, y=1) irrespective of the value of b. For large positive x, y gets very large. Conversely, for large negative x, y gets very small.

(Strictly speaking, the graph is only plotted for rational values of x. The magnitude of y for irrational values of x is found as the limit for closely bounding rational values).

A.4.4 The Logarithmic Function

If $y = b^x$ the value of x which solves this equation (given y and b) is called the *logarithm* of y to base b, and is written

$$x = \log_b y \qquad \text{(A.57)}$$

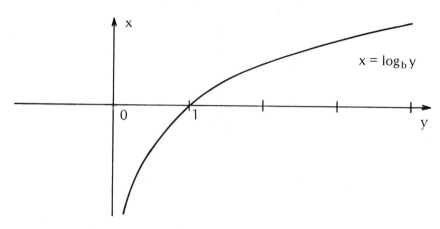

A.19   The Curve $x = \log_b y$

If $y = b^x$ and $z = b^w$ then $yz = b^x b^w = b^{x+w}$, by the law of indices. Hence $x+w = \log_b (yz)$ or, since $x = \log_b y$ and $w = \log_b z$

$$\log_b (yz) = \log_b y + \log_b z. \qquad (A.58)$$

The law of *addition of logarithms* states that the logarithm of a product of factors is the sum of the logarithms of the factors. Clearly, (A.58) can be extended to several terms. Note also,

$$\begin{aligned}
\log (1) &= 0 \\
\log (1/y) &= -\log y \\
\log (y^n) &= n \log y
\end{aligned} \qquad (A.59)$$

These results are true independent of the base, which need not be shown.

The graph of the logarithm is obtained from the graph of $y = b^x$, and is shown in Figure A.19 from which we note that $\log 1 = 0$, $\log (0) = -\infty$, $\log (\infty) = \infty$, and that the logarithm increases *very* slowly to infinity. Thus $N = 10^{30}$ is an *enormous* number, but $\log_{10} N = 30$, a very moderate value only. In almost all mathematical and numerical work, only three bases are used

a) Base 10 for numerical work
b) Base $e = 2.71828 \ldots$ for mathematical formulas
c) Base 2 for (binary) information theory.

Logarithms to base e (the significance of e is considered in detail later in section A.4.5) are called *natural logarithms* and the notation $\log_e$ is sometimes replaced by $\ln$.

It is possible to transform logarithms from one base to another in the following way:

Let $y = b^x$ (so that $x = \log_b y$) and take logarithms of both sides to base a.

$$\log_a y = \log_a b^x = x \log_a b \text{ (by A.59)} = \log_b y \log_a b.$$

Hence

$$\log_a y = \log_b y \log_a b \qquad\qquad (A.60)$$

This relation enables logarithms to base b to be calculated when tables to base a are given. In particular, by taking $y = a$ we get, since $\log_a a = 1$, $\log_b(a) = 1/(\log_a(b))$.

Note that the *positions* of a, b and y in (A.60) are the same as in the formula $\dfrac{y}{a} = \dfrac{y}{b} \cdot \dfrac{b}{a}$. This can be a helpful reminder to (A.60).

A.4.5   The Law of Continuous Growth

The exponential function to base e is best approached from a consideration of the growth of compound interest. Capital X_0, invested for one year at a rate r becomes at the end of the first year,

$$X_1 = X_0 + rX_0 = X_0(1+r) \qquad\qquad (A.61)$$

At the end of the second year, X_1 becomes $X_2 = X_1 + rX_1 = X_0(1+r)^2$. And similarly, at the end of m years, X_0 has grown to $X_m = X_0(1+r)^m$. This is known as the law of compound interest.

Take now the formula (A.61) for the capital at the end of one year, and suppose that instead of rate r per twelve months, the interest is compounded at rate r/2 per six months. At the end of one year we get, instead of (A.61), the value $X_0(1+r/2)^2$. Similarly, if the interest is compounded n times per year at a rate r/n, the capital at the end of the year is $X_1 = X_0(1+r/n)^n$. For *continuous* growth the interval for adding on the earned interest becomes indefinitely small, and n approaches infinity. Thus, for continuous growth,

$$X_1 = \lim_{n \to \infty} X_0 (1+r/n)^n. \tag{A.62}$$

It can be shown that this limit does exist, and that it is finite. If we write $n = rN$ in (A.62) we get

$$X_1 = \lim_{N \to \infty} X_0 (1+1/N)^{rN} = \lim_{N \to \infty} X_0 [(1+1/N)^N]^r \tag{A.63}$$

by (A.56).

The quantity $\lim_{N \to \infty} (1+1/N)^N$ is denoted by e, so that the expression in (A.63) is $X_0 e^r$. Since $(1+1/n)^n$ with $n = 1$ is equal to 2, clearly $e > 2$. With $n = 2$ we get $(1+\frac{1}{2})^2 = 2.25$, and with $n = 3$ we get $(1+1/3)^3 = 2.37$, and so on. The expression increases as n increases, but settles down to the value $2.718\ldots$.

We can use the *binomial theorem* to evaluate e. For all values of x and N we have

$$(1+x)^N = 1 + Nx + \frac{N(N-1)}{1.2} x^2 + \frac{N(N-1)(N-2)}{1.2.3} x^3 + \ldots \tag{A.64}$$

With $x = 1/N$ this gives

$$(1+1/N)^N = 1 + 1 + \frac{(1-1/N)}{1.2} + \frac{(1-1/N)(1-2/N)}{1.2.3} + \ldots$$

Taking N very large, so that 1/N, 2/N etc are negligible, we get

$$e = 1 + 1 + 1/1.2 + 1/1.2.3 + \ldots. \tag{A.65}$$

from which e can be calculated.

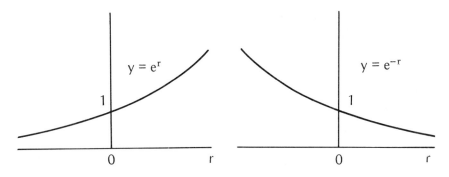

A.20   The Curves $y = e^r$ and $y = e^{-r}$

The product of the first n natural numbers, which occurs here in the denominators of the various terms, is known as *factorial n,* and is written n! Thus, $2! = 1 \cdot 2 = 2$, $3! = 1 \cdot 2 \cdot 3 = 6$, $4! = 1 \cdot 2 \cdot 3 \cdot 4 = 24$, etc. Clearly, for $n > 1$.

$$n! = n \cdot (n-1)! \qquad\qquad\qquad\qquad (A.66)$$

Taking $n = 1$ in this result would give $0! = 1$ as a *self-consistent definition* of 0!, thus extending the range of (A.66) to $n = 1$.

From (A.67) and (A.63), $\underset{n \to \infty}{\text{Lim}} (1+r/n)^n = e^r$, and by using (A.64) with $x = r/N$ we see that

$$e^r \approx 1 + r + r^2/2! + r^3/3! + \ldots \qquad\qquad (A.67)$$

In particular, for small values of r we have the approximation

$$e^r \approx 1 + r \qquad\qquad\qquad\qquad\qquad (A.68)$$

Thus the graph of $y = e^r$, near the origin, looks like a straight line through the point $y = 1$, and with unit slope. It is shown in Figure A.20. The graph of $y = e^{-r}$ is the mirror image of the graph of $y = e^r$ reflected in the line $r = 0$.

The expression $y = e^{-r}$ at $r = 1$ drops to a value $e^{-1} = 0.36788\ldots$ and is a commonly used level at which to assess the decay of an exponential process. Because e enters "naturally" in all proportionate growth and decay calculations, it occurs in most mathematical formulas for these phenomena. As already mentioned, logarithms to base e are called natural logarithms, and are denoted by $\log_e$ or $\ln$.

A.5 DECIBELS AND NEPERS

A.5.1 The Decibel

Power ratios, for example between a transmitter power and receiver power, may involve very large numbers. By taking the logarithm a more manageable number is obtained. Moreover, if the power reduction is due to a number of separable factors, the logarithms of each part can be added to get the final figure. This may make it easier to visualize the relative significance of the various contributions. In these calculations the logarithms are usually taken to base ten.

If P_1 and P_2 are two powers, then $\log_{10}(P_1/P_2)$ is a measure of their ratio in a unit known as a bel. The conventionally used unit is a tenth of a bel, or decibel, symbolized by dB. Thus

$$10\log_{10}(P_1/P_2) = \text{measure of power ratio in decibels} \qquad (A.69)$$

As an example, a path loss of 120 dB corresponds to a power ratio given by $10\log(P_1/P_2) = -120$, or $P_1 = P_2 \times 10^{-12}$.

Since power is proportional to fieldstrength (or voltage or current) *squared,* the decibel measure for these quantities is obtained by taking *twenty* times the logarithm. Thus a path loss of 120 dB corresponds to a fieldstrength ratio given by

$$20\log(E_1/E_2) = -120, \text{ or } E_1 = E_2 \times 10^{-6}.$$

Decibels refer to power *ratios.* Sometimes the *reference level* may be stated explicitly, sometimes it may be implied. The symbol dBW means decibels relative to a power level of 1 watt. The symbol dBmW, often contracted to dBm, means decibels relative to one mW. Decibels of sound are relative to the threshold power level of hearing for the normal ear.

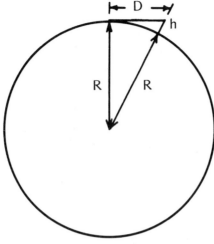

A.21   Horizon Distance

A.5.2   The Neper

Many physical quantities grow or attenuate exponentially. Thus, the decay of the electric field through a lossy material would be given by

$$E_x = F_o e^{-\alpha x} \tag{A.70}$$

where E_x = fieldstrength at depth x (x in meters), E_o = field-strength at the surface (x = 0), and α = attenuation coefficient.

The units for α are neper per meter. At one meter depth $E_x / E_o = e^{-1}$. The decibel measure of this is $20 \log_{10} e^{-1} = -8.686$. This attenuation is also by definition, one neper. Thus 1 neper = 8.686 decibels.

The depth at which the field has attenuated by one neper is known as the *skin depth*, sometimes denoted by δ. Thus $\alpha\delta = 1$ or $\delta = 1/\alpha$: the skin depth is the inverse of the attenuation constant (in nepers per meter).

A.6   GEOMETRICAL FEATURES

A.6.1   Horizon Distance

Figure A.21 depicts a position at height h above the surface of the earth, radius R, with distance D to the horizon. By applying Pythagoras's theorem we get

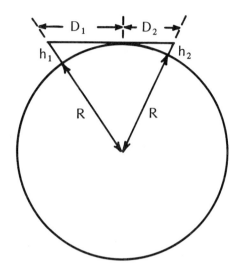

A.22   Grazing Range

$$(h+R)^2 = R^2 + D^2 \text{ or } D^2 = 2h(R+h) \tag{A.71}$$

Usually, h is completely negligible compared to R, and (A.71) is approximated by

$$D = (2Rh)^{\frac{1}{2}} \tag{A.72}$$

This relation can also be solved for h, giving the height necessary for a given horizon distance

$$h = D^2/2R \tag{A.73}$$

In these formulas, all distances should be in the same units, such as meters. However, in engineering applications it is often the case that D is given in miles and h in feet. The value of R is 3960 miles, which number conveniently happens to be 3/4 x 5280; and 5280 is the number of feet in a mile. Hence (A.72) can be re-written

$$D = (3h/2)^{\frac{1}{2}} \tag{A.74}$$

with D in miles and h in feet.

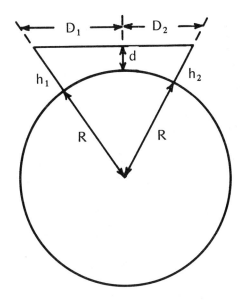

A.23   Clearance Range

When two observation heights are given, as in Figure A.22, the separate line-of-sight distances D_1 and D_2 must each be calculated from the heights h_1 and h_2, using (A.72). The maximum separation distance D is then the sum of the two grazing incidence distances:

$$D = D_1 + D_2 = (2Rh_1)^{1/2} + (2Rh_2)^{1/2} \qquad\qquad (A.75)$$

If a clearance d, as shown in Figure A.23, is needed, it must be subtracted from *each* height before doing the calculation indicated in (A.75)

A.6.2 Atmospheric Refraction of Radio Waves

The preceding results hold for calculations of the *optical* horizon. The radio horizon for line-of-sight links is somewhat different due to atmospheric bending of the rays. This effect is more pronounced for radio waves due to the relatively higher dielectric constant of water vapor at the radio frequencies. It is the gradient of the refractive index of the air, due mainly to variations with height of the water vapor content, that causes the radio waves to be refracted and to follow curved rather than straight lines. It happens that the effect can be allowed for by using a *fictitious* or *effective* earth radius, R_{eff}, in the preceding formulas, where R_{eff} differs from the actual earth radius R by a curvature factor K such that one can write

$$R_{eff} = KR \tag{A.76}$$

The value of K depends on atmospheric conditions, but its *average* value is about 4/3. Hence the radio horizon, *under average conditions* is $(4/3)^{1/2}$, or about 15%, greater than the optical horizon. However, K changes greatly, depending on conditions, and allowance for its range must be made in any practical calculations of clearance or tower height.

A.6.3 Triangles

By inspection of the triangle in Figure A.24 and comparison with the surrounding rectangle it is seen that the area is half that of the rectangle; i.e., the area of a triangle is half its base times its height. From the figure, $h/a = \sin B$ where B is the angle at the apex B. Hence

$$\text{Area} = \tfrac{1}{2}\, ac \sin B \tag{A.77}$$

By symmetry the area also equals $\tfrac{1}{2}\, bc \sin A$ or $\tfrac{1}{2}\, ab \sin C$. Equating these results gives, on re-arrangement

$$\frac{a}{\sin A} = \frac{b}{\sin B} = \frac{c}{\sin C} \tag{A.78}$$

This equation can be used for calculating the remaining sides and angles if two sides and the included angle, *or* two angles and the adjacent side are known. Note that *always* $A + B + C = 180°$ for a triangle. The largest angle is opposite the largest side. Equation (A.78) gives the sine of the angle, leading to *two* possible values between 0 and 180°. The correct value has to be selected from a consideration of the above information. The above area formula

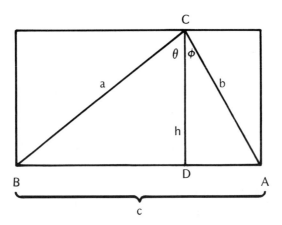

A.24   Triangle and Circumscribing Rectangle

can be used to prove the addition formula for the sine function, as displayed in (A.18). In Figure A.24 $C = \theta + \phi$ and the area formula can be written

Area = ½ ab sin ($\theta + \phi$)

It also equals area BCD + area CDA = ½ ah sin θ + ½ bh sin ϕ. But h = a cos θ and it also equals b cos ϕ. Hence area BCD + area CDA = ½ ab cos ϕ sin θ + ½ ab cos θ sin ϕ. Equating the two results and cancelling ½ ab gives the addition formula for the sine function:

sin ($\theta + \phi$) = sin θ cos ϕ + cos θ sin ϕ

An extension of Pythagoras's theorem to non-right angle triangles is as follows. From triangle BDC, since BD = a cos B,

$$a^2 = h^2 + (a \cos B)^2 \tag{A.79}$$

Similarly, from triangle CDA, since DA = c – BD = c – a cos B

$$b^2 = h^2 + (c - a \cos B)^2 \tag{A.80}$$

Subtracting (A.80) from (A.79) gives

$$a^2 - b^2 = (a \cos B)^2 - (c - a \cos B)^2 = 2 \, ac \cos B - c^2$$

Hence

$$b^2 = a^2 + c^2 - 2\,ac\cos B \tag{A.81}$$

This equation, and the two similar ones obtained by cyclic inter-change of the letters, also enables the angles of a triangle to be calculated when all three sides are given. Since $\cos B = 0$ when $B = 90°$, (A.81) reduces to the usual Pythagorean form for a right-angled triangle.

A.6.4 Areas and Volumes

A few useful results will be quoted here. Most of them require the calculus for their proof, and a further discussion will be given in section A.9.

Areas
a) Circle, radius R. Area $= \pi R^2$
b) Circular sector, radius R, angle of sector α *(in radians)*,
   Area $= \alpha R^2/2$
c) Sphere, radius R. Area of surface $= 4\pi R^2$
d) Cylinder, radius R, height h. The lateral surface area can be calculated by "unrolling" the surface, and gives area $= 2\pi Rh$
e) Cone, base radius R, *slant* height h. Again the lateral surface area can be calculated by unrolling the surface, and gives Area $= \pi Rh$

Volumes
a) Sphere, radius R. Volume $= 4\pi R^3/3$
b) Cylinder, radius R, height h. Volume $= \pi R^2 h$
c) Cone, base radius R, *vertical* height h. Volume $= \pi R^2 h/3$

A.7 COMPLEX NUMBERS

A.7.1 The Quantity j

Since $2\times2 = 4$ and $(-2)\times(-2) = 4$ also, the square root of 4 can be either plus or minus two. The square root of *minus* four can therefore be neither of these, and involves a new quantity denoted by j (or sometimes i) which is equal to the square root of minus one;

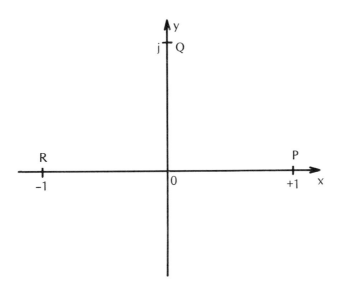

A.25   The Argand Diagram

$j^2 = -1$ or $j = \sqrt{-1} = (-1)^{\frac{1}{2}}$ (A.82)

In terms of it we can write $\sqrt{-4} = \sqrt{-1 \times 4} = \sqrt{-1} \times \sqrt{4} = \pm j2$.

The *real* numbers can be plotted along a single axis, such as the x-axis. *Imaginary* numbers can be plotted along a perpendicular axis, such as the y-axis, as shown in Figure A.25.

To see that this representation is *self-consistent,* we note that multiplying 1 by j rotates the position of the point P, representing 1, to the point Q, representing j, a counter-clockwise rotation of 90°. A further multiplication by j, or a further counter-clockwise rotation by 90° takes P to the point R represented by –1. Thus $j \times j \times 1 = -1$ or $j^2 = -1$.

A.7.2 Complex Numbers

The representation of imaginary numbers in this way is a permissible, self-consistent visual aid, and is called the *Argand diagram.* Points off the axes are called *complex numbers.* Thus both real and imaginary numbers are special cases of complex numbers. A complex number is of the form (a+jb), and is represented by the point x = a, y = b. Thus (a+jb) and (a, b) are equivalent representa-

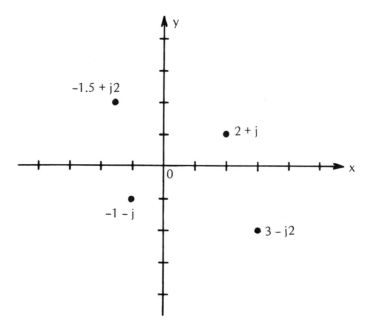

A.26   Illustration of Complex Numbers on the Argand Diagram

tions, but, in the former, the two real numbers a and b are *com-bined together* to form a *single* complex number. Four examples are shown in Figure A.26.

If Z is the complex number a+jb, then we write

Re Z = a

Im X = b (not jb)

(A.83)

where Re and Im stand, respectively, for "the real part of" and "the imaginary part of".

When adding complex numbers, the real parts are added and the imaginary parts are added; $(2+j4) + (1-j2) = (2+1) + j (4-2) = 3+j2$. When multiplying, the expressions are multipled as if they were ordinary numbers, and use is made of $j^2 = -1$. Thus,

$$(5+j7) \cdot (2+j3) = 5×2 + j7×2 + 5×j3 + j^2 7×3$$
$$= 10 + j14 + j15 -21$$
$$= -11 + j29$$

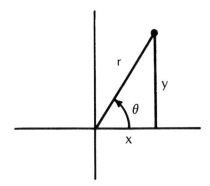

A.27   Polar Representation

Division is more complicated, and is achieved by multiplying both numerator and denominator by the *complex conjugate* of the denominator; i.e. the expression in the denominator *with the sign of j changed.* Thus,

$$\frac{50 + j10}{3 - j4} = \frac{(50+j10)(3+j4)}{(3-j4)(3+j4)} \quad \frac{50\times3 + j10\times3 + j50\times4 + j^2 10\times4}{3\times3 + j3\times4 - j4\times3 - j^2 4\times4}$$

$$= \frac{(150-40) + j(30+200)}{9 + 16} = \frac{110 + j230}{25} = 4.4 + j9.2$$

Note that, since $-j^2 = +1$ the *denominator* will *always* consist of a *sum* of two squares.

A.7.3 Polar Representation

A point can also be represented by the distance r (a positive quantity) from the origin, and the angle θ of a counter-clockwise rotation. Thus we can write x+jy, or (x,y) in the form (r, θ). *This* form does *not* combine r and θ into a single quantity. This will be done later. Meanwhile, we see from Figure A.27 that

$$x = r\cos\theta$$

$$y = r\sin\theta \tag{A.84}$$

This equation gives x and y in terms of r and θ. Correspondingly,

$$r = + \sqrt{x^2 + y^2}$$

$$\theta = \tan^{-1}(y/x)$$

(A.85)

Since the tangent function is periodic with period 180°, the value of θ obtained from (A.85) *has to be checked for the correct quadrant.*

A.7.4 The Complex Exponential

The combination $f(\theta) = \cos\theta + j\sin\theta$ is *very* important in subsequent developments. By direct multiplication, we see that

$$f(\theta) \times f(\phi) = (\cos\theta + j\sin\theta)(\cos\phi + j\sin\phi)$$

$$= (\cos\theta\cos\phi - \sin\theta\sin\phi) + j(\sin\theta\cos\phi + \cos\theta\sin\phi)$$

$$= \cos(\theta+\phi) + j\sin(\theta+\phi)$$

(on using the formulas for cosine and sine of a sum of two angles)

$$= f(\theta+\phi), \text{ from the definition of f.}$$

Since $e^{K\theta} \times e^{K\phi} = e^{K(\theta+\phi)}$, where K is any constant we see that $f(\theta)$ and $e^{K\theta}$ both satisfy the same functional form, namely, the law of indices. This makes it plausible that there is a value of K for which $\cos\theta + j\sin\theta = e^{K\theta}$. To see what value this is we note that, for small θ, $\cos\theta \approx 1$ and $\sin\theta \approx \theta$, while $e^{K\theta} \approx 1+K\theta$. Hence $\cos\theta + j\sin\theta \approx 1+j\theta \approx 1+K\theta$, giving K=j. Hence

$$\cos\theta + j\sin\theta = e^{j\theta}$$

(A.86)

(This is not, of course, a rigorous mathematical proof)

We are now in a position to combine (r, θ) into a single complex form. From (A.84)

$$x+jy = r\cos\theta + jr\sin\theta = r(\cos\theta + j\sin\theta) = re^{j\theta}$$

(A.87)

Thus, if two complex numbers can be represented in polar form, they can be multiplied by multiplying their amplitudes and adding the angles.

$$(r_1 e^{j\theta_1}) \times (r_2 e^{j\theta_2}) = r_1 r_2 e^{j(\theta_1 + \theta_2)} \tag{A.88}$$

A.7.5 Square Roots

Equation (A.88) can be used for finding the square roots of complex numbers. Thus, if

$$(R\ e^{j\theta})^{1/2} = re^{j\phi}, \text{ then } (re^{j\phi})^2 = R\ e^{j\theta} .$$

Using (A.88) we get $r^2 e^{2j\phi} = R\ e^{j\theta}$ so that $r^2 = R$ and $2\phi = \theta$. Hence

$$(R\ e^{j\theta})^{1/2} = R^{1/2} e^{j\theta/2} \tag{A.89}$$

Thus one takes the square root of the amplitude and halves the angle. Since the angle could also have been increased by 360° without altering the original quantity, (A.89) could also be interpreted as involving $e^{j(\theta+360°)/2} = e^{j\theta/2} e^{j180°}$. But $e^{j180°} = \cos 180° + j \sin 180° = -1$. Hence another square root is $-R^{1/2} e^{j\theta/2}$. This gives the familiar ± for square roots. In a physical situation, the physical requirements may determine which root to use. They are both equally valid mathematically.

A.7.6 The Complex Propagation Constant

Since $\cos \theta + j \sin \theta = e^{j\theta}$ we can put

$$\cos \theta = \mathrm{Re}\left(e^{j\theta}\right) \tag{A.90}$$

This important relation enables us to replace trigonometric functions by exponentials. Thus,

$$E_o \cos (\omega t + \phi - 2\pi x/\lambda) = E_o \mathrm{Re}\ e^{j(\omega t + \phi - 2\pi x/\lambda)}$$

$$= \mathrm{Re}\ \left\{ E_o e^{j(\omega t + \phi)} \times e^{-j2\pi x/\lambda} \right\} .$$

In this way the time and distance-dependent terms can be *separated.*

If a wave is attenuated with attenuation constant α, then if E_x is the value of E at depth x we can write, since $E_o e^{-\alpha x}$ is real,

$$E_x = E_o e^{-\alpha x} \cos (\omega t + \phi - 2\pi x/\lambda) \qquad (A.91)$$

$$= \text{Re} \left\{ E_o e^{j(\omega t + \phi)} \times e^{-(\alpha + j 2\pi/\lambda)x} \right\} \qquad (A.92)$$

It is seen that the attenuation and distance-dependent phase terms have been *combined* and also separated from the time-varying term. The quantity $e^{-(\alpha + j 2\pi/\lambda)x}$ contains important valuable information on the behavior of the wave. The *interpretation* is via (A.91) and (A.92), but it is convenient to work with this term isolated.

The quantity $\alpha + j 2\pi/\lambda$ is sometimes denoted by γ and is called the *complex propagation constant.* The *real part* gives the *attenuation,* and the *imaginary part* gives the *velocity of the wave.* Since $\cos(\omega t - 2\pi x/\lambda) = \cos [\omega (t-x/v)]$ we get $\omega/v = 2\pi/\lambda$ or

$$v = \omega/(2\pi/\lambda) = \omega/(\text{Im } \gamma) \qquad (A.93)$$

which, together with

$$\alpha = \text{Re} (\gamma) \qquad (A.94)$$

give important wave properties in terms of γ.

The quantity $2\pi/\lambda$ is sometimes denoted by β or by k, so that we can write

$$\gamma = \alpha + j\beta = \alpha + jk \qquad (A.95)$$

Equation (A.93) can thus also be written $v = \omega/k$.

A.7.7   The Complex Permittivity

Currents in a material involve two aspects, in general. *Conduction Currents* involve flow of electrons, and are associated with the conductivity σ of the material.

Displacement Currents involve *charges* and are associated with the dielectric constant ϵ of the material. Suppose a block of material sandwiched between two parallel plates of area A and separation D. If C is the capacitance of the plates then C = ϵA/D, and the charge for a voltage V is Q = VC = ϵV A/D. The displacement current is the rate of change of charge. If V is sinusoidal, and given by V_0 cos ωt, its rate of change (see later) is $-\omega$V × sin ωt, so that the displacement current is $(V_0 A/D)$ [$-\omega\epsilon$ sin ωt].

The conduction current is V/R = V/(A/σD) = $(V_0 A/D)$ [σ cos ωt]. Hence the *total current* involves σ cos ωt − $\omega\epsilon$ sin ωt = Re[$e^{j\omega t}$ (σ + j$\omega\epsilon$)]. Thus, using complex number notation, the effect of conductivity can be taken into account by augmenting the term j$\omega\epsilon$ by σ, or alternatively, augmenting ϵ by σ/jω. The quantity

$$\epsilon_c = \epsilon + \sigma/j\omega \tag{A.96}$$

is called the *complex permittivity.* It can also be written

$$\epsilon_c = \epsilon_0(\epsilon_r + \sigma/j\omega\epsilon_0) = \epsilon_0(\epsilon_r - j\, 60\lambda_0\sigma) \tag{A.97}$$

where λ_0 is the free-space wavelength and the factor 60 (ohms) comes as $1/\omega\lambda_0\epsilon_0 = 1/[2\pi f\lambda_0/(4\pi\cdot9\cdot10^9)] = 1/[2\pi\cdot3\cdot10^8/(4\pi\cdot9\cdot10^9)] = 1/[1/60] = 60$.

In free space the complex propagation constant is simply j$2\pi/\lambda_0$, but in a medium whose properties are described by ϵ_c, the value of γ is j$(2\pi/\lambda_0)(\epsilon_c/\epsilon_0)^{\frac{1}{2}}$. Hence

$$\alpha + jk = j(2\pi/\lambda_0)(\epsilon_c/\epsilon_0)^{\frac{1}{2}} \tag{A.98}$$

so that

$$k = \text{Re}[(2\pi/\lambda_0)(\epsilon_c/\epsilon_0)^{\frac{1}{2}}] \tag{A.99}$$

$$\alpha = -\,\text{Im}[(2\pi/\lambda_0)(\epsilon_c/\epsilon_0)^{\frac{1}{2}}] \tag{A.100}$$

The last two equations are extremely important for determining velocity and attenuation in a material.

A.7.8   Radio Propagation in Water

As an example of the application of the above results, let us consider the evaluation of the velocity and attenuation of a radio wave in fresh water at a frequency of 300 kHz.

The electrical constant are $\epsilon_r = 80$ and $\sigma = 1$m mho/m. Hence,

$$
\begin{aligned}
\epsilon_c/\epsilon_o &= 80 - j60(3 \cdot 10^8 / 3 \cdot 10^5) \cdot 10^{-3} \\
&= 80 - j60 \\
&= (80^2 + 60^2)^{\frac{1}{2}} \, e^{-j \tan^{-1} (60/80)} \\
&= 100 \, e^{-j 36.8^\circ}
\end{aligned}
$$

$$
\begin{aligned}
(\epsilon_c/\epsilon_o)^{\frac{1}{2}} &= 100^{\frac{1}{2}} \, e^{-j(36.8/2)^\circ} \\
&= 10 \, e^{-j18.4^\circ} = 10[\cos 18.4^\circ - j \sin 18.4^\circ] \\
&= 9.49 - j3.16
\end{aligned}
$$

This gives

$$ \gamma \qquad = (6.28/1000)(9.49 - j3.16) = 0.06 - j0.02 $$

Hence, from (A.99) and (A.100),

$$ k \qquad = 0.06 \text{ radian/m} $$
$$ \alpha \qquad = 0.02 \text{ neper/m} = 0.17 \text{ dB/m} $$

The velocity of the wave is

$$ v = \omega/k = (6.28 \cdot 3 \cdot 10^5)/0.06 = 3.16 \; 10^7 \, \text{m/s} $$

This reduction in velocity relative to free space is largely due to the dielectric constant. However, for appreciably higher conductivities or lower frequencies, the term $60 \lambda_o \sigma$ becomes dominating over ϵ_r, and it is the former that then mainly determines the properties.

Since, for sea water, $\epsilon_r = 81$ and $\sigma = 4$ mho/m, the two terms are equal when $81 = 60 \times 4 \times \lambda_o$ or $\lambda_o = 34$ cm. Thus, for frequencies less than 1 GHz it is the conductivity that mainly determines both the attenuation *and* the velocity in the sea water. The usual expression for refractive index $n = \epsilon_r^{\frac{1}{2}}$ has to be modified to $n = \mathrm{Re}(\epsilon_c/\epsilon_o)^{\frac{1}{2}}$

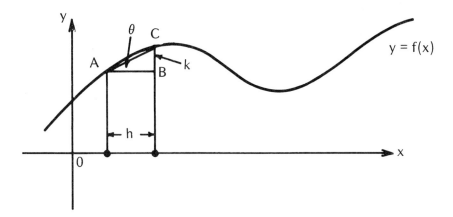

A.28   Illustration of Differentiation

and, in the case of sea water below about 1 GHz, the dielectric constant plays little part. Such a material is said to possess metallic properties, since, in a metal, it is *only* the conductivity which determines the response.

A.8 DIFFERENTIAL AND INTEGRAL CALCULUS

A.8.1 Differentiation

The process of differentiation is concerned with finding the slope or gradient of a curve at a point, defined as the tangent of the angle of slope. It is obtained by approximating an arc of the curve by a chord. As the chord and arc become smaller and smaller they approach equality; in the limit of a vanishingly small arc, the process yields the slope of the curve.

Figure A.28 shows a triangle ABC in which AC is the chord joining the points A and C on the curve. The side AB is denoted by h and BC by k. The tangent of the angle BAC is therefore k/h, and we need to find the limiting value of this as k and h approach zero. If the equation of the curve is y = f(x) then at the point A we have simple y = f(x) where x and y are the coordinates of A. The coordinates of C are x + h and y + k, and these are related by y + k = f(x + h) since C is also on the curve. Hence, from the triangle ABC we get

$$\tan \theta = \frac{BC}{AB} = \frac{k}{h} = \frac{f(x + h) - f(x)}{h} \qquad (A.101)$$

This is true for any sized triangle ABC. If we now take the limit for vanishingly small h, θ becomes the angle of slope of the curve at A and we get

$$\tan \theta = \lim_{h \to 0} \frac{f(x + h) - f(x)}{h} \qquad (A.102)$$

The limiting form of h is usually written dx meaning "a differential increment in x." The d is not to be treated as a multiplier and cannot be separated from the x; dx is an entity. Similarly k is written as dy. Hence (A.101) and (A.102) can be written

$$\tan \theta = \frac{dy}{dx} = \lim_{dx \to 0} \frac{f(x + dx) - f(x)}{dx} \qquad (A.103)$$

A few important exmaples follow.

A.8.2   Examples of Differentiation

By following the process indicated in (A.103) the following results are deduced form first principles:

a) $y = Cx^n$, where C and n are constants
    Then $y + dy = C(x + dx)^n$
    $$= C(x^n + nx^{n-1} dx + \ldots)$$

where the binomial theorem has been used to expand $(x + dx)^n$, and higher order terms in (dx) have not been explicitly shown. The next step is to subtract $y = Cx^n$ from both sides, divide by dx, and take the limit as $dx \to 0$. A term such as $(dx)^2$ will thus be reduced to dx and will vanish in the limit, as will still higher-order terms. Only the term initially proportional to dx will survive the process. Hence we get $dy = C(x^n + nx^{n-1} dx + \ldots) - Cx^n$
$$\frac{dy}{dx} = Cnx^{n-1} + \text{terms vanishing with dx.}$$

On taking the limit as $dx \to 0$ this gives

$$\frac{dy}{dx} = Cnx^{n-1} \qquad (A.104)$$

This method can be applied to a sum of terms, and leads to the formulation that differentiation of a sum equals the sum of the individual differentiated terms.

Note that in the special case n = 0, y = C, a constant and (A.104) then gives dy/dx = 0.

b) $y = Ce^{ax}$; C and a are constants.

$y + dy = Ce^{a(x+dx)} = Ce^{ax} e^{adx}$ from (A.54)

$dy = Ce^{ax} e^{adx} - Ce^{ax}$

$= Ce^{ax} (e^{adx} - 1)$

$= Ce^{ax} (adx + . . .)$ from (A.67)

Hence

$$\frac{dy}{dx} = aCe^{ax} \quad \text{as } dx \to 0 \tag{A.105}$$

c) $y = \sin(\omega x + \theta)$; A, ω and θ are constants.

$y + dy = A \sin(\omega x + \theta + \omega dx)$

$= A \sin(\omega x + \theta) \cos(\omega dx) + A \cos(\omega x + \theta) \sin(\omega dx)$
from (A.18)

$dy = A \sin(\omega x + \theta) [\cos(\omega dx) - 1] + A \cos(\omega x + \theta) \sin(\omega dx)$

Dividing by dx and using the results from section A.2.10 that
$\underset{\phi \to 0}{\text{Lim}} (1 - \cos \phi)/\phi \to 0$ and $\underset{\phi \to 0}{\text{Lim}} \sin(\phi)/\phi = 1$ we get

$$\frac{dy}{dx} = \underset{dx \to 0}{\text{Lim}} A \cos(\omega x + \theta) \frac{\sin(\omega dx)}{\omega dx} \cdot \omega = A\omega \cos(\omega x + \theta) \tag{A.106}$$

In a similar way, or by increasing θ in (A.106) by $\pi/2$ it can be shown that

d) $y = A \cos(\omega x + \theta)$

leads to

$$\frac{dy}{dx} = -A \omega \sin(\omega x + \theta) \tag{A.107}$$

e) $y = \log_e x$

Then $x = e^y$, and by taking C = a = 1 in (A.105) and interchanging the roles of x and y, we get

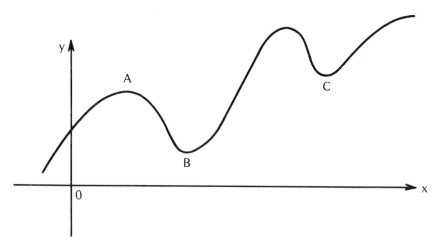

A.29   Maxima and Minima

$$\frac{dx}{dy} = e^y$$

$$\qquad = x, \text{ since } y = \log_e x$$

Inverting this equation gives the important result

$$\frac{dy}{dx} = \frac{1}{x} \qquad\qquad\qquad\qquad\qquad \text{(A.108)}$$

A.8.3 Maxima and Minima

Figure A.29 shows a curve with a maximum at A and a minimum at B. Both positions are known as *turning points,* and at a turning point the gradient is zero. Hence the equation dy/dx = 0 *locates* the values of x for which turning points occur. To find the value of y corresponding to a value of x, i.e. to actually determine the height of the maximum or minimum, the value of x has to be inserted in the equation y = f(x) for the curve. It is usually clear by inspection whether a particular turning point is a maximum but in curves with several turning points it is possible to have a minimum which is actually higher than a maximum. This would be so for the position C, as compared to A, in Figure A.29.

As an example take the curve y = cos x for which dy/dx = -sin x. Since sin x = 0 when x = nπ where n is any integer, including 0, we see that the cosine function has turning points at these posi-

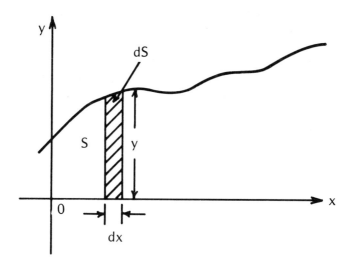

A.30   Area Under a Curve

tions. The value of cos x at one of these positions is $\cos(n\pi) = (-1)^n$. Hence the turning points are alternatively positive and negative, and are numerically of unit magnitude.

A.8.4   Integration

Perhaps the most suitable approach to defining integration is in the determination of the area under a curve. In Figure A.30 the area is denoted by S up to a particular value of x. If x is increased to x + dx, the area increases by an amount dS where dS is represented by the (approximate) rectangle of height y and width dx. As the limit dx → 0 is taken the approximation (due to the curved upper edge of the rectangle) becomes exact, and we get dS = ydx or

$dS/dx = y$ (A.109)

Now S can be thought of as a summation of many strips, each of the form ydx, or $S = \Sigma ydx$ where the summation covers the area of interest. In the limit as dx → 0 the summation sign Σ is replaced by the integral sign $\int$ and the formula for S becomes

$S = \int ydx$ (A.110)

This result should be compared with (A.109), and it shows that if we know the expression which, when differentiated, gives the function y, then this expression represents the area S. Thus integration may be thought of as the inverse process to differentiation, and the results of section A.8.2 may be used to set up a table of results for integration. Note that since the differentiation of a constant gives zero, an arbitrary constant should *always* be added when an integration is performed. Its value can be determined from physical considerations in an actual problem. Sometimes, but not always, its value is zero.

y	Cx^n	$\dfrac{C}{x}$	Ae^{ax}	$A\sin(\omega x+\theta)$	$A\cos(\omega x+\theta)$
$\int y\,dx$	$\dfrac{Cx^{n+1}}{n+1}$	$C\log_e x$	$\dfrac{Ae^{ax}}{a}$	$-\dfrac{A}{\omega}\cos(\omega x+\theta)$	$\dfrac{A}{\omega}\sin(\omega x+\theta)$

Table A.1.   Examples of Integration

In the above table the constant of integration has not been shown explicitly. Note that the integration of x^{-1} is a special case that does not come under the first column but gives $\log_e x$ as in column 2.

A.8.5 Integration Limits

The expression in (A.110) is known as an *indefinite* integral, since the limits for determining the area S are not specified. If the lower limit is L_1 and the upper L_2, then (A.110) would be written

$$S = \int_{L_1}^{L_2} y\,dx \qquad\qquad (A.111)$$

and gives the area between the values L_1 and L_2 for x. If the function which, when differentiated, gives y, is denoted by $F(x)$ then (A.111) gives

$$S = F(L_2) - F(L_1) \qquad\qquad (A.112)$$

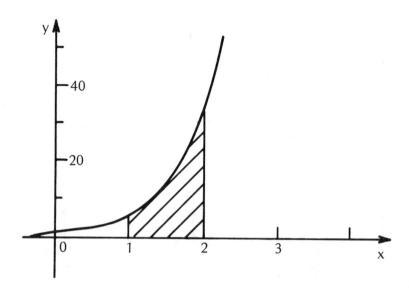

A.31   Area Under the Curve $y = 2 + 4x^3$ between x = 1 and 2

Note that a) the value at the lower limit L_1 is *subtracted* from the
value at the upper limit L_2; and b) it is not necessary to show explicitly the integration constant because it cancels out when the
two functions are subtracted. What has really happened is that
$F(L_1)$ is doing duty as the integration constant. This may be seen
as follows: If we return to (A.110) we could write $S = F(x) + C$,
and hence $S(L_2) = F(L_2) + C$. What value is needed for C? Clearly
$S = 0$ when $X = L_1$ so $S(L_1) = 0 = F(L_1) + C$, or $C = -F(L_1)$.
Hence $S(L_2) = F(L_2) - F(L_1)$, in agreement with (A.112). As an
example, the area under the cubic curve $y = 2 + 4x^3$ between
x = 1 and x = 2, shown in Figure A.31, is obtained from

$$S = \int_1^2 (2 + 4x^3)dx$$

$$= (2x + x^4)\Big|_1^2 \quad \text{from table 1}$$

$$= (4 + 16) - (2 + 1) = 17 \text{ units of area.}$$

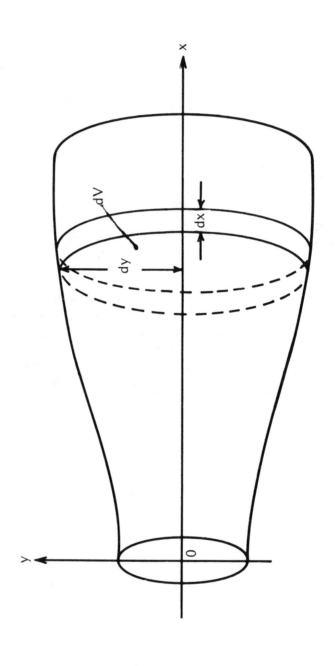

A.32   Volume of Revolution

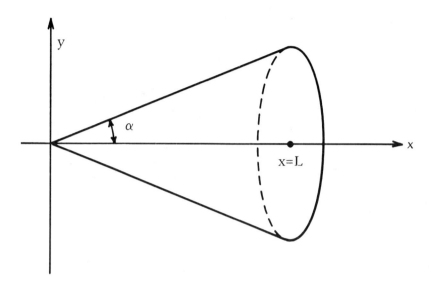

A.33   Cone of Semi-angle α

A.8.6   Volumes of Revolution

In Figure A.32 the differential of volume, dV, is a disc of radius y
and thickness dx. Hence its value is πy^2 dx, and we get

$$V = \int \pi y^2 \, dx \qquad\qquad\qquad (A.113)$$

a) Volume of a cone. The equation for the cone radius, shown in
Figure A.33, is $y = x \tan \alpha$, where α is the half-angle at the cone
apex. Hence

$$V = \int_0^L \pi x^2 \tan^2 \alpha \cdot dx = (\pi/3)x^3 \tan^2 \alpha \Big|_0^L = (\pi/3)L^3 \tan^2 \alpha$$

But $L \tan \alpha$ is the radius at the cone base, whose area is accordingly $A = \pi (L \tan\alpha)^2$. Hence

$$V = (1/3) \text{ (Area of base) x (perpendicular height)} \qquad (A.114)$$

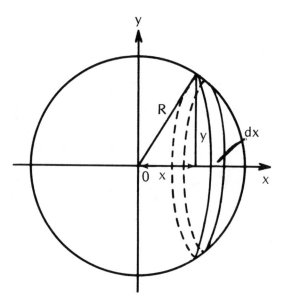

A.34   Calculation of Spherical Volume

b) Volume of a Sphere. The equation of the circle which when rotated determines the spherical shape, is $y^2 = R^2 - x^2$ where R is the radius of the circle or sphere. Hence, as indicated in Figure A.34,

$$V = \int_{-R}^{R} \pi(R^2 - x^2)dx$$

$$= \pi(R^2 x - x^3/3) \Big|_{-R}^{R}$$

$$= \pi(R^3 - R^3/3) - \pi(-R^3 + R^3/3)$$

$$= 4\pi R^3/3 \qquad\qquad\qquad\qquad\qquad (A.115)$$

We can also get the volume of the sphere from its surface, since, if S is the surface, the volume dV of a spherical shell, as shown in Figure A.35, is given by dV = SdR. If we know S as a function of R, we can get V by integrating. Alternatively, knowing V we can calculate S by evaluating dV/dR. Thus, using (A.115) gives

$$S = dV/dR = 4\pi R^2 \qquad\qquad\qquad\qquad (A.116)$$

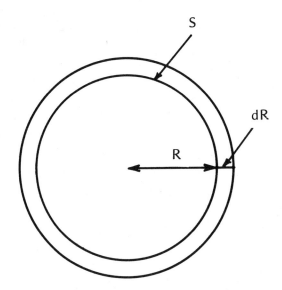

A.35   Calculation of Spherical Surface

A.9   PROBABILITY

A.9.1   Permutations and Combinations

If there are n objects and we wish to select r of them, the *order of selection being important,* we can choose the first object in n ways and, for each of these n ways, we can select the second object in (n – 1) ways, and so on. The total number of ways is n(n – 1)(n – 2) . . . . (n – r + 1). This number is denoted by $^{n}P_{r}$, the number of permutations of n objects r at a time.

$$^{n}P_{r} = n(n - 1) \ldots . . (n - r + 1) = \frac{n\,!}{(n - r)\,!} \qquad (A.117)$$

These r objects can be rearranged among themselves in r! ways, since there is a choice of r for the first position, (r–1) for the second, etc. If we are interested in the number of ways in selecting r objects from n objects, when the *order of selection is of no importance,* then it is clear that (A.117) overestimates this number by a factor r! Denoting by $^{n}C_{r}$ the number of ways of selecting r out of n objects, irrespective of order, then $^{n}C_{r}(r!) = {}^{n}P_{r}$, whence

$$^{n}C_{r} = \frac{n!}{r!\,(n - r)\,!} = \frac{n(n - 1) \ldots . (n - r + 1)}{r!} \qquad (A.118)$$

Note that $^nC_n = 1$, since all n objects can be selected in only one way; i.e. by taking them all. This also follows from the formula, since $0! = 1$.

From (A.118), by replacing r by n – r, we get the relation

$$^nC_r = {}^nC_{n-r} \qquad\qquad (A.119)$$

which expresses a symmetry property of ^nC_r. For reasons that will shortly be apparent, ^nC_r is called a binomial coefficient. Alternative notations are C_r^n and $\binom{n}{r}$, the latter being the most frequent form encountered in mathematical work.

As an example of the application of these results, we shall now prove the binomial theorem for integer powers:

$$(x + y)^n = x^n + {}^nC_1 x^{n-1} y + {}^nC_2 x^{n-2} y^2 + \ldots + y^n \qquad (A.120)$$

We can write $(x + y)^n = (x + y)(x + y)(x + y)\ldots\ldots(x + y)$ (n factors). To get the coefficient of $x^{n-r}y^r$, we have to select (n – r) symbols x and r symbols y when multiplying out the factors. The number of ways of selecting the x's is $^nC_{n-r}$, and the remaining factors automatically give the required number of y's. Hence the required coefficient is $^nC_{n-r}$. From (A.119), $^nC_{n-r} = {}^nC_r$. Hence we have the *binomial theorem* (A.120). As a check, $x^{n-1}y$ comes by selecting one y from any of the n factors, and x's from all of the others. There are n ways of doing this, agreeing with (A.120) since $^nC_1 = n$.

A.9.2 Examples of Permutation Calculations

Not all calculations are simple applications of (A.117) and (A.118). It is often necessary to break a problem up into parts to which the above formulas individually apply, and then combine the results appropriately. This can best be seen from a number of examples.

a) There are 8 boys and 5 girls. How many ways can a group of 6 children be selected when

  i)   there is no restriction on the makeup of the group;
  ii) there must be 4 boys and 2 girls;
  iii) there must be at least 4 boys?

i)   There are 13 children altogether, and the number of ways of selecting any 6 (the *order of selection* being of no consequence) is $^{13}C_6 = \dfrac{13.12.11.10.9.8}{6.5.4.3.2.1} = 1716.$

ii) The number of ways of selecting 4 boys is $^8C_4$ *and* the number of ways of selecting 2 girls is $^5C_2$. These sets are *independent* so that the total number of ways is their product (for *each* of the $^8C_4$ ways of selecting the boys there are $^5C_2$ ways of selecting the girls — hence the numbers are *multiplied*, not added)

$$^8C_4 \times {}^5C_2 = \frac{8.7.6.5}{4.3.2.1} \times \frac{5.4}{2.1} = 700.$$

iii) This can be done by selecting 4 boys and 2 girls, in $^8C_4 \times {}^5C_2$ ways; *or* 5 boys and 1 girl, in $^8C_5 \times {}^5C_1$ ways; *or* 6 boys, in $^8C_6$ ways. These are *mutually exclusive,* so that the total number of ways is their *sum.*

$$^8C_4 \times {}^5C_2 + {}^8C_5 \times {}^5C_1 + {}^8C_6 = 700 + 280 + 28 = 1008$$

Note carefully, that when the sets of numbers are *independent* the total number of ways is the *product* of the sets, but when they are *mutually exclusive* the total number is their sum.

b) How many ways can 4 hands of 13 cards be dealt when i) there are no specifications on the deal; ii) one player (at least) should have all cards of one suit; iii) a specified player should have all the hearts?

i)   The number of ways of dealing 4 hands in a particular order, with the order of the cards in the hands also specified is clearly 52! But each hand can be re-arranged in 13! ways, and the hands re-arranged amongst themselves in 4! ways. Hence the total number of possible deals is $52!/(13!)^4\, 4!$ We take case iii) before ii). Since a specified player has all the hearts, the remaining 39 cards can be dealt to the remaining players (as in case i) in $39!/(13!)^3\, 3!$ ways.

ii) There are more ways of dealing ii) than iii) because any of the 4 suits could have been chosen, increasing the number of ways by 4; and *any* of the players could have

been chosen for the special hand, again increasing the
number of ways by 4. Hence the total number of ways
is $16 \cdot (39!)/(13!)^3 \cdot 3!$.

A.9.3 Probability

If there are n independent ways favorable to the outcome of an
event, and if there are N ways in which things could occur, and
if all of these ways are equi-probable, then the probability of
the event is n/N.

Example 1. What is the probability that at least one player will
have a hand all one suit in a game of bridge? The total number
of deals is as in i) above, and the number favorable deals is as in
ii). Since these deals are all independent and equi-probable, the
required probability is

$$\frac{16(39!)}{(13!)^3 \, 3!} \div \frac{52!}{(13!)^4 \, 4!} = \frac{64(13!)(39!)}{52!}$$

after some cancellation of factors (many more factors would
also cancel). This number works out to about 1 chance in 10^{10}.

Example 2. What is the probability of throwing 10 or more
with a throw of two dice? The total number of ways in which the
dice can appear is 6 x 6 = 36. To get 10 or more we need

12 = (6 + 6)	1 way only
11 = (6 + 5) *or* (5 + 6)	2 ways
10 = (6 + 4) *or* (5 + 5) *or* (4 + 6)	= 3 ways
10 or 11 or 12 1 + 2 + 3 ways	= 6 ways
p(10 or more) = 6/36 = 1/6	

The probability of getting *exactly* 10 is p(10) = 3/36 = 1/12
The probability of *not* getting a ten is (36 – 3)/36 = 11/12
The probability of an event happening is denoted by p.
The probability of an event *not* happening is denoted by q.
Since the event must either happen or not happen, and these are
mutually exclusive possibilities, one of which must happen, we
get p + q = 1 (certainty).

Example 3. What is the probability of getting 3 heads *followed
by* 3 tails in 6 spins of a penny? The probability that the first
spin is heads is ½. The same for the second spin. These are *independent* events. Hence the probability of the first two being

heads is ½ x ½ = ¼. Similarly for each spin, giving $(½)^6$ as the total probability. Note that the same result would have been obtained if the order of heads and tails had been different, so long as it was *specified*.

Example 4. What is the probability of 3 heads and 3 tails in 6 spins of a penny? This differs from the previous example in that the order is not specified and therefore is of no importance. The three heads could have been selected in $^6C_3$ = 20 different ways, all *mutually exclusive*. Therefore we need $(½)^6$ + ... $(½)^6$, with 20 terms, i.e. 20 x $(½)^6$ = $5/16$. (Having got 3 heads the rest *has to be* tails, and therefore does not affect the calculation further.)

Example 5. What is the probability of at least 3 heads in 6 spins of a penny? The *mutually exclusive* possibilities are 3, 4, 5 or 6 heads. Their respective probability is therefore the *sum* of these

$$p(\geqslant 3) = (½)^6 \ [^6C_3 + {}^6C_4 + {}^6C_5 + {}^6C_6] = \frac{20 + 15 + 6 + 1}{64} = \frac{21}{32}$$

In some cases it is easier to work with the non-occurence of an event, and to deduce its probability from p = 1 − q.

Example 6. What is the probability of getting at least one 6 in a throw of 4 dice? The probability of *not* getting a six in one throw is 5/6. Since the throws are independent, the probability of *no* sixes in 4 throws is $(5/6)^4$. Hence the probability of *at least one six* (the mutually exclusive *and* exhaustive alternative to *no* sixes) is $1 − (5/6)^4$ = 671/1296.

A.9.4 Venn Diagrams

Probabilities can be represented as a ratio of two areas, a favorable area to a total area. The shape of the area is irrelevant. Favorable areas are shaded in the following.

Figure A.36 shows eight examples of the representation of different types of events. One could imagine in the first example that the rectangle represents a dart board onto which darts are thrown in a uniform and random manner. Some of them will land in the shaded area A. The probability of this happening is proportional to the area of A, and is equal to the area of A to the total area. Examples of the use of this method of representation follow.

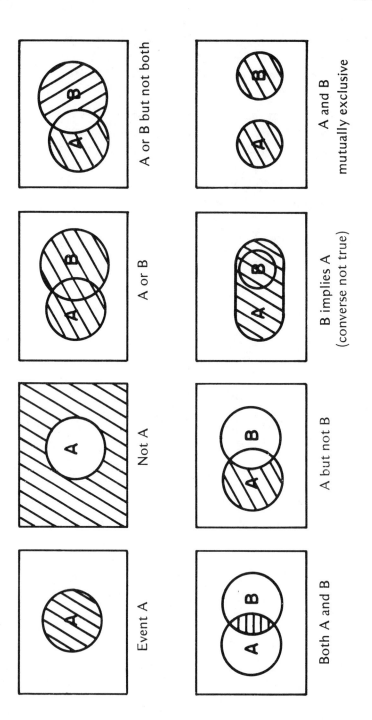

A.36 *Venn Diagrams*

A.9.5   Joint Probability

We use P(A or B) to mean "the probability of event A *or* B," and P(A,B) to mean "the probability of both A *and* B."

From the third example in Figure A.36

$$P(A \text{ or } B) = \frac{(\text{area A only}) + (\text{area B only}) - \text{overlap area}}{S = \text{total area of rectangle}}$$

The overlap area is subtracted in order *not to count it twice* in calculating the areas of A and B. Hence

$$P(A \text{ or } B) = P(A) + P(B) - P(A,B) \tag{A.121}$$

If the events are mutually exclusive or incompatible, they cannot occur together and P(A,B) would be zero. Hence, for *mutually exclusive events* (only)

$$P(A,B) = 0 \text{ and } P(A \text{ or } B) = P(A) + P(B) \tag{A.122}$$

Note that this is the particular case of (A.121) when the events are mutually exclusive.

From the fourth example in Figure A.36 we deduce that

P(A or B but not both)

$$= \frac{(\text{Area A only}) + (\text{Area B only}) - 2 \times \text{overlap area}}{S = \text{Total area of rectangle}}$$

Hence

$$P(A \text{ or } B \text{ but not both}) = P(A) + P(B) - 2P(A,B) \tag{A.123}$$

Note that the overlap is now subtracted twice because it has to be excluded from both area A and area B. Compare to equation (A.121).

A.9.6   Conditional Probability

If two events A and B are related in some way, e.g. total equipment failure, and failure of one component, we can write $P(A|B)$ to represent the *conditional* probability of A happening, *given* that B has happened. Thus one can ask what is the probability that a given transistor has failed if it is known that a piece of equipment has failed (for any one of a number of reasons including, but not limited to, a transistor failure).

The Venn diagram pertinent to this type of situation is shown in Figure A.37. The first diagram shows P(A) when nothing in known about B. But if B is *known to have occurred,* the area favorable to event A is reduced to the overlap area of B. Also the field of possible events has shrunk to the area of B only (since B is *known* to have occurred). Hence

$$P(A|B) = \frac{\text{area } (A,B)}{\text{area } B} = \frac{\text{area } (A,B)}{\text{area } S} \div \frac{\text{area } B}{\text{area } S}$$

$$= \frac{P(A,B)}{P(B)}$$

This yields the theorem of *joint probability:*

$$P(A,B) = P(A|B)P(B) \tag{A.124}$$

A similar argument gives

$$P(B,A) = P(B|A)P(A) \tag{A.125}$$

But P(A,B) = P(B,A) since both are the probability of A *and* B. Hence

$$P(A|B)P(B) = P(B|A)P(A) \tag{A.126}$$

This equation connects P(A|B) with P(B|A). Note that these probabilities are not the same thing. For example, if A stands for the event of a premature failure of a tube, and B for the event that its filament is defective, then P(A|B) is the probability of a premature failure given tht the filament is defective, whereas P(B|A) is the probability that the filament is defective, given that the tube has prematurely failed. The first probability would be of interest for investigating the effects of a certain defect, the second for investigating the causes of a certain failure.

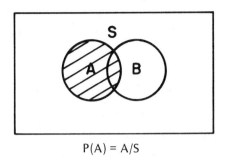

$P(A) = A/S$

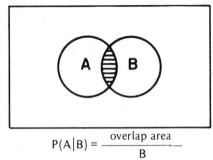

$P(A|B) = \dfrac{\text{overlap area}}{B}$

A.37   Venn Diagram Illustrating Conditional Probability

If A and B are *independent,* then A does not depend on B, so $P(A|B)$ is simply $P(A)$. Hence (A.124) becomes

$$P(A,B) = P(A)P(B) \tag{A.127}$$

This is the formula for *multiplication* of probabilities for *independent* events.

If the event B can take place only in conjunction with n mutually exclusive events $A_1, A_2 \ldots A_n$ with conditional probabilities $P(B|A_r)$ for the rth event, then

$$P(B) = \sum_{1}^{n} P(B,A_r)$$
$$= \sum_{1}^{n} P(B|A_r)P(A_r) \tag{A.128}$$

This is called the theorem of *total probability.*

In many applications we know the value of $P(B|A)$ and need to get $P(A|B)$. This comes from (A.126), since

$$P(A|B) = \frac{P(B|A)\,P(A)}{P(B)} \tag{A.129}$$

Using (A.128) to eliminate $P(B)$ we get

$$P(A|B) = \frac{P(B|A)P(A)}{\sum\limits_{1}^{n} P(B|A_r)P(A_r)} \tag{A.130}$$

This extremely important theorem is known as Baye's theorem.

A.9.7   Continuous Variables and Probability Density

A continuous random variable X may fall in the range $x < X < x + dx$ with a probability proportional to dx. The proportionality factor depends on x, and is known as a *probability density*. It is defined by

$$P(x < X < x + dx) = p(x)dx \tag{A.131}$$

Since X must take some value between $-\infty$ and $+\infty$ (in special cases within a more restricted range), we have

$$\int_{-\infty}^{\infty} p(x)dx = 1 \tag{A.132}$$

The *distribution function* F(x) is defined by

$$F(x) = P(X < x) = \int_{-\infty}^{x} p(x)dx \tag{A.133}$$

The probability that X will fall in the finite range $a < x < b$ is accordingly

$$P(a < x < b) = \int_{a}^{b} p(x)dx = F(b) - F(a) \tag{A.134}$$

The joint probability density for two continuous random variables x and y is defined by

$$P(x < X < x + dx , y < Y < y + dy) = p(x,y)dx\,dy \tag{A.135}$$

In (A.130) the events may be associated with continuous variables X and Y rather than discrete sets A_r. In this case the summations are replaced by integrations, the range covering all permissible values of the variables. In the most general case this would be from

$-\infty$ to $+\infty$, though in special cases the range would be restricted for physical reasons. The form taken by (A.130) for the continuous case is

$$P(x|y) = \frac{P(y|x)P(x)}{\int_{-\infty}^{\infty} P(y|x)P(x)dx} \qquad (A.136)$$

A.9.8 Mean Value and Variance

The average value, or mean value, of a function $f(X)$ of a discrete random variable X is defined by

$$<f(X)> = \sum_{1}^{n} f(X_r)\, P(X_r) \qquad (A.137)$$

where X takes the values X_r, $r = 1$ to n, with probabilities $P(X_r)$. If X is a continuous random variable the sum is replaced by an integral to give

$$f(X) = \int_{-\infty}^{\infty} f(x)p(x)dx \qquad (A.138)$$

In particular, the average value $<X>$ is given by

$$<X> = \int_{-\infty}^{\infty} xp(x)dx \qquad (A.139)$$

As before, physical restriction may reduce the integration range. For example, if x can take only positive values, the lower limit is zero instead of $-\infty$.

The *variance* $D(X)$ is defined as the difference between the average value of the square of X and the square of its average value:

$$D(X) = <X^2> - (<X>)^2 \qquad (A.140)$$

If X represents a noise signal, its average value is zero and the variance reduces to $<X^2>$, proportional to the power in the noise.

A.9.9 Reliability

Suppose that the probability that a component, having lasted a time t, has the probability $\lambda\, dt$ of failing in the interval dt. The quantity λ is a measure of the *quality* of the component and may itself vary with the component's age. But we shall here assume it is constant. This is approximately so in many cases. It exactly represents the situation in radioactive decay, and in many other examples where random processes are responsible for failure.

What is the probability that the component will reach at least an age t before failing?

Let A = component will reach at least age t

   B = component will fail in the interval t to t + dt

   $\overline{B}$ = component will *not* fail in this interval

Then $P(A,\overline{B})$ is the probability that the component will reach age t *and* will survive the next interval dt. If P(t) is the probability that the component survives at least to age t, then P(A) = P(t) and $P(A,\overline{B})$ = P (t + dt). Also P(B) = λdt and $P(\overline{B})$ = 1 – λdt. Hence

$$P(t + dt) = P(t) (1 - \lambda dt) \tag{A.141}$$

But from the definition of differentiation we have

$$\frac{P(t + dt) - P(t)}{dt} = \frac{dP}{dt} \quad \text{whence}$$

$$P(t + dt) = P(t) + \frac{dP}{dt} dt \tag{A.142}$$

Combining this with (A.141) gives $P(t) + \frac{dP}{dt} dt = P(t) - P(t)\lambda dt$, so that

$$\frac{dP}{dt} = -\lambda P \tag{A.143}$$

This is a differential equation for P, and is of a form satisfied by the exponential function. Hence

$$P = Ke^{-\lambda t} \tag{A.144}$$

where K is a constant. Now when t = o, (A.144) gives P(t = o) = K. But P(t = o) = 1 since there is certainty that the component will survive zero time. Hence K = 1 and (A.144) reduces to

$$P(t) = e^{-\lambda t} \tag{A.145}$$

Many physical decay processes, e.g. radioactive decay, mortality statistics, etc., follow this relation either exactly (λ independent of age) or approximately. λ is a measure of *quality* in a component; small λ means long life or higher quality.

The *average life expectance* T_{av} is given by $\int_0^\infty e^{-\lambda t} dt = 1/\lambda$. It is different from the *half-life* $T_{1/2}$, which is the time to produce a probability of ½ of failure. $T_{1/2}$ is accordingly determined by ½ = $e^{-\lambda T_{1/2}}$, or $\lambda T_{1/2} = 0.6931$. Since the average life is $T_{av} = 1/\lambda$ the relation between the two is

$$T_{1/2} = 0.6931\, T_{av} \qquad\qquad (A.146)$$

T_{av} is also called the *mean time between failures*, MTBF.

A.10 LIST OF ACRONYMS*

ACK	Acknowledgement
ADCCP	Advanced Communication Control Procedures
ADP	Automated Data Processing
ADU	Accumulation and Distribution Unit
ANK	Alphanumeric Keyboard
AO	Abort Output
APR	Active Page Register
ARPANET	Advanced Research Projects Agency Network
ARQ	Automatic Repeat Request
ASC	Automatic Switching Centers
ASCII	American Standard Code for Information Interchange
AUTOVON	Automatic Voice Network
AYT	Are You There
B1	Bulk 1 Transfer
B2	Bulk 2 Transfer
BCC	Block Check Character
BEC	Buffer and Executive Control
BER	Bit Error Rate
BISYNC	Binary Synchronous
BKB	Bookkeeping Block
BOP	Bit Oriented Protocol
BOT	Beginning of Tape
BP	Block Parity
bps	Bits per Second

*This list was kindly provided by Ford Aerospace and Communications Corporation.

BSC	Binary Synchronous Communication
BSL	Binary Segment Leader
CAN	Cancel Character
CAU	Crypto Ancillary Unit
CBEC	Cumulative Block Error Count
CCB	Configuration Control Board
CCP	Control Character Protocols
CCSL	Character Canned Segment Leader
CCU	Channel Control Unit
CDRL	Contract Data Requirements List
CFF	Critical Flicker Frequency
CHOM	Channel Output Module
CI	Configuration Item
CLS	Communications Line Switch
CM	Configuration Management
CMB	Connection Management Block
CMK	Clock Modification Kit
COI	Community of Interest
COMP	Computer (PS)
COMSEC	Communications Security
CONUS	Continental United States
CPC	Computer Program Components
CPCI	Computer Program Configuration Item
CPM	Computer Program Module
CPU	Central Processing Unit
CRC	Cyclic Redundancy Check
CREP	Computer Resource Evaluation Program
CRT	Cathode Ray Tube
CS	Clear-to-Send
CSA	Communications Service Authorization
CSC	Computer Sciences Corporation
CSL	Character Segment Leader
CSR	Control and Status Registers
CUSL	Character Unclassified Segment Leader
CXA	Channel Extractor
DAA	Data Access Arrangement
DCAOC	Defense Communications Agency Operations Center
DCE	Data Communication Equipment
DCEC	Defense Communication Engineering Center
DEC	Digital Equipment Corporation
DECCO	Defense Commercial Communication Office

DEL	Delete Character
DEMUX	Demultiplexer
DEST	Destination Address
DIU	Dial-In User
DIV	Data In Voice
DLE	Data Link Escape
DLE CAN	Data Link Escape Cancel
DLE DC1	Data Link Escape Device Control One
DLE DC2	Data Link Escape Device Control Two
DLE DC3	Data Link Escape Device Control Three
DLE DC4	Data Link Escape Device Control Four
DLE EM	Data Link Escape — End of Medium
DM	Data Mode
DMA	Direct Memory Access
DSN	Display Selection Number
DSR	Digital Service Rate
DSU	Data Service Units
DTE	Data Terminal Equipment.
DTMF	Dual Tone Multifrequency
DTR	Digital Transmission Rate
EBCDIC	Extended Binary Coded Decimal Interchange Code
EC	Erase Character
ECP	Engineering Change Proposal
EEL	External Event List
ELA	Electronics Industries Association
EL	Erase Line
EM	End of Medium
EMC	Electromagnetic Compatibility
EMI	Electromagnetic Interference
EOB	End of Block
EOM	End of Message
ESTD	Established
ETB	End of Transmission Block
ETX	End of Text
FACC	Ford Aerospace and Communications Corp.
FCS	Frame Check Sequence
FDX	Full Duplex
FIN	Finish
FIPS	Federal Information Processing Standards
FISH	First in Still Here
FOC	Final Operational Capability

FSW	Frame Sync Word
FTS	Federal Telecommunications System
GFE	Government-Furnished Equipment
GMT (ZULU)	Greenwich Mean Time
HCU	Host Control Unit
HDT	Host Channel Description Table
HDX	Half Duplex
HSI	Host Specific Interface
HSLA	High Speed Line Adapter
I/A	Interactive
ICU	Interface Control Unit
IDP	Integrated Data Processing
IIE	Input Interrupt Enable
I/O	Input/Output
ILS	Integrated Logistics Support
INPRO	Input Process Output
IOC	Initial Operational Capability
IP	Interrupt Process
ISP	Integrated Support Plan
ITP	Induced Transient Pulse
kbps	Kilobits Per Second
KIPS	Kilo Instructions Per Second
LCM	Line Control Module
LCU	Line Control Unit
LIB	Line Interface Base
LM	Loop Module
LRC	Longitudinal Redundancy Checks
LTU	Line Termination Unit
Mbps	Megabits Per Second
MCCU	Multiple Channel Control Unit
MDT	Mean Down Time
MOS	Metallic Oxide Semiconductor
MRF	Module Release Form
MTBF	Mean Time Between Failures
MTTR	Mean Time to Repair
MUX	Multiplexer
N	Narrative
NCC	Network Control Center
NETINT	Network Interface
NOP	No Operation
NPR	Non-Processor Request
NRZI	Non-Return to Zero Inverted

NSTRAP	Node Sizing and Traffic Analysis Program
NVT	Network Virtual Terminal
OCU	Office Channel Unit
OIE	Output Interrupt Enable
OGFE	Optional Government-Furnished Equipment
OLU	Office Line Unit
OPINT	Operator Interface
PAD	Pad Character
PAR	Page Address Register
PC	Printed Circuit
PCL	Parallel Communications Link
PDR	Page Descriptor Register
PFPT	Password Function Permission Table
PPS	Packets Per Second
PRF	Program Release Form
PS	Packet Switch
PSL	Program Support Library
PSN	Packet Switch Node
PT	Password Table
PTC	Programmable Terminal Controller
PTF	Patch and Test Facility
PTFSS	Patch and Test Facility Subsystem
QA	Quality Assurance
QOS	Quality of Service
Q/R	Query/Response
RAB	Received Abort
RACT	Received Active
RAM	Random Access Memory
RBKB	Receive Bookkeeping Block
RD	Receive Data
RDA	Receive Data Available
RDB	Receiver Data Buffer
RDP	Receiver Data Path
RDYI	Ready In
RDYO	Ready Out
REJ	Reject
RENA	Receive Enable
REOM	Received End of Message
REP	Reply Character
RFI	Radio Frequency Interference
RFN(s)	Ready for Next Segment(s)
RFO	Reason for Outage

RGA	Received Go-Ahead
RJE	Remote Job Entry
RM	Record Marker
RMA	Reliability, Maintainability, Availability
RNR	Receiver Not Ready
ROR	Receive Overrun
ROM	Read-Only Memory
RR	Receiver Ready
RSA	Receive Status Available
RSI	Receive Shift-In
RTAC	Remote Terminal Access Controller
RQI	Request In
RQO	Request Out
RSOM	Received Start of Message
SCM	Switch Control Module
SCN	Specification Change Notice
SCCU	Single Channel Control Unit
SD	Source/Destination
SDL	Software Development Laboratory
SDLC	Synchronous Data Link Control
SEC	Security
SECACT	Security Accountant
SEL	Select Character
S/F	Sync/Flag
SI	Serial In
SIP	Segment Interface Protocol
SLU	Station Line Unit
SMO	System Management Office
S/N	Signal to Noise
SO	Serial Out
SOH	Start of Heading
SOM	Start of Message
SOW	Statement of Work
SREJ	Selective Reject
SSCM	Supervisory Switch Control Module
SSTF	Switch Software Task File
SST	Segment Status Table
STF	System Test Facility
STR	Sender to Receiver
STX	Start of Text
SYN	Synchronous Idle Character
SYNC	Synchronous

TAC	Terminal Access Controller
TAM	Task Modeling Matrix
TATR	Task/Traffic Matrix
TBKB	Transmit Bookkeeping Block
TBMT	Transmitter Buffer Empty Bit
TC	Time-Controller
TCB	Transmission Control Block
TCC	Transmission Control Code
TCF	Technical Control Facility
TCMS	Tone Control Modem Selector
TCP	Transmission Control Program
TDB	Transmitter Data Buffer
TDM	Time Division Multiplex
TDMI	Time Division Multiplex Interface
TDSR	Transmitter Data Shift Register
TEC	Technical Control
TENA	Transmitter Enable
TEOM	Transmit End of Message
TF	Test Facility
TH	Terminal Handler
THP	Terminal-to-Host Protocol
TIM	Traffic Instruction Mix
TSA	Transmitter Status Available
TSOM	Transmit Start of Message
TTY	Teletype
TXAB	Transmit Abort
TXGA	Transmit Go Ahead
USYRT	Universal Synchronous Reciever Transmitter
UMB	User Multiplexing Block
VCO	Voltage Controlled Oscillator
VF	Voice Frequency
VRC	Vertical Redundancy Check
WBS	Work Breakdown Structure
WU	Western Union
WWOLS	Worldwide Online System
ZULU	See GMT

A

H

I

R

T

Y

Z